SPACE TELESCOPE SCIENCE INSTITUTE

SYMPOSIUM SERIES: 5
Series Editor S. Michael Fall, Space Telescope Science Institute

MASSIVE STARS IN STARBURSTS

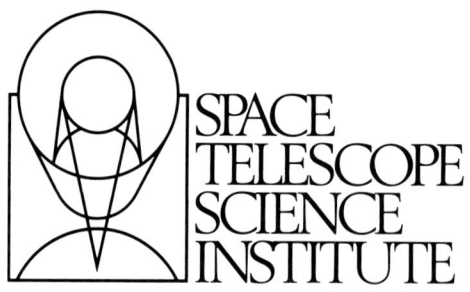

Titles in the Space Telescope Science Institute Symposium Series

1 Stellar Populations
   Edited by C.A. Norman, A. Renzini and, M. Tosi 1987 0 521 33380 6

2 Quasar Absorption Lines
   Edited by C. Blades, C.A. Norman and, D. Turnshek 1988 0 521 34561 8

3 The Formation and Evolution of Planetary Systems
   Edited by H.A. Weaver and L. Danly 1989 0 521 36633 X

4 Clusters of Galaxies
   Edited by W.R. Oegerle, M.J. Fitchett and, L. Danly 1990 0 521 38462 1

5 Massive Stars in Starbursts
   Edited by C. Leitherer, N.R. Walborn, T.M. Heckman and, C.A. Norman
     1991 0 521 40465 7

MASSIVE STARS IN STARBURSTS

Proceedings of the Massive Stars in Starbursts Meeting
Baltimore, 1990 May 15–17

Edited by

CLAUS LEITHERER
Space Telescope Science Institute

NOLAN R. WALBORN
Space Telescope Science Institute

TIMOTHY M. HECKMAN
Space Telescope Science Institute

COLIN A. NORMAN
Space Telescope Science Institute

Published for the
Space Telescope Science Institute

CAMBRIDGE UNIVERSITY PRESS

Cambridge
New York   Port Chester
Melbourne   Sydney

Published by the Press Syndicate of the University of Cambridge
The Pitt Building, Trumpington Street, Cambridge CB2 1RP
40 West 20th Street, New York, NY 10011-4211, USA
10 Stamford Road, Oakleigh, Melbourne 3166, Australia

© Cambridge University Press 1991

First published 1991

Printed in Great Britain at the University Press, Cambridge

British Library cataloguing in publication data available

Library of Congress cataloguing in publication data available

ISBN 0 521 40465 7 hardback

CONTENTS

Frontispiece x

Preface xiii

Participants xv

Chapter 1

Massive Stars—From the Solar Neighborhood to Starburst Galaxies 1
C. LEITHERER
Definition: How Massive are Massive Stars? 1
Direct and Indirect Observations of Massive Stars 3
Models for Massive Stars 4
The Stellar Census of Massive Stars 7
Spectral Signatures of Massive Stars 10
The Formation of Massive Stars in Starbursts 13
References 15
Discussion — D. Hummer, Chair 17

Chapter 2

Spectra of Massive Blue Stars 21
P. S. CONTI
Classical Spectroscopy 21
Spectrophotometry 23
Indirect Signatures 31
Wolf-Rayet Galaxies 36
Present and Future Directions 38
References 39
Discussion — D. Hummer, Chair 40

Chapter 3

Massive Red Stars in Galaxies 45
R. M. HUMPHREYS
Red Supergiants in the HR Diagram 45
The Luminosity/Stability Limit and the Most Luminous Red
Supergiants—Their Use as Distance Indicators 46
Identification and Measurement of Red Supergiants via Their Spectroscopic
and Photometric Signatures at Optical and Infrared Wavelengths 49
Massive Red Stars in Starburst Galaxies 53
Massive Red Stars in More Distant Galaxies—Observations
With the Hubble Space Telescope 53
References 54
Discussion — D. Hummer, Chair 56

Chapter 4

Model Atmospheres of Massive Hot Stars — 59
R. KUDRITZKI, R. GABLER, D. KUNZE, A. W. A. PAULDRACH, J. PULS

Introduction	59
Model Atmospheres—The State of the Art	59
Ionizing Fluxes	65
Winds and Metallicity	69
Detailed Diagnostics	77
References	91
Discussion — D. Hummer, Chair	93

Chapter 5

Massive Star Evolution — 97
A. MAEDER

Introduction: Are Massive Stars the Same Throughout the Universe?	97
Mass and Z Effects in the Course of Evolution	98
Lifetimes as Supergiants and Questions Related to SN1987A	100
W-R Stars: Properties and Numbers in Galaxies	102
Final Masses, Chemical Yields and Mass Limits	105
References	108
Discussion — N. Langer, Chair	109

Chapter 6

The Observational H-R Diagram and IMF of Massive Stars — 115
C. D. GARMANY

Introduction: Stars and the Starburst Phenomena	115
IMF for Massive Stars: How it's Done Directly	115
Magellanic Clouds	121
How are the Masses of the Stars Determined?	122
The Most Massive Individual Stars	124
Relation of Nebular $H\alpha$ Flux to Stellar Content	125
Conclusions	126
References	126
Discussion — N. Langer, Chair	128

Chapter 7

IR- and mm-Observations of Regions of Massive Star Formation — 133
I. GATLEY, K. M. MERRILL, A. M. FOWLER

Introduction	134
Global Star Formation in the L1630 Cloud	134
Star Formation in M17	135
Star Formation in NGC 2023	138
Star Formation in the Orion Nebula	139
Conclusions	141
References	141
Discussion — N. Langer, Chair	141

Chapter 8

30 Doradus, Starburst Rosetta — 145
N. R. WALBORN

Introduction	145
Scales and Populations	146
Optical Spectroscopy and Photometry: The Current Generation	148
Infrared: New Beginnings	149
High Energy: Last Things	150
References	153
Discussion — G. Shields, Chair	154

Chapter 9

Properties of Giant H II Regions — 157
R. C. KENNICUTT, JR

Introduction	157
Physical Characteristics of Giant H II Regions	158
Systematics of H II Region Populations in Galaxies	160
Stellar Content and Initial Mass Function	161
Kinematics, Dynamics, Evolution	163
References	165
Discussion — G. Shields, Chair	167

Chapter 10

Supernovae and Supernova Remnants in Starbursts — 169
R. A. CHEVALIER

Introduction	169
Supernovae and Young Supernova Remnants	170
The Surroundings of Supernovae	172
Supernovae in Starburst Galaxies	173
Large-Scale Flows	175
References	178
Discussion — G. Shields, Chair	179

Chapter 11

Observations and Models of Blue Compact Dwarf Galaxies — 183
T. X. THUAN

Introduction	183
The Evidence for Starbursts in BCDs	184
The Star Formation History in BCDs	187
Comparison with Massive Starburst Galaxies	193
References	199
Discussion — E. Bica, Chair	201

Chapter 12

The Initial Mass Function in M82 — 205
G. RIEKE

Introduction — 205
Starburst Modeling — 206
Observable Parameters — 206
Comparison with Models — 212
References — 213
Discussion — E. Bica, Chair — 214

Chapter 13

Population Synthesis Models of Starbursts — 217
B. ROCCA-VOLMERANGE

Introduction — 217
Typical Spectral Features — 218
Models — 220
Comparison with Observations and Discussion — 222
Conclusion — 227
References — 228
Discussion — E. Bica, Chair — 230

Chapter 14

H_2 and Infrared in Global Starburst Galaxies — 233
N. SCOVILLE, B. T. SOIFER

Starbursts — 234
Luminous Infrared Galaxies — 234
An Example—Arp 220 — 242
Starburst Models — 244
Merger-Induced Starbursts — 247
Summary — 250
References — 251
Discussion — K. Freeman, Chair — 252

Chapter 15

Stellar Content of Starburst Galaxies — 259
R. JOSEPH

Introduction — 259
Infrared Techniques for Measuring Extinction — 259
Starburst Model — 261
The Lower Mass Cutoff — 261
The Upper Mass Cutoff — 262
Population of Evolved Massive Stars — 263
NGC 3256 — 264
Conclusions — 265
References — 265
Discussion — K. Freeman, Chair — 266

Chapter 16

Models of Starburst Galaxies 271
C. NORMAN
Introduction 271
Dynamical Studies 273
Interstellar Medium 276
Star Formation 277
Making the Burst 278
The Post-Starburst Phase 278
Face-On Starbursts 279
Galaxy Formation 280
References 280
Discussion — K. Freeman, Chair 282

Chapter 17

The Starburst–AGN Connection 289
T. M. HECKMAN
Introduction 289
The Energy Input From Massive Stars vs. AGN's in the Universe 290
AGN's Without Black Holes 291
Far-Infrared Galaxies: Starbursts or 'Buried' Quasars? 297
Star-Formation in Active Versus Normal Galaxies 303
References 308
Discussion — R. Griffiths, Chair 313

Chapter 18

Cosmological Consequences of Starbursts 317
D. WEEDMAN
Introduction 317
Luminosity Functions for Star Forming Galaxies 318
Predictions for Limiting Observations 324
Summary 328
References 328
Discussion — R. Griffiths, Chair 329

Chapter 19

Final Discussion 331

The ionizing cluster of the 30 Doradus Nebula in the Large Magellanic Cloud, as seen in the historic, first scientifically significant image obtained by the Hubble Space Telescope. It was acquired by the Goddard High Resolution Spectrograph team with the Wide Field Camera on August 3, 1990, in a 40-second exposure through a 230 Å-wide filter centered at 3680 Å. North is at top and east to the left; the extent in declination is $2'$ and the resolution is $0''\!.1$–$0''\!.2$. A 7×7-pixel median-filtered image has been subtracted from the original raw image to qualitatively suppress the effects of spherical aberration. Note the resolution of the concentrated central object R136, and compare the best ground-based image on p. 147. Courtesy of Jack Brandt and Sally Heap, with technical assistance by Eliot Malumuth, Richard White, Robert Hanisch, Zoltan Levay, and Skip Westphal.

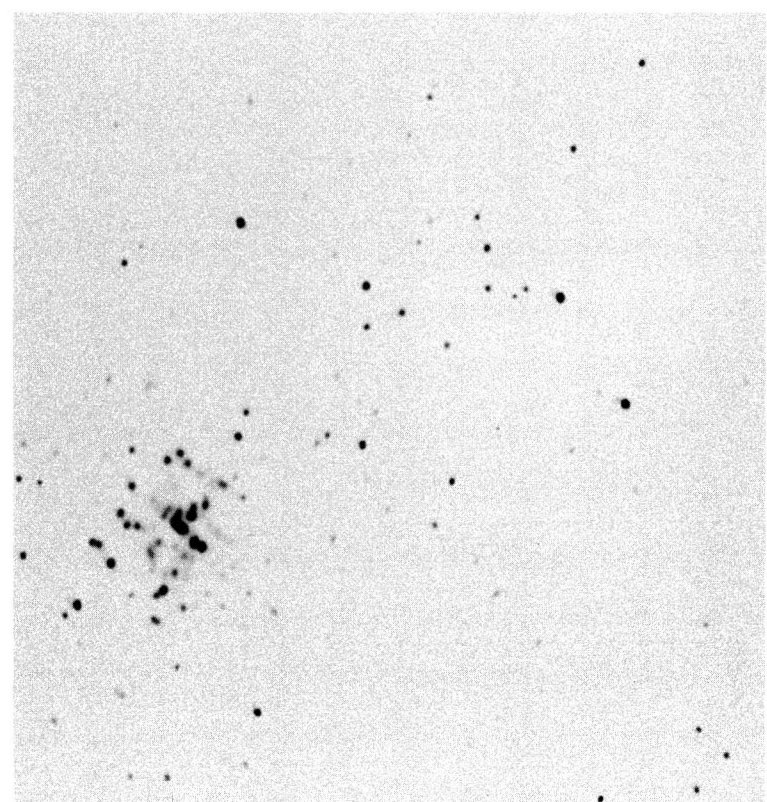

The central object R136 in 30 Doradus, as shown by a Lucy restoration with factor-of-two oversampling from the preceding (raw) HST/WFC image, performed by Richard L. White. North is at top and east to the left; the extent in declination is $18''$ and the resolution is $0\rlap{.}''1$–$0\rlap{.}''2$. The ground-based speckle interferometric components within R136 can be seen by inspection. The three brightest isolated stars to the north and west are of types O3 If and WN, and are probably similar to the brightest R136 components; the fact that they remain unresolved is consistent with very high individual luminosities and masses of order 100–200 $M_\odot$ (see p. 145).*

PREFACE

The premises underlying this Workshop were that, although their subjects are clearly related, the massive-star and extragalactic-starburst communities have been practically disjoint, and that a meeting bringing the two together might provide a significant stimulus to interrelated research on starburst phenomena. An attempt was made to focus the presentations and discussions on the interaction between the two fields, rather than on either per se. It is probable that future resolution of ambiguities in the interpretation of starburst regions will depend upon improved observations and understanding of specific stellar signatures in their spectra. It will be useful for starburst specialists to be aware of both the considerable recent advances and uncertainties (the latter emphasized by SN 1987A) concerning massive stellar atmospheres, winds, and evolution as functions of mass and metallicity. Conversely, exposure of the stellar specialists to the physical parameters of starbursts could promote relevant observational and theoretical studies. Hence, the Workshop was conceived not merely as a review of the pertinent fields, but as a dynamical contribution toward future research. Although the organizers felt occasional trepidation about the prospects of this rather ambitious objective, there was ample evidence during the meeting that the desired interaction substantially succeeded, with both stellar and extragalactic participants frequently expressing interest and even amazement at the results and dilemmas of their counterparts. It is hoped that this volume will preserve both the substance and the spirit of the interaction for the benefit of others who could not attend as well as the participants themselves, thus contributing to the longer-term objective of constructively influencing research directions. This volume records the eighteen invited review talks and associated discussions; the contributed poster papers have been recorded separately and are available from the publications office of the STScI.

Many of the participants were saddened by the untimely passing of Nick Sanduleak the week prior to the Workshop, on May 7, 1990, and the assembly stood for a moment of silence in honor of his life and work on the morning of May 16. Nick was one of the quiet foundation builders of astronomy, and his name was frequently heard during the meeting, particularly in relation to his fundamental spectroscopic surveys of the Magellanic Clouds. Nevertheless, he also gave his name to two of the most famous massive stars in our universe, namely Stephenson-Sanduleak (SS) 433, and Sanduleak (Sk) $-69°202$, the progenitor of SN 1987A. His work will retain a permanent place in the continuing progress of our science.

At the time the Workshop was proposed, there was a reasonable prospect that the first relevant images from the Hubble Space Telescope might be available in time to present there. Such was not to be the case, but in the interval before completion of this publication, they did materialize, and the initial discouragement produced by the unanticipated spherical aberration problem has given way to qualified optimism about the unique capabilities remaining nevertheless, and the prospects for eventual correction of the problem. Happily, the first image of scientific interest obtained was of the 30 Doradus cluster in the Large Magellanic Cloud, and two representations of it are presented as a frontispiece to the volume. They indicate clearly the remarkable contributions to the fields of the Workshop which HST will provide in both the near and long terms.

Nolan R. Walborn
Claus Leitherer
Timothy M. Heckman
Colin A. Norman

PARTICIPANTS

B. Altner	R. Giacconi	M. Plavec
P. Appleton	A. Grandchamps	R. Pogge
L. Armus	R. Griffiths	G. Rieke
V. Balzano	J. Haller	C. Robert
K. Bernlohr	J. Hawthorn-Bland	V. Robledo-Rella
E. Bica	S. Heap	B. Rocca-Volmerange
C. Blaha	T. Heckman	N. Scoville
B. Bohannan	G. Hensler	G. Shields
F. Bruhweiler	J. Hillier	J. Shields
G. Bruzual	D. Hummer	E. Smith
A. Campbell	R. Humphreys	P. Solomon
P. Carrall	C. Jog	L. Spight
J. Cassinelli	R. Joseph	S. Sreenivasan
H. Casteneda	R. Kennicutt	R. Stanga
R. Chevalier	G. Koenigsberger	G. Stasinska
P. Conti	E. Kontizas	R. Stencel
N. Devereux	R. Kudritzki	C. Struck-Marcell
A. Dey	S. Lamb	N. Sykes
B. Draine	N. Langer	T. Thuan
L. Dressel	C. Leitherer	K. Truong
L. Drissen	C. Lonsdale	J. Turner
D. Ebbets	A. Maeder	W. Vacca
M. Fanelli	D. Massa	S. Vogel
E. Fitzpatrick	V. McIntyre	N. Walborn
K. Freeman	I. Mirabel	R. Walterbos
K. Fricke	A. Moffat	Z. Wang
A. Fullerton	P. Morris	D. Weedman
K. Garmany	D. Neufield	A. Wilson
D. Garnett	C. Norman	J. Young
I. Gatley	J. Parker	

MASSIVE STARS—FROM THE SOLAR NEIGHBORHOOD TO STARBURST GALAXIES

Claus Leitherer[*]
Space Telescope Science Institute
3700 San Martin Drive
Baltimore, MD 21218
USA

Abstract. Significant progress has been achieved in massive-star research in recent years. Theoretical modeling and high-quality observations of massive stars have improved our understanding of the formation, structure, atmosphere, and the evolution of these objects. Massive stars are formed in violent bursts of star formation and are probably related to phenomena observed in active galaxies. Although both topics—*stellar* astrophysics of massive stars and research on *extragalactic* starburst regions—receive wide attention individually, the interaction between the two fields is still rather weak. This paper reviews various results and open questions related to massive stars, both considering these stars individually and as a whole population when observed in distant starbursts.

1. DEFINITION: HOW MASSIVE ARE MASSIVE STARS?

Stars are known to be forming with a mass spectrum extending from about 0.1 $M_\odot$ up to 100 $M_\odot$ (Scalo 1986). The lower limit is imposed observationally by the intrinsic faintness of stars with $M_v > 16^m$. The upper limit is defined by very few stars only and is thus quite uncertain. Spectroscopic analyses of several hot, luminous stars indicate masses around 100 $M_\odot$ (Kudritzki 1988).

What is a reasonable *working definition* for the mass range of *massive* stars? An astrophysically significant dividing line exists at ~ 10 $M_\odot$. Main-sequence stars with $M \gtrsim 10$ $M_\odot$ (corresponding to a luminosity of $M_{Bol} \lesssim -6^m$) have a photospheric radiation field around the Lyman edge strong enough to initiate massive stellar winds via radiation pressure (see Lamers 1988 for the observations, and Kudritzki 1988 for the theory). Since the stellar luminosity in the far-UV decreases with decreasing main-sequence mass (*e.g.*, Panagia 1973), stellar winds become unimportant for stars with $M < 10$ $M_\odot$ (cf. Figure 1). Above ~ 10 $M_\odot$, the stellar mass-loss rates are high enough to affect stellar evolution profoundly; the evolutionary relation between early and late stellar

---

[*] Affiliated with the Astrophysics Division of the Space Science Department of the European Space Agency

stages (such as, e.g., supernovae) critically depends on mass loss. The return of mass and energy from stars to the interstellar medium is dominated by stars above ~ 10 $M_\odot$ (Abbott 1982). Finally, the output of Lyman-continuum quanta from stars with $M < 10$ $M_\odot$ is too low to ionize an observationally significant quantity of the ambient interstellar hydrogen so that observed H II regions are produced by higher mass stars. Hence, it is physically plausible to define massive stars as stars in the mass regime 10 $M_\odot < M < 100$ $M_\odot$.

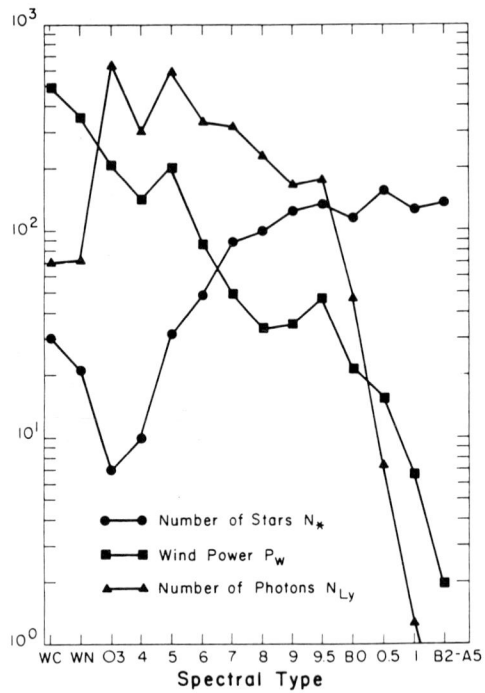

Figure 1. *Distribution with spectral type of the number of stars per kpc^2 in the solar neighborhood (taken from Abbott 1982); number of ionizing photons contributed by these stars; and the wind power. The relatively rare Wolf-Rayet- and O-stars dominate the return of wind and radiant energy to the interstellar medium.*

Since massive stars experience strong mass loss, their mass decreases with time. The descendants of initially very massive stars (such as, e.g., Wolf-Rayet stars) may have masses as low as 5 $M_\odot$ (Cherepashchuk 1990). Because of their evolutionary relationship with originally massive stars it is reasonable to include these objects in our definition of massive stars. Evolved descendants of massive stars are observed in the close vicinity of their progenitors (due to the short lifetimes of massive stars) and are fundamental for our understanding of the starburst phenomenon. Note that the preceding discussion of the dependence of the mass loss and the output of ionizing photons refers to stars on the main sequence only. Wolf-Rayet stars in particular have a very different mass-luminosity relation and chemical composition as compared to OB stars (Langer 1989) and do have high mass-loss rates although their masses may be low.

Using our working definition for massive stars as stellar objects with initial masses 10 $M_\odot < M < 100$ $M_\odot$ including their evolved descendants, *massive stars comprise O-*

and early B-stars on the main sequence, supergiants of all spectral types, Wolf-Rayet stars, and supernovae.

2. DIRECT AND INDIRECT OBSERVATIONS OF MASSIVE STARS

In comparison with other stellar groups, massive stars are easily detectable at large—even extragalactic—distances. Since stars closely follow a mass-luminosity relation $L \sim M^\alpha$ with $\alpha > 1$ (Maeder 1987), the most massive stars are also the most luminous stars. The census of massive stars is complete up to a distance from the sun which is larger than for any other stellar class. Individual massive stars are observable in a number of Local Group galaxies. Table 1 is a summary of the magnitudes of the brightest blue supergiants in several nearby galaxies (partly from Humphreys 1983). Blue supergiants have been individually detected up to a distance modulus of 27.6^m (M81). The apparent magnitude for the brightest blue supergiants in the Virgo group has been calculated assuming $M_V = -10^m$. It is to be expected that individual blue supergiants will actually by detected in the Virgo cluster with the next generation of telescopes, including the *Hubble Space Telescope*.

Table 1. *Apparent visual magnitude of the brightest blue supergiants in several nearby galaxies. The magnitudes are observations taken from Humphreys (1983) except for Virgo, where the empirically expected magnitude is given.*

Galaxy	$(V - M_V)_0$	V(Blue SG)
LMC	18.3	8.5
SMC	19.0	9.8
NGC6822	23.2	14.7
M33	24.2	14.7
M31	24.3	14.3:
IC1613	24.3	16.1
M81	27.6	17.7:
Virgo	31.7	21.7:

The magnitudes listed in Table 1 demonstrate that massive stars can be studied in detail in several galaxies of the Local Group (Massey 1985, Humphreys 1987). Such studies can test the variation of the formation and evolution of massive stars in different galactic environments. Local Group galaxies cover a variety of morphological types, have different gas content, and vary widely in their chemical composition.

The brightest red supergiants are typically two magnitudes fainter than their blue counterparts ($M_V \approx -8^m$). Unlike the brightest blue supergiants, however, the average absolute magnitude of the brightest red supergiants is independent from the absolute magnitude of the parent galaxy (Humphreys 1983). Sandage and Tammann (1974) drew attention to the use of red supergiants as important primary extragalactic distance indicators.

For most of their lifetime massive stars are hot enough to emit a significant fraction of their radiation below 912 Å. The fractional energy emitted in the Lyman continuum relative to the total energy output $(\int_{\nu_{Ly}}^\infty F_\nu d\nu)/(\int_0^\infty F_\nu d\nu)$ typically amounts to $\sim 50\%$ for a core-hydrogen burning star with $M > 10\ M_\odot$ (Panagia 1973). Massive stars are formed in Giant Molecular Clouds (Scoville and Good 1987). These regions are mostly optically thick in the Lyman continuum. Stellar radiation is trapped, reprocessed, and reemitted at longer wavelengths so that optical recombination radiation of hydrogen (*e.g.*, Hα, Kennicutt 1984) or thermal bremsstrahlung at cm-wavelengths (*e.g.*, Klein and Gräve 1986) are detectable. Observations of such H II regions allow one to probe the underlying massive star population in star forming regions where direct observations of the stars are infeasible due to their distance and/or dust absorption. Such *indirect* observations of massive stars greatly enhance our capabilities to investigate the stellar content of star forming regions.

Radiation emitted by hot massive stars can be absorbed by circum- and interstellar dust which is heated to an equilibrium temperature of $\sim 40\ K$. Thermal radiation from dust heated by massive stars has been detected by *IRAS* in a large number of galaxies including our own. Wood and Churchwell (1989) detected *individual* O stars embedded in galactic molecular clouds from their radio- and $100\mu m$-flux, and thus obtained a *direct* measure of their population in the Galaxy. On the other hand, far-IR measurements of spiral galaxies (de Jong *et al.* 1984) probe the *integrated* total stellar content of these galaxies. Except for very powerful starbursts, massive stars contribute only less than $\sim 50\%$ to the total far-IR flux (Lequeux 1989) and detailed modeling is required to infer the underlying massive star population. Using the far-IR flux as a tracer of thermal dust radiation it is possible to detect massive stars in ultraluminous infrared galaxies up to a distance of $z \approx 0.1$ (Sanders *et al.* 1987).

Massive stars with mass-loss rates of $10^{-6} - 10^{-5}\ M_\odot yr^{-1}$ and terminal wind velocities of $\sim 2000\ kms^{-1}$ convert about 10% of their radiative luminosity into mechanical power, which is transferred to the surrounding interstellar gas. As a consequence, wind-blown shells resulting from the interaction of the stellar wind with the interstellar gas are formed (Weaver *et al.* 1977). Figure 2 is an example of a stellar-wind bubble in the irregular galaxy NGC6822. It has most probably been generated by one or more massive stars inside the bubble. No spectroscopic confirmation of these stars has yet been obtained but the presence of the wind-blown bubble can be taken as a fairly reliable tracer of massive stars. Galaxies undergoing intense star formation, like M33, display a large number of wind bubbles (Courtès *et al.* 1987) indicating the sites where large numbers of massive stars are being formed.

A variety of direct and indirect techniques exist to detect and study massive stars. Traditionally, most of the work using direct methods has been done in our Galaxy. Conversely, indirect techniques have been employed in distant galaxies. In recent years observations of individual massive stars together with their integrated properties have become available. The Magellanic Clouds and other Local Group galaxies are ideal laboratories for these observations (Leitherer 1990). *Local* starburst regions like 30 Doradus in the LMC (Walborn, this meeting) will be important calibrators for indirect massive star tracers, which are used to investigate *distant* starbursts.

3. MODELS FOR MASSIVE STARS

Figure 3 is a schematic outline of the structure of a typical 60 $M_\odot$ star on the main sequence. The star has a hydrogen abundance of 70% (by mass) and a helium abundance of 28%. All other elements contribute 2% to the composition. The energy production rate ϵ in the central 20 $M_\odot$ exceeds $10^3 ergs^{-1}g^{-1}$. We define this as the stellar core where hydrogen is burnt. The central 45 $M_\odot$ of the star are fully convective whereas energy is transported by radiation in the outer stellar regions. The lack of outer

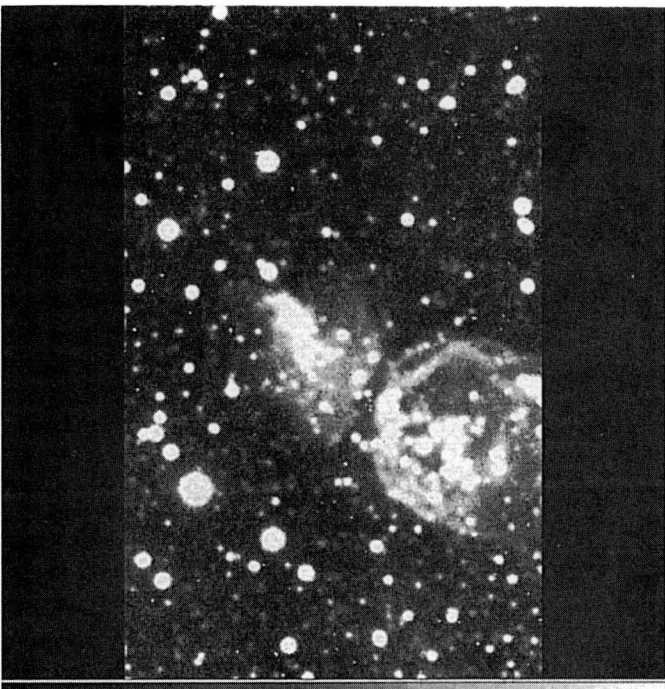

Figure 2. *H II regions No. 2 and 4 (Hodge 1977) in the Local Group galaxy NGC6822. CCD frame obtained at the ESO 2.2m telescope using an R filter. North is up and east to the right. Field size is 2 × 3 arcmin corresponding to a linear size of 240 × 360 pc. Hodge 4 (off-center) has a prominent ring-like structure due the interaction of stellar winds with the surrounding interstellar gas.*

convection zones in hot massive stars may be responsible for the absence of coronae in these stars. Deep in the stellar interior the radiation field is *nearly* completely isotropic. It is not quite isotropic, though because we *see* the star, i.e. $dT/dr \neq 0$. Further out, temperature and density become lower, and the chance for a photon to escape increases. At $\tau_R = 2/3$ the photon has a 50% probability of escaping after the emission process. This defines the stellar photosphere. At the photosphere and above, the radiation field strongly deviates from isotropy. Absorption of photons therefore transfers momentum and a net outward force is exerted. The force is negligible in the continuum but very strong in a large number of UV absorption lines. An outflow of matter is initiated which develops supersonic velocities within 1.5 stellar radii. The net force on the outflowing matter becomes very small beyond ∼10 stellar radii, and a constant outflow velocity is maintained. Eventually the stellar wind is decelerated due to the interaction with the ambient interstellar medium, and a stellar-wind blown bubble may be formed. The interstellar medium around the star is ionized by stellar photons emitted in the Lyman continuum.

This model for a massive star has intentionally been constructed in a 'unified' way, i.e. emphasizing the continuous transition from the stellar interior to the interstellar medium. In practice, models for massive stars are not obtained in such a way. Separate models exist for the stellar interior (*e.g.*, Chiosi and Maeder 1986), the photosphere (*e.g.*, Hubeny 1988), the wind (*e.g.*, Pauldrach *et al.* 1990), and the surrounding interstellar medium (e.g., Clegg et al. 1987). Interactions between these individual stellar and circumstellar layers are only taken into account as second-order corrections.

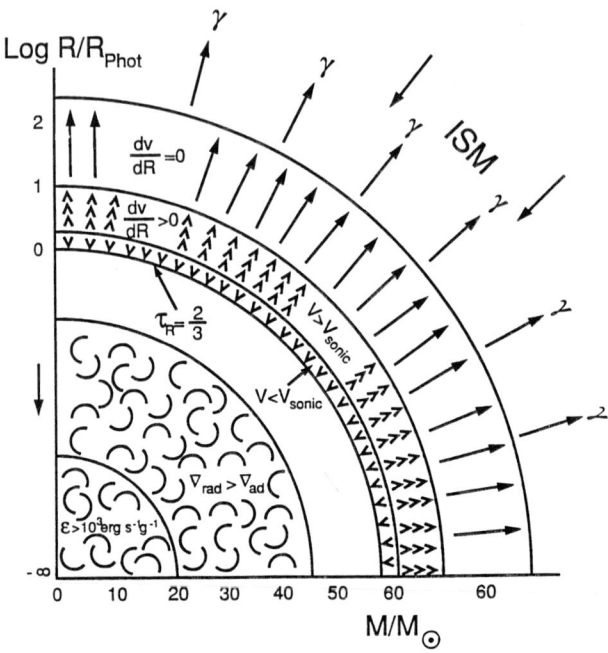

Figure 3. *Structure of a 60 $M_\odot$ star on the main sequence. Notice the different quantities defining abscissa and ordinate. Both axes are on a non-linear scale. The transition from the stellar interior to the wind and to the interstellar medium is emphasized. Understanding the structure and evolution of a massive star requires inclusion of all aspects illustrated in this figure.*

Models for the stellar structure and evolution assume a gray atmosphere as an upper boundary condition. Significant errors are introduced by this assumption, e.g., for the computed position of Wolf-Rayet stars in the HRD (Langer 1989). This point is of relevance for models of violent star formation processes (Warmers; Terlevich and Melnick 1985).

Stellar winds are incorporated in evolutionary models—but only as an empirical correction to decrease the stellar mass with time. However, evolutionary models predict different stellar parameters if mass loss is included, which in turn influences the stellar wind models so that an iterative process is required to achieve self-consistency (Leitherer and Langer 1990).

With respect to massive stars in extragalactic starburst regions, the dependence of the stellar properties on chemical composition is crucial. Recent results imply that stellar evolution is significantly different in environments with non-solar chemical composition—mainly due to different mass-loss properties of the stars (Maeder 1990a, and this meeting). Generally, massive stars evolve from the main sequence towards cooler temperatures at nearly constant luminosity. Depending on their initial mass they either become red supergiants and may again evolve back towards hotter temperatures and form Wolf-Rayet stars, or they never enter the red supergiant stage but become Wolf-Rayet stars directly after passing through the blue supergiant phase.

Quantitative predictions of that scenario are still uncertain. However, such models are a fundamental ingredient for population synthesis models of starbursts (*e.g.*, Arnault, Kunth, and Schild 1989). Currently, considerable progress is being achieved in constraining models for the stellar evolution and mass loss. Observational studies of the massive stellar content of several galaxies of the Local Group allow one to recognize similarities and differences in the formation and evolution of stars in different environments (Humphreys 1987). Such observations of *stellar populations* may be used to test predictions of models for *stellar atmospheres* via population synthesis models (Maeder 1990b).

Massive stars contribute to the chemical enrichment of galaxies by returning processed material to the interstellar medium via mass loss (by stellar winds and supernova explosions). The mass fractions of various elements released to the interstellar matter as a function of initial stellar mass are shown in Figure 4 (from Chiosi and Maeder 1986). After weighting the stellar yields with an appropriate initial mass function the results can be adopted as an input for the chemical evolution of galaxies (Matteucci 1989). In principle, models for the evolution of galaxies can then be used to predict the variation of chemical composition with time, which in turn is input for stellar evolutionary models.

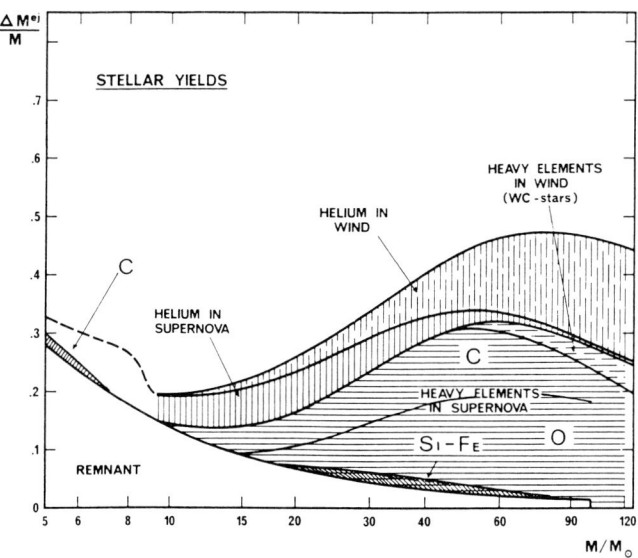

Figure 4. *Mass fraction of helium and heavy elements ejected as a function of the initial stellar mass (from Chiosi and Maeder 1986).*

4. THE STELLAR CENSUS OF MASSIVE STARS

Our knowledge of the population of massive stars as a whole rests on the stellar census within $2-3$ kpc of the solar neighborhood. Such a volume is large enough to contain a significant number of very massive stars, which are very rare. Conversely, stellar surveys are complete down to about 10 $M_\odot$ in the solar neighborhood. This point is further illustrated in Figure 5. The upper section of this figure gives the apparent magnitude of an average 10 $M_\odot$ star on the main sequence as a function of distance from the sun. Two cases are shown: no interstellar extinction and a uniform extinction of 1 magnitude. Spectral classification surveys become incomplete at about 11th magnitude so that the upper limit for the completeness of the massive star content of the solar neighborhood is at ~ 4 kpc.

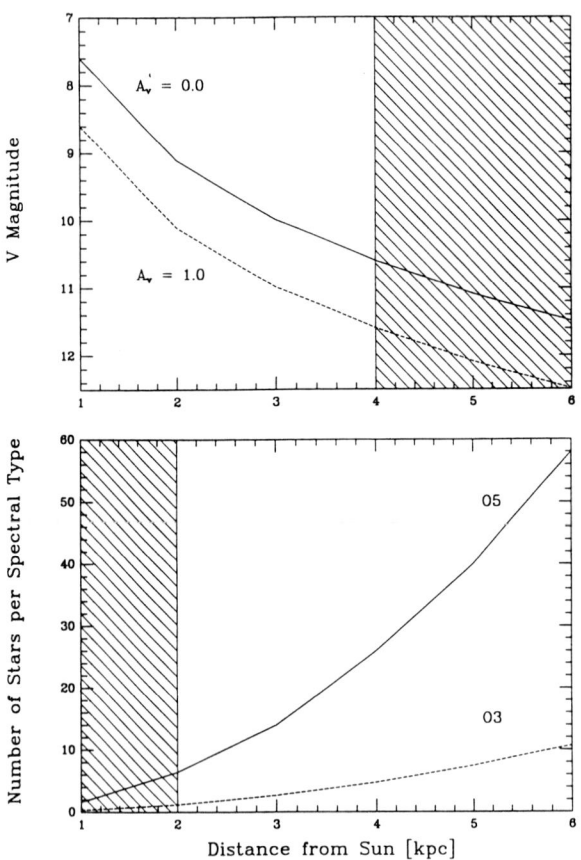

Figure 5. *Upper part: apparent magnitude of a 10 $M_\odot$ star on the main sequence as a function of distance. Lower part: cumulative number of O3- and O5-stars versus distance. The shaded areas indicate distances where the stellar census is affected by incompleteness and small number statistics, respectively.*

The lower section of Figure 5 addresses the statistics of the most massive stars. Using the initial mass function and star formation rate derived by Garmany, Conti, and Chiosi (1982), the average (cumulative) number of O3- and O5-stars as a function of distance from the sun has been calculated. The steepness of the initial mass function makes it very unlikely to detect a significant number of very massive stars within 2 kpc of the sun.

Combining both parts of Figure 5 it is obvious that a volume with a radius of 3 kpc around the sun contains an satisfactory sample of massive stars. It is large enough to include a statistically significant number of stars at the high-mass end, and on the other hand, stellar surveys are still complete for less massive stars.

Various authors studied the initial mass function of massive stars in the solar neighborhood (*e.g.*, Garmany, Conti, and Chiosi 1982, Humphreys and McElroy 1984, Van

Buren 1985, Blaha and Humphreys 1989). The mean of their results indicates a stellar mass spectrum $\psi \sim M^{-2.5}$, which is slightly steeper than Salpeter's (1955) IMF slope of 2.35. However, the scatter of the individual results around the mean value is considerable—although the massive stellar population in the solar neighborhood is the best-studied sample of all galactic and extragalactic massive stars. In fact, it is as large as the IMF variations found between different starburst galaxies, which have been interpreted as real differences in the star-formation process (see Scalo 1986).

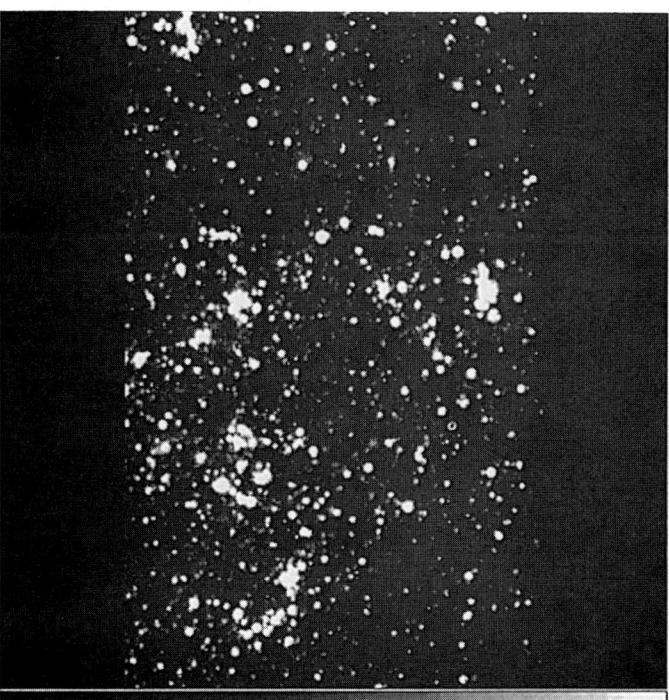

Figure 6. *Region of strong star formation activity northeast of the central bar of the irregular galaxy IC1613. Technical data are the same as in Figure 2, except for the passband, which is V, and the linear size, which is 400×600 pc.*

The surface area of the solar neighborhood with a reliable stellar census projected on the galactic disk covers less than 2% of our Galaxy. Our knowledge of the massive stellar content of our Galaxy is based on an extrapolation of what is known from the solar neighborhood. Indirect means to probe the population of massive stars in the entire Galaxy, such as the statistics of giant H II regions (Smith, Biermann, and Mezger 1977) and ultracompact H II regions (Wood and Churchwell 1989) can complement results derived in the solar neighborhood. However, relatively uncertain assumptions on the absorption properties of gas and dust with respect to the incident stellar radiation have to be made for a quantitative assessment of the population of massive stars.

Recent progress in observational techniques, such as two-dimensional CCD photometry, opened up the possibility to obtain a census of massive stars in nearby galaxies. Such studies are underway (*e.g.,* Massey *et al.* 1989) and will produce an unbiased view of the massive stellar content of an entire galaxy. However, even at the distance of the closest irregular galaxies like the Magellanic Clouds, NGC6822, or IC1613 crowding becomes a problem. Figure 6 is an example of a region with ongoing star formation activity in the irregular galaxy IC1613. At a distance of ~0.7 *Mpc* a typical OB association extends over only a few arcseconds. There is considerable loss of spatial resolution as compared with galactic OB associations so that studies of *individual* stars are degraded. On the other hand, extragalactic star formation regions provide the opportunity to study the formation of massive stars *as a whole* and their interaction with the interstellar medium.

5. SPECTRAL SIGNATURES OF MASSIVE STARS

Massive stars in starburst regions beyond the Local Group of galaxies can no longer be detected individually. Their presence can be inferred from the integrated energy distribution of large stellar and interstellar complexes. Although thermal and non-thermal radio radiation, nebular Hα emission, and thermal emission from dust are reliable indicators of massive stars, the most powerful tracers are direct spectral signatures of the stars themselves. Spectral features unambiguously proof the presence of individual stellar types. *They can be calculated by theoretical stellar models and calibrated locally via direct observations.* In contrast, secondary massive-star tracers, like thermal radio radiation, need additional assumptions and modeling of the interstellar medium.

Spectral features of massive stars on the main sequence are rather weak as compared with those of stars in later evolutionary stages. Moreover, spectral lines which are fairly strong, such as hydrogen and helium lines, are blended with nebular emission from interstellar gas. This makes the direct detection of massive main-sequence stars in distant starburst regions rather difficult (Melnick, Moles, and Terlevich 1985). The situation improves if ultraviolet data are available. Hot, massive stars—when they are on the main sequence and the more so when they have already evolved into the blue-supergiant phase—exhibit characteristic ultraviolet lines such as Si IV λ1400 and C IV λ1550 which are formed in stellar winds. These lines have been observed in starburst galaxies (*e.g.*, NGC7714, cf. Figure 7) and interpreted as direct proof for the presence of massive stars (Weedman *et al.* 1981). However, detailed modeling of the wind properties of massive, hot stars is required before these lines can be used to synthesize the underlying stellar population quantitatively (Leitherer and Lamers 1990).

Given a typical IMF, the observed flux even at $\sim$ 1500 Å contains a significant contribution from lower-mass stars below 10 $M_\odot$ (O'Connell 1990) so that spectral-line features from very massive stars are difficult to detect in the composite energy distribution. Starburst regions contain large amounts of dust so that the ultraviolet spectral region suffers from considerable interstellar extinction, and massive stars are selectively obscured. As a consequence, ultraviolet observations of only a relatively small number of starburst galaxies with spectral features from massive stars on or close to the main sequence exist (*e.g.*, Joseph, Wright, and Prestwich 1986).

Massive stars with masses around $\sim$ 40 $M_\odot$ evolve into red supergiants about $4\,10^6$ *yr* after their formation (Maeder 1990a). The most prominent absorption features in the near-IR spectrum of red supergiants are the Ca II triplet lines at $\lambda\lambda 8500, 8540, 8660$ Å. Terlevich *et al.* (1990) detected the Ca II triplet in an H II region of the starburst galaxy NGC3310 (lower part of Figure 8). For comparison, the upper part of Figure 8 is a reproduction of the spectrum of the nucleus of NGC3310, where the Ca II triplet is due to an old population of lower-mass stars in the galactic bulge. The occurrence of these lines in a giant H II region can be understood if a population of massive stars is observed about 4 million years after the onset of the burst of star formation.

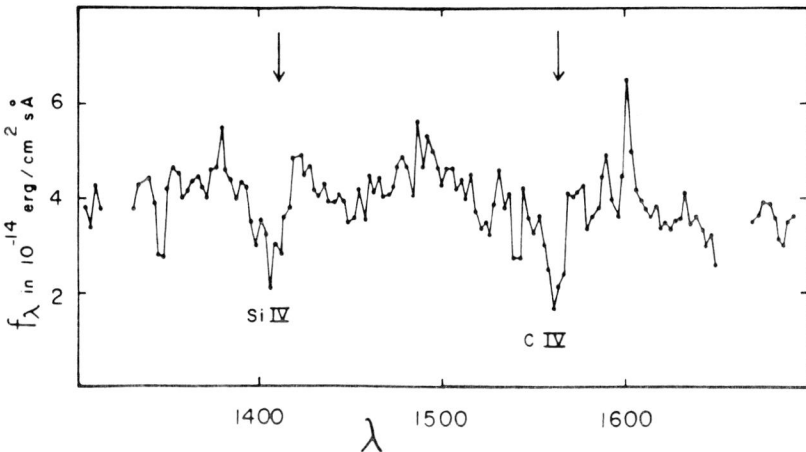

Figure 7. *Ultraviolet spectrum of the starburst galaxy NGC7714 obtained with the IUE satellite (taken from Weedman et al. 1981). Absorption lines due to Si IV and C IV indicate the presence of hot, massive stars.*

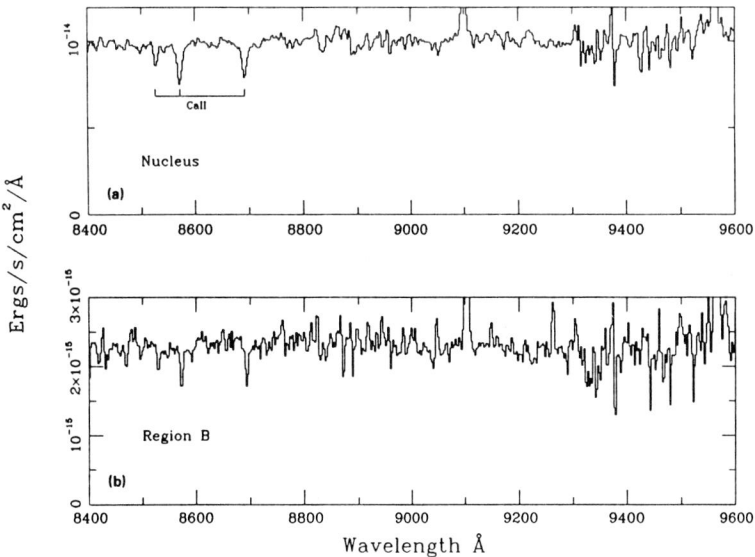

Figure 8. *Near-infrared spectrum of a giant H II region in NGC3310 obtained by Terlevich et al. (1990). Absorption lines due to the Ca II triplet hint at an evolved population of massive stars in their red-supergiant phase.*

Broad He II $\lambda 4686$ emission has been detected in some galaxies with strong starburst activity (*e.g.,* Kunth and Sargent 1981, Armus, Heckman, and Miley 1988). Figure 9 gives an example of such observations. Broad $\lambda 4686$ is clearly present in the Blue Compact Dwarf galaxy Tol3. In contrast, IIZw40 and Pox4 show *narrow*, unresolved He II $\lambda 4686$ originating from ionized interstellar material. Broad He II $\lambda 4686$ is a characteristic emission feature in the spectra of galactic WN stars. The strength of the line can be used to infer large numbers of W-R stars in these galaxies. Since W-R stars are believed to be the low-mass descendants of previously massive stars (Lamers *et al.*

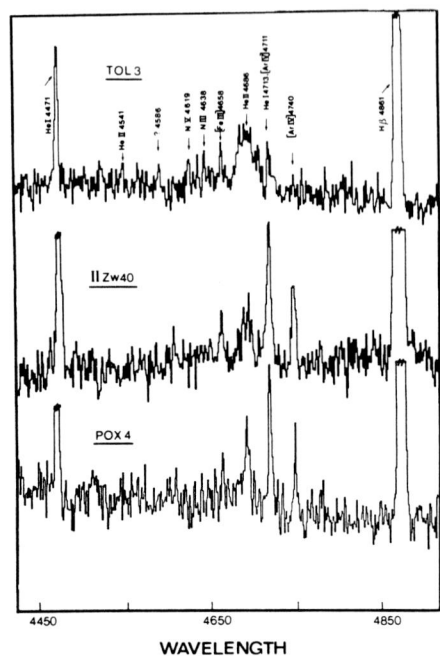

Figure 9. *Emission due to He II λ4686 in the Blue Compact Dwarf galaxies Tol3, IIZw40, and Pox4 (Kunth and Schild 1981). Large numbers of W-R stars have been inferred from the presence of this line.*

1990), the λ4686 line is expected if massive stars are still being formed or a burst of star formation has ended recently. Population synthesis models can predict the star formation rate and the initial mass function of the starbursts (Arnault, Kunth, and Schild 1989).

In most cases, spectral signatures of massive stars detected in distant extragalactic starbursts arise from *evolved* stars, for which theoretical stellar models are rather uncertain—not from core-hydrogen burning stars still on the main sequence. This is due to the scarcity and weakness of absorption lines in massive main-sequence stars, which are difficult to detect in the integrated spectrum with contributions from less massive stars and H II regions. Certain features (like He II λ4686, Si IV λ1400) become more prominent as the stars evolve from the main sequence because of increasing mass-loss rates and wind densities. These lines are not formed in *hydrostatic* photospheres but in the *hydrodynamic* outer atmospheres. Detailed modeling of the radiative transfer in the stellar wind and subsequent implementation in evolutionary models is required before such features can be interpreted quantitatively in distant starbursts.

An example for the uncertainties of post-main-sequence evolution is given in Figure 10. The figure shows the location of two well-known LMC stars in the HRD: Sk −69°202 is the progenitor of Supernova 1987A (Arnett et al. 1989), and R84 is a late WN star recently analyzed by Schmutz et al. (1990). Unlike R84, Sk −69°202 never evolved into the Wolf-Rayet phase but exploded as a supernova. This indicates that a critical mass exists between 20 $M_\odot$ and 30 $M_\odot$ below which no Wolf-Rayet stars are formed. Population synthesis models reproducing the occurrence of W-R stars in starbursts are rather sensitive on the exact limiting mass for the Wolf-Rayet channel. However, there is general agreement that the presence of W-R stars indicates that stars with masses above ∼25 $M_\odot$ must originally have been formed.

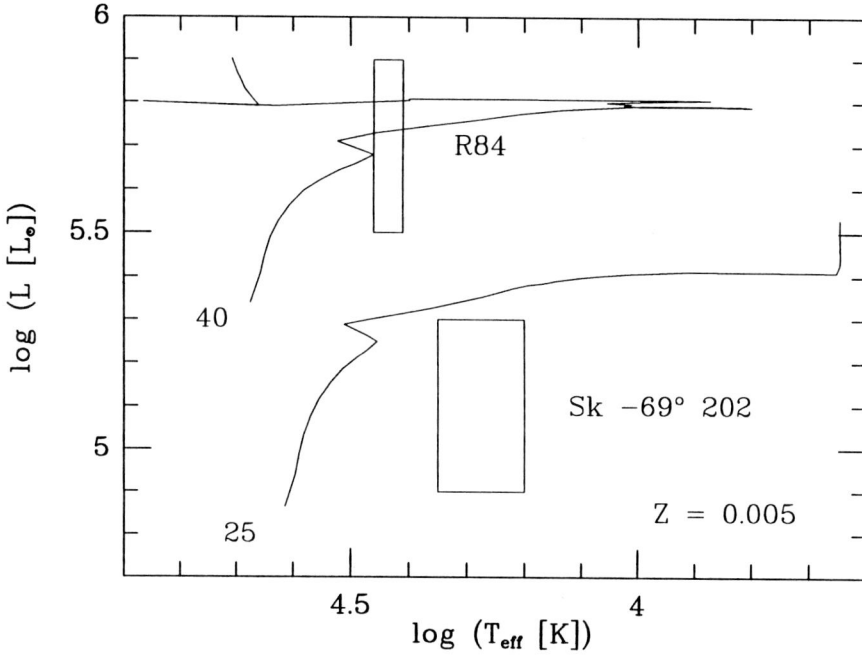

Figure 10. *Location of the LMC stars R84 and Sk −69°202 in the HRD. Evolutionary models for 25 $M_\odot$ and 40 $M_\odot$ stars are from Maeder (1990a). These two stars are examples for two vastly different late stages of stellar evolution despite similar initial conditions.*

6. THE FORMATION OF MASSIVE STARS IN STARBURSTS

Current ideas for the formation and evolution of massive stars are largely based on information obtained *locally*. Note that *locally* may mean something different to different groups of astronomers. The 'optical' astronomer who is limited by the number of photons counted in a spectrograph or photometer may have a more restricted definition of 'local' than do 'infrared' or 'radio' astronomers whose restriction is mainly the scarcity of massive star-formation regions in the immediate vicinity of the sun. In any case, keeping in mind the *caveat* expressed in the previous sections, a fairly detailed picture has emerged for local regions of star formation (see Myers 1988 for a recent summary).

May we simply apply what has been found in our own Galaxy to more distant starbursts? Table 2 (based on values published by Kennicutt 1984) serves as a reminder that regions of star formation in our Galaxy are quite small in comparison with some more distant starbursts. I am implicitly assuming that sites of star formation like the Orion nebula are 'starbursts'. A starburst is simply an event in space and time where temporarily the star-formation rate is significantly higher.

Classic H II regions like M42 or M8 have no *very* massive stars at all. Even regions like NGC3372, which contains most of the very massive galactic O3 stars known (Walborn and Hesser 1982) are dwarfed by extragalactic starburst regions like 30 Doradus. NGC3603 and W49 are among the most massive galactic starburst regions. Due to their distance and location in the galactic plane they suffer from high interstellar extinction so that their stellar content is only vaguely known. NGC3603 and W49 are similar in

Table 2. *Properties of several galactic and extragalactic H II regions (data from Kennicutt 1984). N(O5V) is the equivalent number of ionizing stars to account for the Hα flux if a Salpeter-type IMF with a mass range of 10 $M_\odot < M < 100\ M_\odot$ is assumed.*

Object	Galaxy	Diameter [pc]	$M(H^+)$ [$M_\odot$]	$N(O5V)$
M42	Galaxy	5	50	< 1
M8	Galaxy	25	900	< 1
NGC2244	Galaxy	50	2000	1
NGC3372	Galaxy	200	44000	9
NGC3603	Galaxy	100	39000	20
W49	Galaxy	150	45000	27
30 Dor	LMC	370	600000	230
NGC604	M33	400	700000	65
A	M81	450	400000	25
A	NGC2403	600	550000	210
A	M82	450	≥250000	600
NGC5461	M101	1000	6000000	1000

their properties to the most massive giant H II regions in M31, M81, and other Sb and Sbc galaxies (Kennicutt 1984) but are still small when compared to 30 Doradus. This may indicate that the difference in the H II region population of Local Group galaxies is a morphological-type effect (Kennicutt, Edgar, and Hodge 1989).

Although there is some overlap in size and massive-star content between galactic and extragalactic H II regions, distinct differences with respect to their density structure (Kennicutt 1984) and kinematics (Melnick et al. 1987) have been found. As a consequence, conditions for star formation may differ from regions of star formation in the solar neighborhood—a point well worth remembering when applying theoretical models to extragalactic starbursts.

Supernova 1987A in the Large Magellanic Cloud demonstrated impressively the shortcomings of theoretical models of massive stars (Arnett et al. 1989). Canonical models relating early and late stages of massive-star evolution did not predict the progenitor star Sk −69°202 to explode as a supernova. The failure may be due to incompleteness in the models and/or due to different properties of extragalactic massive stars. The importance of the chemical composition of the stellar birthplace on the evolution of massive stars has only recently been addressed in a quantitative way (*e.g.*, Maeder 1990a, b). Observed differences in the massive stellar content in distant starbursts have often been interpreted in terms of variations of the initial mass function (*e.g.*, Sekiguchi and Anderson 1987), i.e. indicating differences in the conditions of *star formation*. However, differences in *stellar evolution* may also be able to account for the observed properties of distant starbursts.

The number of massive stars formed in several well-studied starburst galaxies exceeds those found in even the most luminous giant H II regions of Table 2. Weedman et al. (1981) estimate a total ∼$10^8\ M_\odot$ which is locked into stars after the burst of star formation in NGC7714. Although a direct comparison with the corresponding masses in giant H II regions is not straightforward because of possible differences in the stellar mass spectrum, the value derived for NGC7714 is roughly 3–4 orders of magnitude higher than what is observed in any H II region of Table 2. The total mass of stars formed in the exceedingly luminous starburst galaxies Arp 220 and NGC 6240 is even higher (Rieke et al. 1985). Are these events simply scaled-up versions of nearby starburst regions or do we observe a phenomenon which has an entirely different triggering mechanism, and which has produced a population of massive stars with different properties?

Finally, the question arises to what extent the starburst phenomenon is responsible for the properties of *all* morphologically peculiar types of galaxies. Terlevich and Melnick (1985) suggested that large nuclear starbursts followed by supernova explosions and supernova remnants can account for the observed properties of ordinary active galactic nuclei. This suggestion is under debate (see Heckman, this meeting). Norman and Scoville (1988) and Scoville and Norman (1988) developed a hybrid model for AGN's invoking a nuclear, non-thermal source *and* a burst of star formation in the immediate vicinity of the nucleus. Although large uncertainties still exist—both on the observational and the theoretical side—these models demonstrate the importance of massive stars for our understanding of even the most peculiar galaxies.

REFERENCES

Abbott, D. C. 1982, *Ap. J.*, **263**, 723.
Armus, L., Heckman, T. M., Miley, G. K. 1988, *Ap. J.*, **326**, L45.
Arnault, P., Kunth, D., and Schild, H. 1989, *Astr. Ap.*, **224**, 73.
Arnett, W. D., Bahcall J. N., Kirshner, R. P., and Woosley, S. E. 1989, *Ann. Rev. Astr. Ap.,* **27**, 629.
Blaha, C., and Humphreys, R. M. 1989, *A. J.*, **98**, 1598.
Cherepashchuk, A. M. 1990, in *IAU Symposium 143, Wolf-Rayet Stars and Interrelations with Other Massive Stars in Galaxies*, ed. K. van der Hucht and B. Hidayat (Dordrecht: Kluwer), in press.
Chiosi, C., and Maeder, A. 1986, *Ann. Rev. Astr. Ap., ***24**, 329.
Clegg, R. E. S., Harrington, J. P., Barlow, M. J., and Walsh, J. R. 1987, *Ap. J.*, **314**, 551.
Courtès, G., Petit, H., Sivan, J.-P., Dodonov, S., and Petit, M. 1987, *Astr. Ap.*, **174**, 28.
de Jong, T., Clegg, P. E., Soifer, B. T., Rowan-Robinson, M., Habing, H. J., Houck, J. R., Aumann, H. H., and Raimond, E. 1984, *Ap. J.*, **278**, L67.
Garmany, C. D., Conti, P. S., and Chiosi, C. 1982, *Ap. J.*, **263**, 277.
Hodge, P. W. 1977, *Ap. J. Suppl.*, **33**, 69.
Hubeny, I. 1988, *Comp. Physics Comm.*, **52**, 103.
Humphreys, R. M. 1983, *Ap. J.*, **269**, 335.
_____. 1987, *Pub. A. S. P.*, **99**, 5.
Humphreys, R. M., and McElroy, D. B. 1984, *Ap. J.*, **284**, 565.
Joseph, R. D., Wright, G. D., and Prestwich, A. H. 1986, in *New Insights in Astrophysics*, ed. E. J. Rolfe (ESA SP-263), p. 597.
Kennicutt, R. C. 1984, *Ap. J.*, **287**, 116.
Kennicutt, R. C., Edgar, P. K., and Hodge, P. W. 1989, *Ap. J.*, **337**, 761.
Klein, U., and Gräve, R. 1986, *Astr. Ap.*, **161**, 155.
Kudritzki, R. P. 1988, in *18th Advanced Course, Swiss Society of Astronomy and Astrophysics*, ed. Y. Chmielewski and T. Lanz (Sauverny-Versoix: Geneva Observatory), p.3.
Kunth, D., and Sargent, W. L. W. 1981, *Astr. Ap.*, **101**, L5.

Lamers, H. J. G. L. M. 1988, in *Mass Outflows from Stars and Galactic Nuclei*, ed. L. Bianchi and R. Gilmozzi (Dordrecht: Reidel), p. 39.
Lamers, H. J. G. L. M., Maeder, A., Schmutz, W., and Cassinelli, J. P. 1990, *Ap. J.*, in press.
Langer, N. 1989, *Astr. Ap.*, **210**, 93.
Leitherer, C. 1990, *Ap. J. Suppl.*, **73**, 1.
Leitherer, C., and Lamers, H. J. G. L. M. 1990, *Ap. J.*, submitted.
Leitherer, C., and Langer, N. 1990, in *IAU Symposium 148, The Magellanic Clouds*, ed. R. F. Haynes, and D. K. Milne (Dordrecht: Kluwer), in press.
Lequeux. J. 1989, in *Evolution of Galaxies—Astronomical Observations*, ed. I. Appenzeller, H. J. Habing, and P. Léna (Berlin: Springer), p. 147.
Maeder, A. 1987, *Astr. Ap.*, **173**, 247.
_____. 1990a, *Astr. Ap. Suppl.*, **84**, 139.
_____. 1990b, *Astr. Ap.*, in press.
Massey, P. 1985, *Pub. A. S. P.*, **97**, 5.
Massey, P., Garmany, C. D., Silkey, M., and Degioia-Eastwood, K. 1989, *A. J.*, **97**, 107.
Matteucci, F. 1989, in *Evolutionary Phenomena in Galaxies*, ed. J. Beckman and B. E. J. Pagel (Cambridge: Cambridge University Press), p. 297.
Melnick, J., Moles, M., and Terlevich, R. 1985, *Astr. Ap.*, **149**, L24.
Melnick, J., Moles, M., Terlevich, R., and Garcia-Pelayo, J.-M. 1987, *M. N. R. A. S.*, **226**, 849.
Myers, P. C. 1988, in *Galactic and Extragalactic Star Formation*, ed. R. E. Pudritz and M. Fich (Dordrecht: Kluwer), p. 331.
Norman, C., and Scoville, N. Z. 1988, *Ap. J.*, **332**, 124.
O'Connell, R. W. 1990, in *Windows on Galaxies*, ed. G. Fabbiano, J. S. Gallagher, and A. Renzini (Dordrecht: Kluwer), p. 39.
Panagia, N. 1973, *A. J.*, **78**, 929.
Pauldrach, A. W. A., Puls, J., Gabler, R., and Gabler, A. 1990, in *Properties of Hot Luminous Stars*, ed. C. D. Garmany (Provo: Brigham Young University), p. 171.
Rieke, G. H., Cutri, R. M., Black, J. H., Kailey, W. F., McAlaray, W. F., Lebofsky, M. J., Elston, R. 1985, *Ap. J.*, **290**, 116.
Salpeter, E. E. 1955, *Ap. J.*, **121**, 161.
Sandage, A., and Tammann, G. A. 1974, *Ap. J.*, **191**, 603.
Sanders, D. B., Soifer, B. T., Neugebauer, G., Scoville, N. Z., Madore, B. F., Danielson, G. E., Elias, J. H., Matthews, K., Persson, C. J., and Persson, S. E. 1987, in *Star Formation in Galaxies*, ed. C. J. L. Persson (NASA CP-2466), p. 411.
Scalo, J. M. 1986, *Fund Cosmic Phys.*, **11**, 1.
Scoville, N. Z., and Good, J. C. 1987, in *Star Formation in Galaxies*, ed. C. J. L. Persson (NASA CP-2466), p. 3.
Scoville, N. Z., and Norman, C. 1988, *Ap. J.*, **332**, 163.
Schmutz, W., Leitherer, C., Hubeny, I., Vogel, M., Hamann, W.-R., and Wessolowski, U. 1990, *Ap. J.*, in press.
Sekiguchi, K., and Anderson, K. S. 1987, *A. J.*, **94**, 644.
Smith, L. F., Biermann, P., and Mezger, P. G. 1977, *Astr. Ap.*, **66**, 65.
Terlevich, E., Díaz, A. I., Pastoriza, M. G., Terlevich, R., and Dottori, H. 1990, *M. N. R. A. S.*, **242**, 48p.
Terlevich, R., and Melnick, J. 1985, *M. N. R. A. S.*, **213**, 841.
Van Buren, D. 1985, *Ap. J.*, **294**, 567.

Walborn, N. R., and Hesser, J. E. 1982, *Ap. J.*, **252**, 156.

Weaver, R., McCray, R., Castor, J., Shapiro, P., and Moore, R. 1977, *Ap. J.*, **218**, 377.

Weedman, D. A., Feldman, F. R., Balzano, V. A., Ramsey, L. W., Sramek, R. A., and Wu, C. C. 1981, *Ap. J.*, **248**, 105.

Wood, D. O. S., and Churchwell, E. 1989, *Ap. J.*, **340**, 265.

DISCUSSION

R. Joseph: I think there are some direct tracers of massive stars that one can get from infrared diagnostic tools of various sorts and I will be talking about that on Thursday.

Leitherer: You are talking of spectral lines or on infrared coutinua?

R. Joseph: Both of them I will be talking about. Some are indirect and some are more direct tracers.

Leitherer: There have been recent developments which are really very promising and we can hope that we will get some new tracers opening an entire new field of studying these massive stars.

A. Moffat: You mentioned "direct" observations of masses based on luminosities. Actually, this is still a theoretical approach depending on the M-L relation. If you look at truly direct observations, namely binaries with massive stars in them, there is no system where you find a star above 60 solar masses.

Leitherer: What I was referring to were a few stars which have been studied in detail by spectroscopic techniques. The masses which have been derived this way were on the order of 100 solar masses but, of course, admittedly with relatively large error bars. So probably it would also be consistent with 50–60 solar masses.

A. Moffat: I would like to make reference to our poster contribution in which we find evidence for the most massive star in a binary system from eclipses. The upper mass limit for this star is about 60 solar masses. So if there is a massive star in the galaxy that's one of the most massive ones.

P. Conti: I haven't seen the posters so I can't answer that but I have always thought of Plaskett's star which has a minimum of 60 solar masses. Now I don't want to push that up. But there are stars that are hotter and brighter, so I think that 100 is not out of the question. Are you suggesting that 100 is out of the question?

R. Kudritzki: I will show in my talk results of spectroscopic analyses of hot stars having zero-age main-sequence masses considerably larger than 100 solar masses. I will present new methods to determine the masses very precisely. We get these very high masses. Why they are not found in binaries might be another problem. You cannot expect to find each type of a star in a binary system because there could be several other reasons why the most massive stars are single.

A. Moffat: Yes, I agree that there are selection effects. Let me answer to the issue of Plaskett's star. There is now a more recent paper that says that this is a single-line binary and the previous mass estimates are not reliable (Stickland et al. 1988).

J. Bland Hawthorne: Supernova rates have been claimed, at least in starburst galaxies, from the detection of compact, non-thermal radio sources. But the anticipated new sources have not been observed over a ten year baseline, as far as I am aware. Please review the best current estimators, both observational and theoretical, for galactic supernova rates.

Leitherer: Let me first give you my opinion on theoretical supernova rates. In principle, of course, you can make very fair predictions from evolutionary models on supernova rates and actually, in principle, supernova rates would be a very good indicator, e.g., for star-formation rates and for the initial mass function of massive stars. Unfortunately, such theoretical supernova rates are very sensitive on the assumption of which stars actually explode as a supernova. The lesson which supernova 1987A taught us should be kept in mind. If you applied stellar evolutionary models which do not predict a supernova explosion for star in the evolutionary state of $Sk-69°202$ to distant starburst galaxies, you will obtain completely wrong answers. So that is the first part of my answer: on the theoretical side, in principle, you are able to make very precise predictions but on the other hand, the systematic uncertainties are so large that using the supernova rate to draw *quantitative* conclusions for the massive-star population is probably not advisable. Of course we know that if we detect non-thermal radio radiation, and if we can relate that to the presence of a large number of supernova explosions which must have occurred, it is usually a *qualitatively* safe indicator for the presence of a massive burst of star formation. However, using these numbers to infer actual quantities like the IMF or whatever, that's probably rather uncertain.

As for *observed* supernova rates I would feel more comfortable if I could pass the second part of the answer to somebody who is actually working on this field—deriving supernova rates in galaxies. If somebody is here who would like to give the answer I would gladly pass the question on to him or her.

D. Weedman: I am not going to answer your questions but I do want to reiterate that the primary radio indicator of a starburst region is a very, very non-thermal spectrum in the radio. We see very little indication of the flat thermal spectrum that you would expect and the non-thermal radial luminosity is much greater than we would predict on the basis the known non-thermal radial luminosity of known supernova remnants. A former student of mine, Peter Stine, who just did a dissertation about a year ago, tried to find some luminosity indicator amongst starburst galaxies. He found that the radio and infrared properties of starburst galaxies do not correlate with luminosity. They remain the same over a factor of 10^5 in luminosity. When you are modeling starburst regions, and if you are trying to understand the radio non-thermal luminosity, we could find no indicator scaling with luminosity. It's as if the same thing is happening no matter how luminous the starburst is. The same ratio of radio non-thermal to infrared. But more importantly the radio spectral index also appears to stay the same. So this is not a very sensitive indicator of the initial mass function, it is only weakly dependent on the initial mass function. At least from the standpoint of the radio non-thermal you cannot find any evidence that the initial mass function varies from starburst to starburst. And this is over a luminosity range of at least 10^6 because he deliberately picked things with a wide range of luminosities. Pete Stine is not here but I did want to mention that in the radio at least we don't seem to be able to go any further in disentangling the supernova rate.

C. Norman: I would like to raise a couple additional questions. We hear that in starbursts the mass function is truncated at the lower end, that is, there are no low-mass stars forming. And I would like this conference in the next three days to address this question, is this true or not? And the other question that I would also like to raise is if this is true, can we try to understand why do massive stars form from starbursts? We have lots of observations on massive star formation close by and we need to look at the environmental physics of those regions and try to understand why massive stars form there. Then we have to consider the question why they actually form in starburst galaxies. It seems to me, those are the questions I'd like to ask.

S. Lamb: I have one question going back to one of your viewgraphs where you showed some evolutionary tracks for stars having masses of 25 and 40 $M_\odot$. I was curious as to why the 40 solar mass star went back and I wanted to ask you specifically what the treatment of mass loss and the semiconvection is in those two models? Those are the two things which seem to have the most effect upon that kind of path and I was curious if you knew?

Leitherer: The evolutionary tracks you see here are from André Maeder's paper on evolutionary tracks for various metallicities, and these are the tracks for the LMC. There is one fundamental property of the 25 $M_\odot$ track: unlike the 40 $M_\odot$ track it does not go back from the red point of the HRD to the blue. I think that it's a well known property of these tracks. They fail to reproduce certain aspects of stellar evolution (cf. SN1987A) unless other parameters are varied. You are completely right, you can actually play around with parameters like mass-loss rates, overshooting, convection to actually make the tracks go from the red supergiant phase back to the blue and vice-versa but the point, I think, is that nobody has yet succeeded in producing a self-consistent set of tracks which actually reproduces both the high-mass and the low-mass end of stellar evolution. The behavior of Supernova 1987A is certainly one of our cornerstones for understanding stellar evolution. But it is very clear those evolutionary tracks are highly sensitive to the adopted mass-loss rate because they depend on how much mass you can peel off from the core and this then determines if and when a star will go back to the blue side of the Hertzsprung-Russell diagram.

SPECTRA OF MASSIVE BLUE STARS

Peter S. Conti
Joint Institute for Laboratory Astrophysics
University of Colorado and
National Institute of Standards and Technology
Boulder, Colorado 80309
USA

Abstract. In this paper I shall talk primarily about the SIGNATURES of hot, luminous, and massive stars, not only *direct* spectroscopy but also *indirect* means for discovering and studying them. I will consider the observations of radiation from various wavelengths and what may be inferred from stars in our Galaxy, in nearby systems where individual stars have been or potentially may still be identified, and in even more distant galaxies where only integrated properties of such objects may be observed. I will point out the opportunities that await us at wavelengths other than optical and the openings of new windows on understanding massive-star evolution in galaxies by the Hubble Space Telescope. It is important to keep in mind that these stars are very luminous, particularly in the UV regions, and have total lifetimes of only a few million years. Their presence serves to notify us of star formation in the very recent past.

1. CLASSICAL SPECTROSCOPY

Here I refer to the utilization of the stellar absorption lines in the "blue" part of the optical spectrum. This wavelength region was initially chosen as it was nominally the most sensitive for photographic emulsions; fortunately, it also included some of the Balmer lines which are prevalent in so many stars. With the advent of new electronic detectors, we are no longer limited to this narrow window of stellar spectroscopy, but it still affords us a well-calibrated region to begin our observations. The stellar absorption-line spectrum arises from the photosphere, which for most stars can be taken to be plane parallel; furthermore, there is a close and generally well-understood relationship between the properties of this photosphere and that of the underlying star. In other words, one can derive effective temperature, T_{eff}, and luminosity, L, from observations of the stellar spectrum once the "calibration" problem has been addressed.

Massive blue stars are those of spectral type O, with the addition of the somewhat more evolved B and A supergiants. The lowest mass of stars called "massive" could be that for which the final episode is a SN explosion, say 10 solar masses. A more readily accessible lower limit might be that of the boundary between O and B main-sequence stars, or about 15 solar masses. The latest evolutionary phases are represented by the W-R objects for the *most* massive stars, more than, say, 35 solar masses. K and M supergiants, which are the subject of the following talk, generally correspond to the core He-burning phases of somewhat less massive objects. I shall concentrate my remarks on only the most massive stars, which produce most of the Lyman-continuum radiation in galaxies and whose winds are important contributors to the dynamics of the surrounding interstellar medium.

O stars can be readily identified by the presence of He II absorption lines in their spectra; the feature at 4541 Å and its comparison to a He I line at 4471 Å give the numerical subtypes. A subset of O stars, called Of, have the He II $\lambda 4686$ line in emission, a point to which I shall return later. B supergiants are distinguished from O types by the absence of He II; the prominent lines are those of He I and the Balmer series. The line widths are narrow and several line ratios (*e.g.*, He I $\lambda 4471$ vs. Mg II $\lambda 4481$) lead to the numerical subtypes. In A supergiants, the He I lines are absent, the Balmer series lines are narrow but at maximum strength, and the numerical subtypes are distinguished by the strength of the K line of Ca II compared to the nearby hydrogen features.

A recent paper by Walborn and Fitzpatrick (1990) shows in great detail the spectral sequences of Galactic O and B stars in the blue region of the spectrum. These "linearized" (normalized) data were derived from electronic detectors which will be the standard scheme by which future classifications of individual stars will be done. Spectroscopy is under way by these authors for analogous objects in the Large and Small Magellanic Clouds. The stars in these galaxies, being somewhat metal deficient with respect to the solar vicinity, show differences in spectral appearance and in their classification parameters which are rather subtle for the LMC but pronounced for the SMC.

Wolf-Rayet (W-R) stars, by contrast, evidence a predominant emission-line spectrum in the optical, and in other wavelength regions. This is caused by their strong stellar winds, due, in turn, to their high luminosities and advanced evolutionary state (Abbott and Conti 1987). The relationship between the spectroscopic properties of the wind and the underlying star is not yet all that well understood, but in the last few years there has been great progress in modeling such objects (*e.g.*, Hillier 1989; Schmutz *et al.* 1989). While the presence of emission features has delayed our understanding of the physics of the spectra, such lines appear well above the continuum of the star, thus enabling us to detect and classify them to much fainter magnitudes than stars with only absorption lines.

The W-R classification types are primarily two, the WN and WC. Strong helium and nitrogen lines are found in the former, and helium, carbon, and oxygen ions in the latter (*e.g.*, Smith 1968). A sparsely populated subset of W-R stars with strong oxygen lines in the optical is called type WO (Barlow and Hummer 1982). These types represent objects in which the products of core H-burning (WN) and core He-burning (WC, WO) have been revealed at the stellar surfaces due to previous and continuing mass loss by stellar winds, along with internal mixing of material.

The strongest line in WN stars in the optical is the (4-3) transition of He II at 4686 Å; those of WC stars include C III $\lambda 4650$, along with C III $\lambda 5696$ and C IV $\lambda 5808$. In WO stars, the O VI feature at 3820 Å is prominent. Various numerical subtypes, with corresponding terms WNE (early), WNL (late), WCE, and WCL can be found from the line ratios of various ions in W-R winds.

2. SPECTROPHOTOMETRY

With the advent of electronic detectors has come the ability to acquire observations over wavelengths not readily accessible to photographic emulsions. One also finds an opportunity to obtain the monochromatic energy distribution, and line fluxes. In other wavelength bands, for example the far-UV regions, OB stars show evidence of their own stellar winds from the presence of P-Cygni profiles in several of the resonance lines, even while photospheric absorption features are still found. In W-R stars, the resonance transitions also typically show P-Cygni profiles, while the subordinate UV lines are in emission due to the strong stellar wind.

In several figures to follow, I will show some examples of newly reduced data from the International Ultraviolet Explorer satellite, combined with optical spectrophotometry adapted from Conti and Massey (1989) and from the near-infrared (NIR) regions (Conti et al. 1990). These enable one to get a feeling for the spectral signatures of O and W-R stars over nearly a decade of wavelength. The data have been fluxed and corrected for interstellar extinction using the normal prescriptions for OB stars by Savage and Mathis (1979) and for W-R stars by Vacca and Torres-Dodgen (1990). For hot stars such as these, the continuum fluxes between 1200 Å (the short-wavelength IUE cutoff) and about 1 micron (the long-wavelength CCD limit) change by a factor of about 1000. This is only the "tail" of the total energy emitted by these stars; the maximum of the emergent energy distribution is below the IUE limit for many of the O and W-R stars. As I will discuss later, the radiation from below the Lyman limit manifests itself in other ways.

Figure 1 gives the ultraviolet and optical spectra of a typical early-type Of star (the NIR data are not yet available). The strong absorption plus emission features at 1400 Å and 1550 Å are the P-Cygni profiles of the Si IV and C IV resonance transitions, respectively; N IV at 1718 Å is also seen. Since this is an Of star, emission features due to He II at 1640 Å, 4686 Å, and 6560 Å are found, along with a N III blend at 4640 Å. Careful inspection will reveal the presence of absorption along higher members of the Balmer series and in some other He II transitions. The flux scale used enables one to plot the entire spectrum, but most spectral absorption features are very difficult to see. A typical O star would appear much like this star, with the exception that the He II $\lambda\lambda$1640, 4686, and 6560 lines would be in absorption, the N III lines absent, and the UV resonance lines less prominent.

Figure 2 shows the UV, optical, and NIR spectra of a typical WN7 (WNL) star. Here the overall spectral features are in emission, with the exception of the P-Cygni profile for the C IV resonance line. The leading emission features are at 4686 Å due to He II, and the N III blend at 4640 Å (also found in Of stars). Nearly all the other lines are due to helium and nitrogen ions. Hydrogen is present in this WN7 star as may be seen by comparing the strength of the purely Pickering series at 5411 Å, 4541 Å with the stronger Balmer/Pickering lines at 6560 Å, 4860 Å, 4340 Å. A Paschen line is also found near the He II feature at 10124 Å.

Figure 3 shows the spectrum of a typical WN5 (WNE) star (one without hydrogen). The lines are stronger, and broader, than in the WN7 star of Fig. 2. One may also notice that the line widths appear to increase with wavelength, but this is an artifact of plotting in these units rather than in velocity width. The strongest line in the spectrum is the (4-3) transition of He II, and several transitions of N IV appear quite prominent. It must be obvious that the classification of a W-R star such as this, with its emission-line spectrum, is much easier than for an absorption-line star such as illustrated in

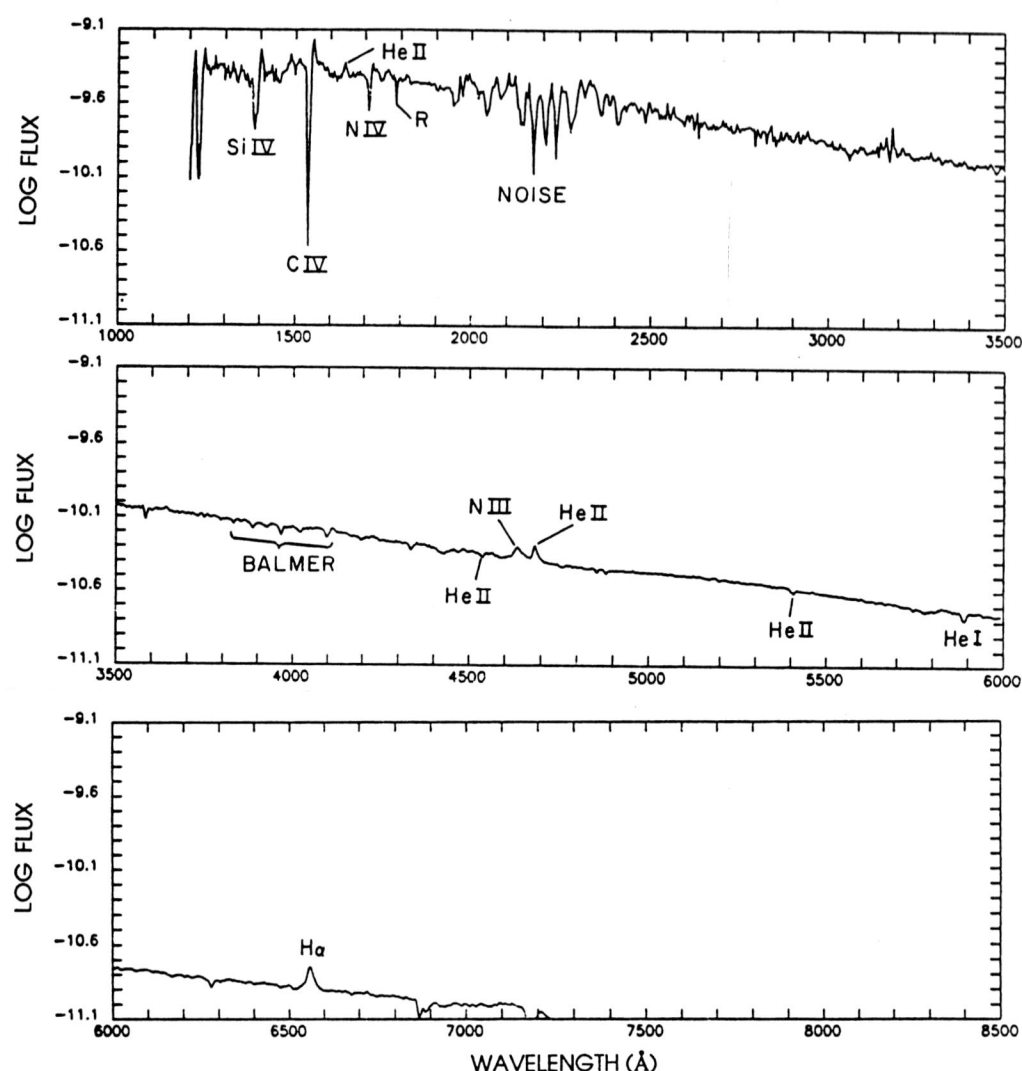

Figure 1. *Spectrophotometry of HD 14947, type O4f. The P-Cygni lines in the UV, and the emission lines of N III and He II in the optical, indicate the presence of a moderately strong stellar wind. The absorption features come from the stellar photosphere.*

Fig. 1. It is also a relatively straightforward procedure to distinguish among the WN subtypes, at least between WNE and WNL.

Before turning to the WC class, I would like to show some spectra of a WN star from the infrared (IR) regions at a few microns. Spectroscopy at these wavelengths for massive blue stars is in its infancy. In fact, I am not aware of any systematic studies of O or Of stars beyond 10830 Å, where a prominent He I line is found. Figure 4, adapted from Hillier et al. (1983), shows the spectrum of another typical WN5 star. Here the data have not been fluxed. The prominent lines are nearly all due to high-level transitions of He II. The strongest feature is the (7–6) transition at 3.09 microns; it has a larger equivalent width than 4686 Å but, of course, its line flux is much smaller. This line, or other He II transitions which do not blend with hydrogen, would be good indicators of the presence of WN stars in IR spectral regions.

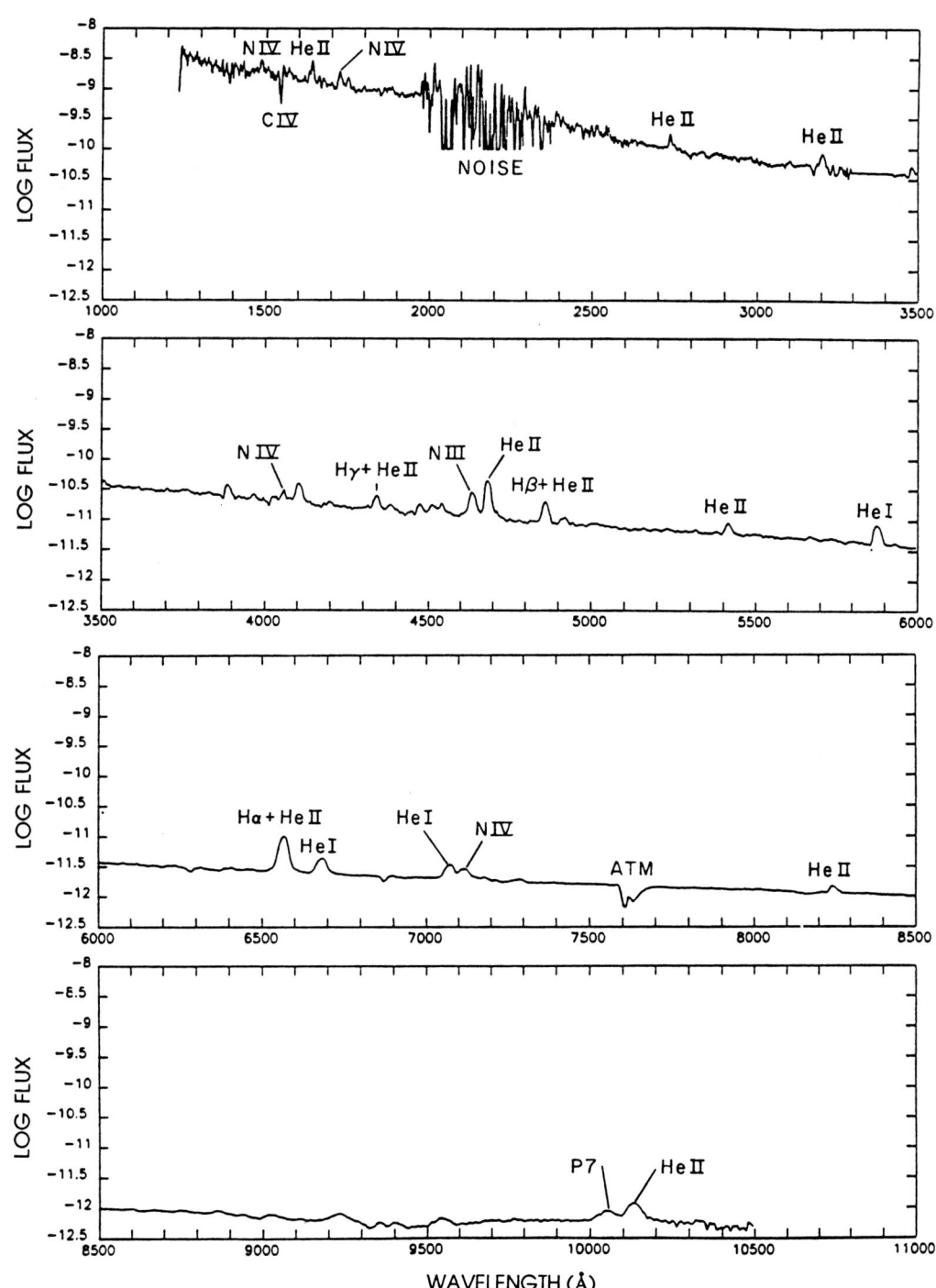

Figure 2. *Spectrophotometry of WR12, a typical WN7 (WNL) star. Nearly all lines are in emission, indicating a very strong stellar wind. The spectral subtype is given by the appearance of the N III and N IV features in the optical.*

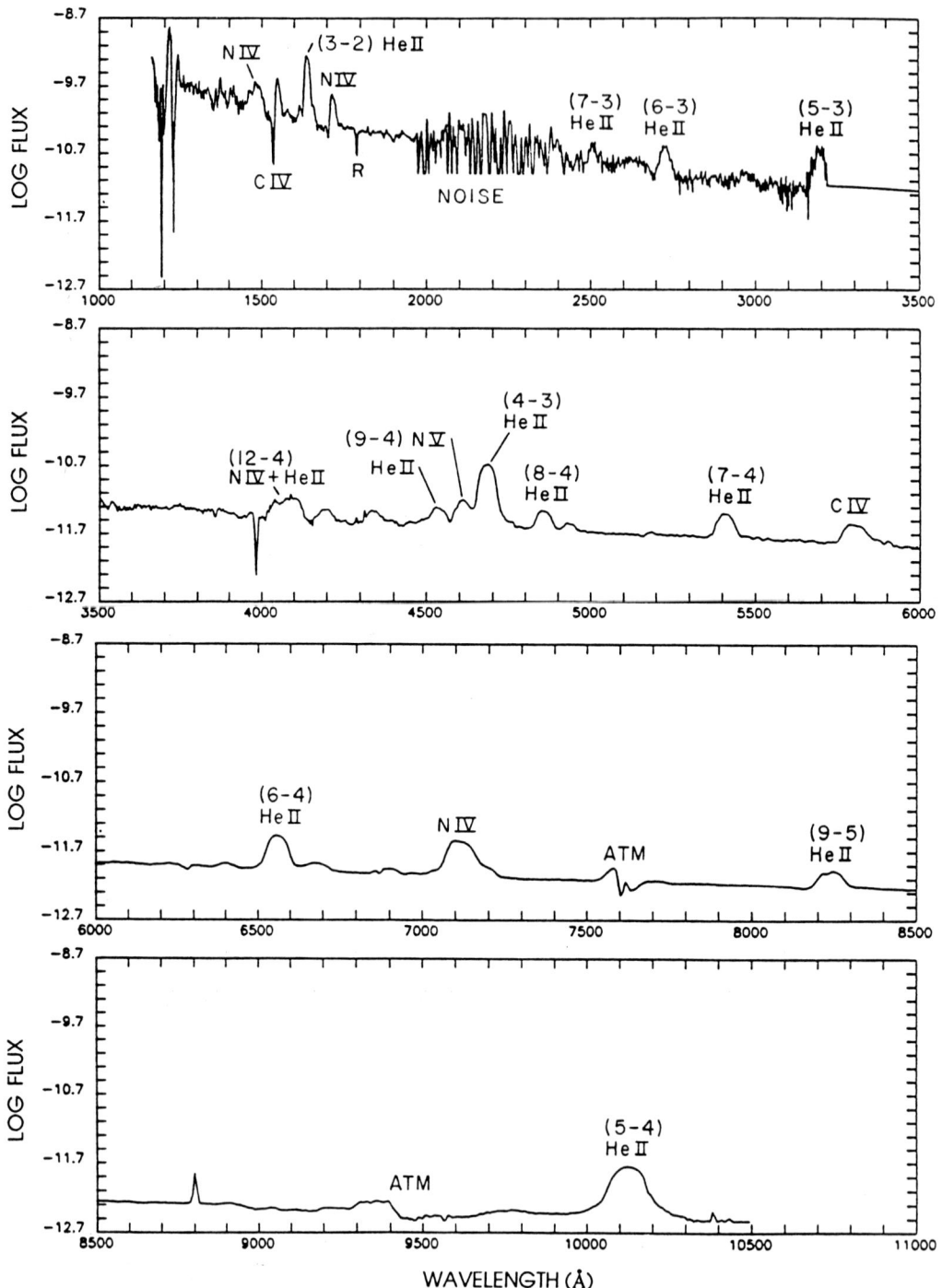

Figure 3. *Spectrophotometry of HD 89358, a typical WN5 (WNE) star. Nearly all lines are in emission, indicating a very strong stellar wind. The spectral subtype is given by the appearance of the N IV and N V features in the optical. Some of the He II transitions are labeled.*

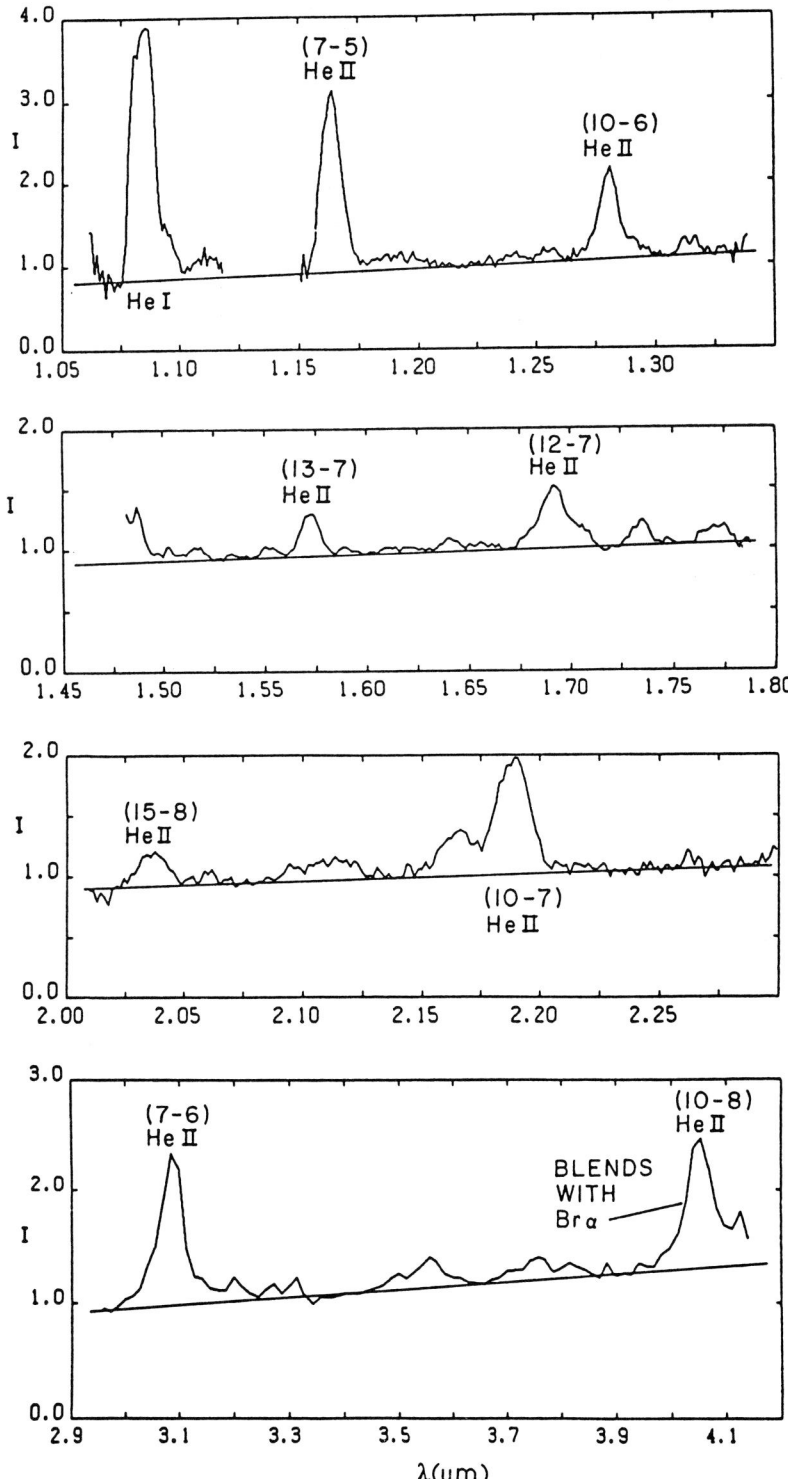

Figure 4. *IR spectroscopy of HD 50896, a typical WN5 (WNE) star. Most of the emission features are due to recombination transitions in He II. The (7–6) line at 3.09 microns is the strongest in the spectrum, rivaling 4686 Å in equivalent width.*

Figure 5 shows the spectrum of a typical WC8 (WCL) star. The most prominent lines are the classification pair at C III $\lambda 5696$ and C IV $\lambda 5808$, along with the C III $\lambda 4650$ feature and 9711 Å also due to C III. At the log flux wavelength scale shown here, the UV features at C IV $\lambda 1550$ and C III $\lambda 1909$ are relatively weak in equivalent width compared to the lines in the optical or NIR but their fluxes are, of course, larger.

Shown in Figure 6 is the spectrum of a typical WC6 (WCE) star. There are some overall differences from the WC8 star of Fig. 5, but the most important change is the ratio of the classification lines: C III $\lambda 5696$/C IV $\lambda 5808$. Here the ratio is less than unity; in WCL stars it is greater than 1. In WC stars, the line width also becomes larger with earlier spectral subtype; in Figs. 5 and 6 the increase in line width with increasing wavelength, in wavelength units, is obvious but is an artifact of plotting with this unit rather than velocity width. It is quite easy to distinguish between WN and WC stars in any of the spectral regions: UV, optical, or NIR. There is little difference in the continua of these objects as we are observing out on the "tail" of the emergent energy distribution.

Infrared spectra of several WC stars have been presented by Smith and Hummer (1988), but their figures do not easily lend themselves to reproduction for this talk. Instead, I show in Figure 7 a spectrum of a WC5 star, HD 193793, which while not typical, nevertheless shows a very strong C IV emission-line feature at 2.09 microns (adapted from Lambert and Hinkle 1984). This line is present only in WC types; conversely, the (10–7) He II transition at 2.19 microns that is very strong in WN types is not present to any great extent in WC stars. By detecting either of these emission features one would be able to ascertain the presence of WN or WC stars in the 2-micron window.

Why would IR spectroscopy of W-R stars be useful? There are *splendid opportunities* to detect WN and WC stars in heavily obscured regions. This may have application to the study of galactic structure, or to the investigation of the massive-star population of heavily reddened IRAS bright galaxies. For example, I have recently been curious about the location of spiral arms in our own Galaxy. We (Conti and Vacca 1990) have plotted the locations of the known W-R star population (which is a highly non-uniform and incomplete sample) in galacto-centric coordinates. We find a vertical scale height of 45 pc as befits an extreme Population I sample. However, the segregation of stars in longitude, while suggestive of "arms" or regions with concentrations of these massive stars, does not show any "spiral" structure. Probably this is due to incompleteness, and this problem is dominated by the extinction in the galactic plane.

A typical O or W-R star has an M_V of -5. At a distance of 10 kpc (a little beyond the galactic center) the distance modulus is 15. Without extinction, such a star would have an apparent magnitude of $+10$, well within the reach of very modest telescopes. As an illustration of the opportunities afforded by the study of blue stars in the infrared, Table 1 is instructive. I have illustrated the expected apparent magnitudes at several bands for an O star at 10 kpc with an assumed 25 magnitudes of extinction in V.

TABLE 1. *Extinction at various wavelengths*

Band (λ):	V	R	I	J	H	K	L
A_λ:	25	18.8	12.0	7.0	4.7	2.8	1.4
$V - \lambda$:	0	-0.1	-0.4	-0.7	-0.8	-0.9	-0.9
m_λ:	35	28.9	22.4	17.7	15.5	13.7	12.3

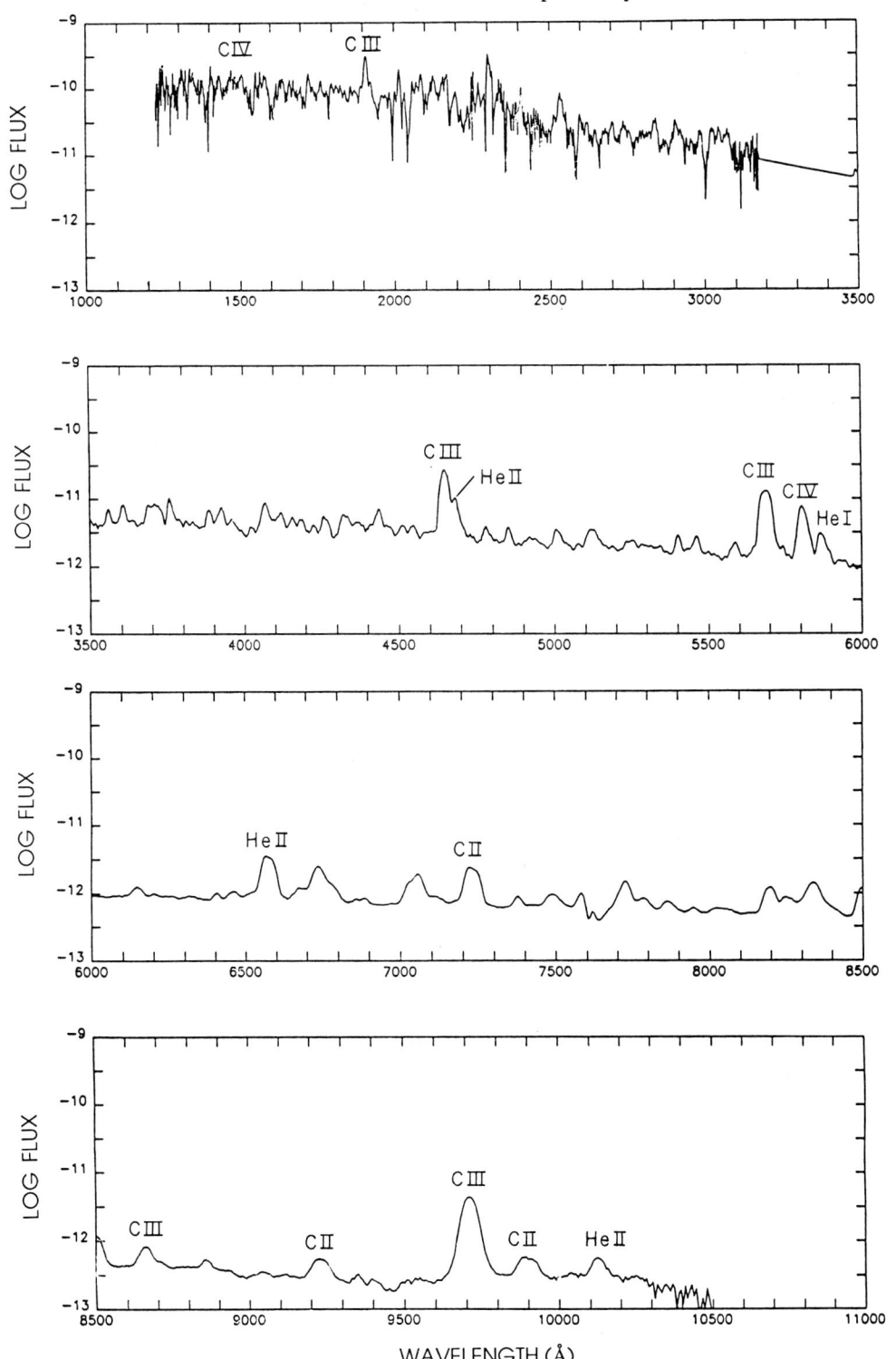

Figure 5. *Spectrophotometry of HD 117297, a typical WC8 (WCL) star. Nearly all lines are in emission, indicating a very strong stellar wind. The spectral subtype is given by the appearance of C III and C IV features in the yellow part of the spectrum.*

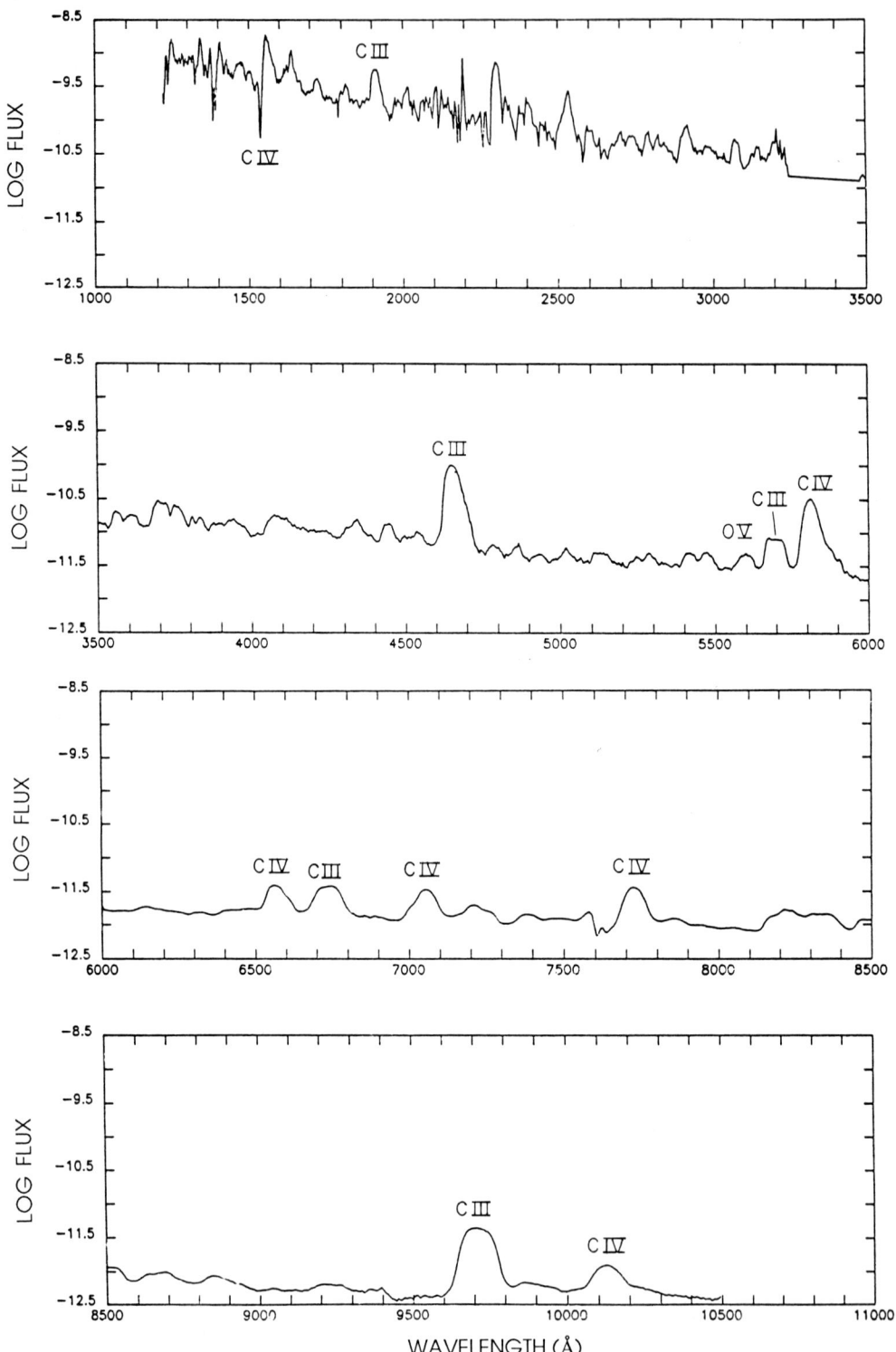

Figure 6. *Spectrophotometry of HD 92809, a typical WC6 (WCE) star. Nearly all lines are in emission, indicating a very strong stellar wind. The spectral subtype is given by the appearance of C III, C IV, and O V features in the yellow part of the spectrum.*

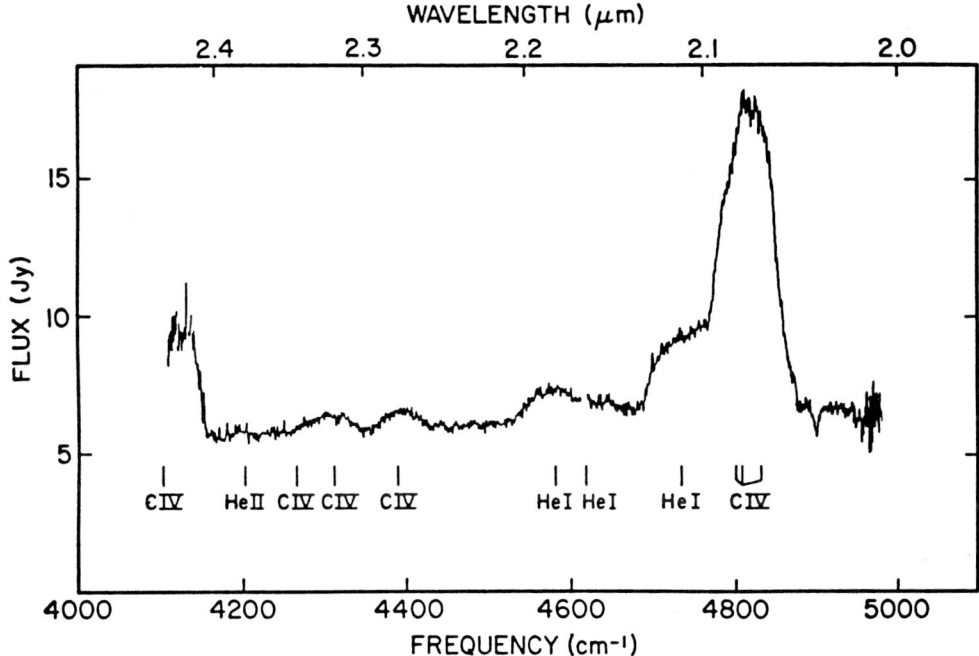

Figure 7. *IR spectroscopy of HD 193793, a WC5 star. Although this object is not typical of the class, the presence of a strong C IV feature at 2.09 microns is found throughout most of the WC sequence.*

The point of this table is two-fold: (1) to remind all of us that as one goes into the infrared wavelength regions (*JHKL*), the extinction drops dramatically; and (2) there is not all that much difference in the intrinsic magnitudes in these wavelength bands, as related to V, for blue stars (in comparison to type A).

With the advent of IR area detectors, it is now possible to image at apparent magnitudes which are similar to those in the last row of Table 1. An intriguing project would be to devise narrow-band filters which isolate, say, the He II 2.19μ and C IV 2.09μ emission-line features in WN and WC stars and attempt to detect W-R stars at much deeper locations in our Galaxy than has heretofore been possible. Unfortunately, the spatial coverage of the current generation of IR detectors is too small to just blindly point along the Galactic equator. A pilot program to investigate the W-R population of obscured giant H II regions is under way by myself, Taft Armandroff, and Phil Massey. We are working at 1 micron, where large-scale CCD detectors are still sensitive, and using line filters which isolate the He II 10124 Å line (Figs. 2 and 3) and C III 9711 Å (Figs. 5 and 6), along with an intermediate continuum point. This may not be at a long-enough wavelength to sufficiently penetrate the dust in and near these massive star-forming regions, but it is a start in a new wavelength regime.

3. INDIRECT SIGNATURES

Massive blue stars are typically associated with surrounding gas and dust. In some cases, the very youngest O and B stars may be completely obscured by their initial birth clouds and not visible at all at optical wavelengths. However, the radiation from such stars, and W-R objects, may be detected indirectly, from their *Lyman continuum radiation* which both excites the gas and heats the dust.

Individual hot stars, or associations of them, give rise to ionized "Strömgren spheres", or H II regions, in the surrounding ISM. The excited gas may be detected by the presence of nebular emission lines of various ionic species and from radio recombination lines of hydrogen (and certain molecules). Radio observations also lend themselves to the detection of the CO molecule, which is typically associated with molecular hydrogen and is often found concentrated in Giant Molecular Clouds (GMCs). These are the sites of massive-star formation (*e.g.,* Myers *et al.* 1986).

In some cases, particularly the "youngest" H II regions, the parent dust cloud may completely shroud the exciting stars. The emergent stellar radiation will heat the dust, which will then radiate in the IR where it can be readily detected through the intervening IS medium as a "point" or slightly extended source. Wood and Churchwell (1989 and references therein) have studied such so-called ultra-compact H II (UCHII) regions. It is possible to make estimates of the number of exciting stars from the integrated luminosity in the infrared. It will be interesting to compare the expected numbers in these regions with actual counts of O stars once IR spatial photometry begins to be applied (*e.g.,* Little *et al.* 1989).

The radio fluxes and the nebular recombination lines may be used to estimate, among other things, the quantity and quality of the Lyman-continuum radiation emitted by the star or stars within an H II region. This step requires a *calibration* of the emergent Lyman-continuum predictions from stellar models with properties of real stars. Nearly all investigators have used Panagia (1973) to derive from the observed total of Lyman-continuum photons emitted an appraisal of the number of "equivalent" main-sequence O stars present.

The most energetic H II regions have come to be called giant H II (GHII) regions; the word supergiant H II region has also appeared in the literature. I have been unable to find precise definitions of the boundaries between these designations but there seems to be general agreement among investigators as to which regions are which. See also Kennicutt's review in this volume. In Table 2 I have adapted from Shields (1990) and Kennicutt (1984) some observed and inferred properties of selected H II and GHII regions in our own and other galaxies, along with two so-called W-R galaxies (from Osterbrock and Cohen 1982). W-R galaxies are a subset of "starburst" galaxies in which the presence of W-R stars is inferred from the appearance of a He II $\lambda 4686$ stellar emission line above the integrated continuum. Bill Vacca, a student at Boulder, and I have been involved in some detailed analysis of about ten of these objects (see our poster paper).

I have labeled Table 2 as "Massive Star-Forming Regions" to emphasize the *continuum of properties of the blue stellar population within them.*

Readily *observed* properties are the number of Lyman-continuum photons, N(Lyc) (photons/sec), and the line luminosity at Hα (ergs/sec). A value of log N(Lyc) of 51.0 appears to be a useful dividing line between H II and GHII regions: those less energetic have typically been given the former label and those more energetic the latter. This is as good a place as any to suggest such a value as a defining property.

An important *derived* property is the equivalent number of main-sequence O5 V stars producing the observed Lyman-continuum radiation. The calibration of Panagia (1973) was based on the Auer-Mihalas (1972) non-LTE, *unblanketed* hot-star models and the Conti (1973) empirical calibration of spectral types. More recent observations and improved models, which incorporate the effects of wind line blanketing upon the emergent continuum, suggest the calibration of Panagia is reasonably accurate for later

TABLE 2. *Massive star-forming regions (examples)*

	From Nebular-Line Analyses			From λ4686	
Object	Log N (Lyc) (photons/sec)	Log L (Hα) (ergs/sec)	#O5 Vs	#W-Rs	Category
Orion Nebula	48.8	37.0	(0.2)	0	H II
Rosette Nebula	49.8	37.9	1	0	H II
Carina Nebula	50.6	38.8	9	3	H II
NGC 3603	51.0	39.2	20	>1	Giant H II
W49	51.2	39.3	27	?	Giant H II
NGC 604 (M33)	51.5	39.6	65	> 3	Giant H II
30 Doradus (LMC)	52.0	40.2	230	> 15	Giant H II
NGC 5471 (M101)	52.5	40.7	750	?	Giant H II
NGC 6764	52.5	40.7	660	1000	W-R galaxy
Mrk 309	53.5	41.7	7100	20,000	W-R galaxy

O and B main-sequence-type stars, but for the earliest O classes it is about one subtype too hot (Bohannan *et al.* 1990). In other words, the column headed O5 V really refers to O4 V stars under the modern spectral-type—temperature calibration. I need to note here that the Panagia calibration was derived for stars of solar composition and may not apply as well for those with substantially different abundances. This latter issue is currently being addressed by Kudritzki and associates.

The number of O5 V stars (I will continue this usage for ease of comparison with the literature) is a single-point estimate of the number of O stars present; if an IMF is specified one could then infer the numbers of O stars at other spectral types. This has sometimes been done using a "standard" IMF. Note that such an analysis from H II regions is *insufficient* to *derive* an IMF. This would require an analysis of the *quality* of the Lyman-continuum radiation, that is, the piecewise contributions over wavelengths below 912 Å. Such a step is usually not taken but could in principle be accomplished by detailed modeling of line strengths of ions other than hydrogen; this is hard to disentangle from potential composition differences and the physically non-uniform spatial distribution of density among various excited regions. For a thorough discussion of the prospects and difficulties in deriving an IMF for massive stars in "starburst" galaxies, see Scalo (1989).

In (so far) a few cases, actual counts of O stars and O5 V "equivalents" have been made for some H II regions and compared to those expected from the Lyman-continuum estimates (Massey *et al.* 1989a,b). The agreement is within a factor 2 and thus gives us confidence in the overall appropriateness of such indirect measurements.

The penultimate column of Table 2 gives either the counted number of W-R stars, or in the W-R galaxies an estimate based upon the measured integrated flux of He II λ4686. The O5 V/W-R number ratio in H II and GHII regions is of the order of 10 which is about what would be expected if the former and latter numbers represent the core-H-burning and He-burning lifetimes, respectively, and *all* (equivalent) O5 V stars become W-R types. In the W-R galaxies, the O5 V/W-R number is smaller; while this might be interpreted as indicating anomalous IMFs, I would rather wait until our analysis of such systems is complete before asserting such a strong conclusion.

The main point of Table 2 is to show some well-known examples of H II and GHII regions in terms of their hot-star content and to illustrate the connection to "starburst" galaxies, of which the W-R galaxies are not yet a well-studied subset. As Kennicutt (1984) has stressed, H II and GHII regions should not be considered to form a *single-parameter* family of increasing energy. There are substantial morphological distinctions among them, for example, the surface brightness and the stellar densities (not illustrated here). On the other hand, as far as the numbers of hot stars are concerned, it is of interest to compare such massive star-forming regions. There are nearly five orders of magnitude in Lyman-continuum luminosity in passing from Orion to the brighter W-R galaxies (at least one "IRAS galaxy", discovered by Armus et al. (1988), is a W-R type and may be somewhat more luminous than Mrk 309). It seems to me that we may be able to understand the massive-star content of "starbursts" by analogy with the better-known GHII regions. NCG 5471 in M101 is similar in its integrated hot-star properties to the W-R galaxy NGC 6764, for example.

There are some intriguing questions raised by Table 2. For example, the GHII region W49 is highly obscured by surrounding dust and cannot be observed in the optical. Some 27 O5 V-star "equivalents" are present from the Lyman-continuum analysis; are several W-R stars also there? It may be possible to address such questions by IR narrow-band photometry utilizing the W-R emission features at 2.09 and 2.19 microns, as I have indicated previously. Are there really *more* W-R stars in NGC 6764 and Mrk 309 than O5 V equivalents? This would imply a very flat IMF or a very short "starburst" interval. NGC 5471 has 750 O5 V-star equivalents. Does it have *any* W-R star population?

Let me discuss the spectral signatures and hot-star content of several of the GHII regions given in Table 2. Here we are dealing with objects for which at least some of the individual stars have been identified and studied spectroscopically. If we were to compare complete stellar data with integrated spectra of these nearby regions, then we could better understand more distant GHII clouds where the individual stars cannot be resolved and we can only observe their integrated properties. This intercomparison has not yet been addressed quantitatively in the literature, due both to the incompleteness in the stellar spectroscopy and the lack of integrated spectrophotometry for the regions in question.

NGC 3603 (the core is catalogued as HD 97950) is a relatively unobscured GHII region in our own Galaxy. The core, about 1 pc in size, contains several unresolved hot stars, at least one of which is a WN star from the optical spectrum (*e.g.*, Moffat et al. 1985). Some 20 O5 V-star equivalents should be present from the nebular-line analysis, but presumably would be found within a more extended 5-pc region centered on HD 97950; Figure 8 illustrates its integrated UV and optical spectrum. The line widths of the emission features indicate the presence of a WN star; O stars are also present as indicated by the Balmer, He I, and He II absorption lines. Such a spectrum is *typical* of what is seen in some much more distant GHII regions and W-R galaxies (see below).

Moffat et al. (1985, 1987), Melnick (1985), and Walborn (1986) have discussed the hot-star content of the core (R136) and vicinity of 30 Dor in the LMC. R136 is of a few pc in size and the brightest component R136a has been resolved by speckle techniques into four brighter and four fainter components (Weigelt and Baier 1985), one of the former of which is a WN type. An integrated optical spectrum of R136 is given, for example, by Moffat et al. (1985) and an ultraviolet one by Cassinelli et al. (1981). One finds the presence of broadened He II $\lambda\lambda 4686$ and 1640 emission features in these spectra. Over 200 O5 V-star equivalents are predicted to be present in the

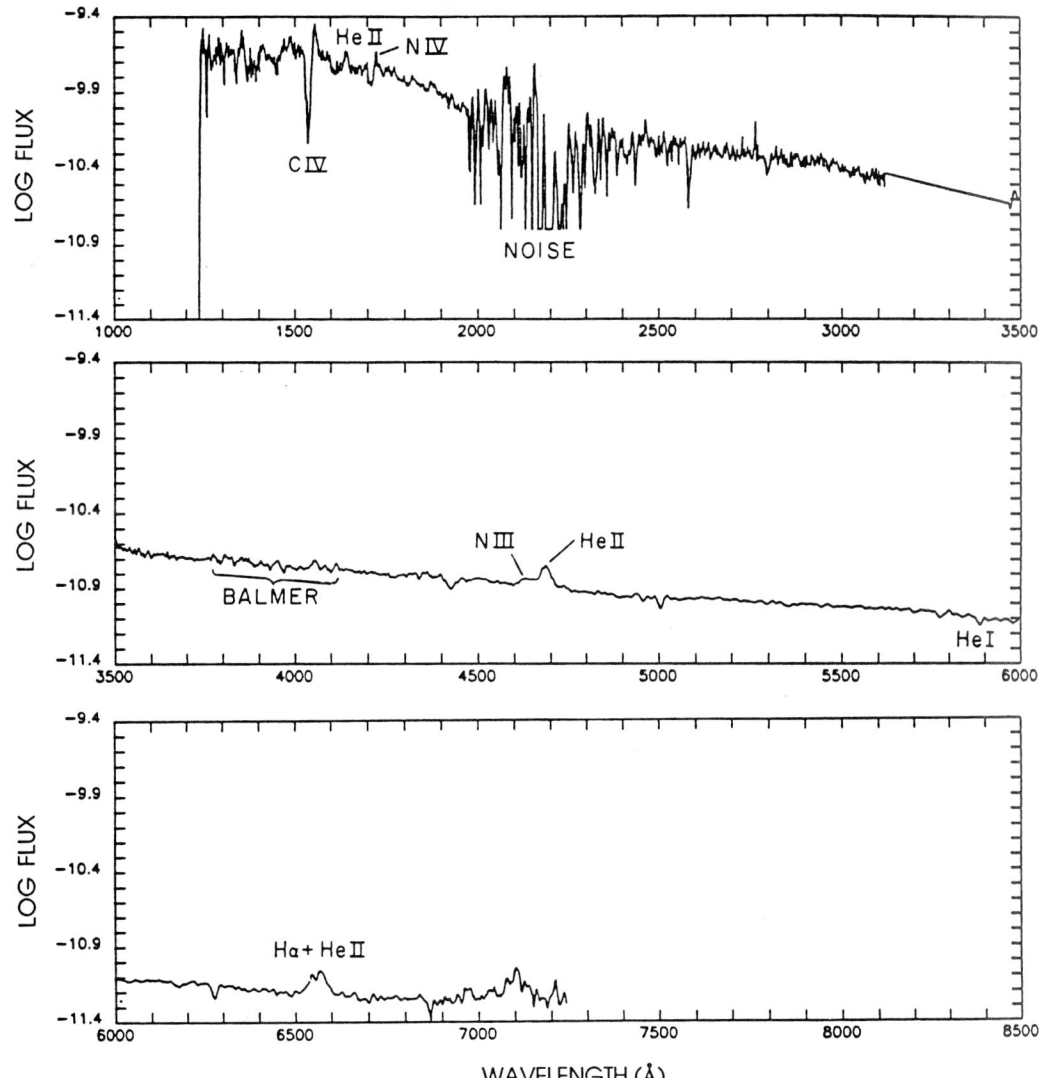

Figure 8. *Spectrophotometry of HD 97950, the unresolved "core" of the GHII region NGC 3603. Although the spectrum is composite, due to the presence of WN-type and O-type stars, it appears superficially similar to the Of star pictured in Fig. 1.*

extended 30 Dor GHII region from the nebular-line analyses. Ongoing identification and classification of the stars in the 30 Dor vicinity is being carried out by Garmany, Massey, Parker, and Walborn (private communication), who have identified at least 100 O types (see the review by Walborn in this volume). At least 15 W-R stars have been identified in this region according to Moffat et al. (1987). While as yet there exists no integrated spectrum of the *nebulosity* of 30 Dor which extends over several arc minutes (about 50 pc), it will be important to compare the final O-star statistics with the predictions from such recombination spectra.

Optical and UV spectra of the GHII region NGC 604 have been given by Rosa (1980), Conti and Massey (1981), and Benvenuti (1983). W-R emission features at 4686 Å and 1640 Å due to He II are found in the integrated spectrum. At least three W-R stars have been identified within NGC 604, which is about 100 pc in size, comparable to the 30 Dor Nebula; more may be present. It will be difficult to attempt spectroscopy of the O-star population of NGC 604 given the distance modulus of 24.3 for M33, but 65 O5 V-star equivalents are predicted. I would like to stress that the *integrated* spectrum of NGC 604 looks very much like R136 and HD 97950 in having O-type and W-R features.

Finally, integrated ultraviolet and optical spectra of the GHII region NGC 5471 have been given by D'Odorico et al. (1983) and Rosa (1980). Here there is certain evidence for O stars but that for W-R types is ambiguous. A narrow and possibly nebular He II $\lambda 4686$ emission line is found in the optical; the IUE data have too low a signal/noise to be useful for the detection of He II $\lambda 1640$. Given the results for the previous three GHII regions, I would be surprised *not* to have W-R stars present in NGC 5471. Spectral observations of this distant GHII region need better S/N and higher spectral resolution than currently exist. Over 750 O5 V-star equivalents are expected; optical spectroscopy is probably not possible with the sensitivity of current ground-based instruments, but might be feasible with the new generation of 8-m telescopes.

It would be important to obtain UV spectra of the individual exciting stars of GHII regions such as those listed above. The crowding problem, and UV brightness, could be overcome by the Hubble Space Telescope observational performance. Guaranteed-Time-Observer approved programs will consider the nearest of the two GHII regions discussed above.

4. WOLF-RAYET GALAXIES

I would now like to take a large step outward in distance (and interpretation) to those galaxies (or condensations within) in which only integrated spectra can be obtained and which show evidence of W-R stars from the presence of He II $\lambda 4686$ emission in their integrated optical spectra. These W-R *galaxies* overlap but extend another decade in luminosity from GHII regions as indicated in Table 2 (and another decade brighter still as inferred from the luminous IRAS galaxy found by Armus et al. 1988). Kunth and Schild (1986) have discussed some properties of 11 of these objects and note that the flux of the He II $\lambda 4686$ feature scales very roughly with the integrated magnitude of the galaxy. It seems to me that these objects form a continuum in their hot-star properties with the GHII regions in nearby galaxies. Some 30 W-R galaxies are known; they have been found serendipitously by those studying emission-line galaxies. While it is possible to initially confuse W-R galaxies with Seyferts having a He II $\lambda 4686$ emission feature, the *nebular* lines in the former are *narrow* and their excitation is *stellar*.

As I have already noted, Vacca and I are involved in a detailed analysis of a subset of W-R galaxies for which we have been able to obtain 4-m time at CTIO. I would like to show, in Figures 9 and 10, examples of spectra of the W-R galaxy He2-10, the "first" known W-R galaxy (Allen et al. 1976). Figure 9, adapted from our 2D-Frutti observations, shows the strong nebular emission-line spectrum excited by O-type stars; the continuum is very "blue" and a weak Balmer series is found in absorption. On the scale shown here, a broad He II $\lambda 4686$ emission feature is not very obvious but is present. Aside from the nebular lines, the overall spectrum is similar to that shown in Fig. 1 and Fig. 8. Among questions we would like to answer from spectra such as these are: numbers (and types?) of exciting stars inferred from the recombination lines; composition of excited gas; numbers and types of stars contributing to the continuum and to the absorption-line spectra; age and duration of the starburst; etc.

Figure 10 is an expanded version of Fig. 9 near the vicinity of the He II $\lambda 4686$

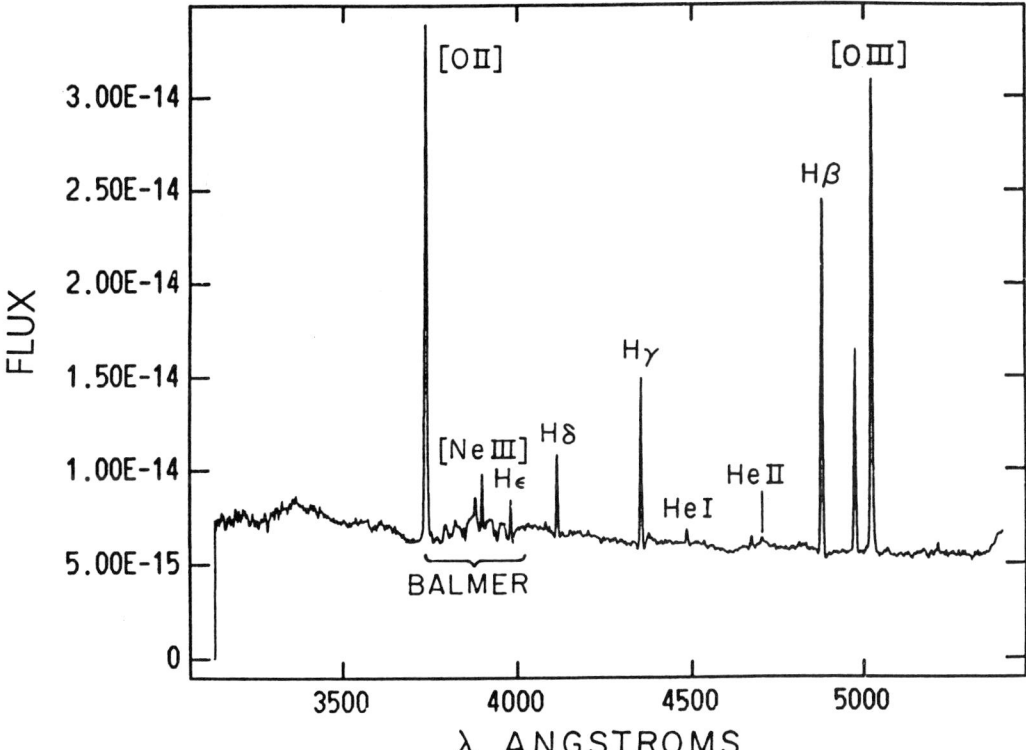

Figure 9. *Spectroscopy of He2-10, a W-R galaxy. The nebular lines are excited by O-type stars; the continuum is relatively "blue"; weak upper Balmer series lines and a Balmer jump due to A-type stars can be seen. A broadened emission feature at He II λ4686 in the integrated spectrum arises from the presence of WN stars in this "starburst" galaxy. Aside from the nebular recombination lines, this object has a spectrum much like that of HD 97950, the core of NGC 3603 (Fig. 8).*

feature; its broadened nature is apparent. The nebular lines are unresolved at the resolution of 4 Å used with the 2D-Frutti detector at the 4-m. Several "artifacts" of this instrument remain in the data. The narrow emission line just shortward of 4686 Å is the [Fe III] line at 4658 Å. It is also present in many (all?) of the W-R galaxies known to me, and also in a few blue compact galaxies with only O stars identified. The overall emission-line spectra of W-R galaxies look very much like those of GHII regions. I would like to stress that for studies of galaxies such as these, the highest feasible S/N and spectral-resolution data are required.

In our (Vacca and Conti) preliminary analysis we also have identified the N III λ4640 emission-line blend in several W-R galaxies. We are thus reasonably certain that these broadened features are due to WN-type stars. We have no certain evidence for C III λ4650, nor for lines at C III λ5696 and C IV λ5808, which in WC stars are as strong as or stronger than He II λ4686 in WN-type stars. Such lines are found in *some* GHII regions containing individual WC stars. It appears that W-R galaxies *may* have relatively *more* WN than WC subtypes, as contrasted to the solar vicinity where the ratio is near unity. The Magellanic Clouds contain more WN than WC stars, which has been attributed to their lower "metal" abundance (*e.g.*, Maeder 1990). W-R galaxies *might* be like the MCs in their composition *and* W-R content but this potential connec-

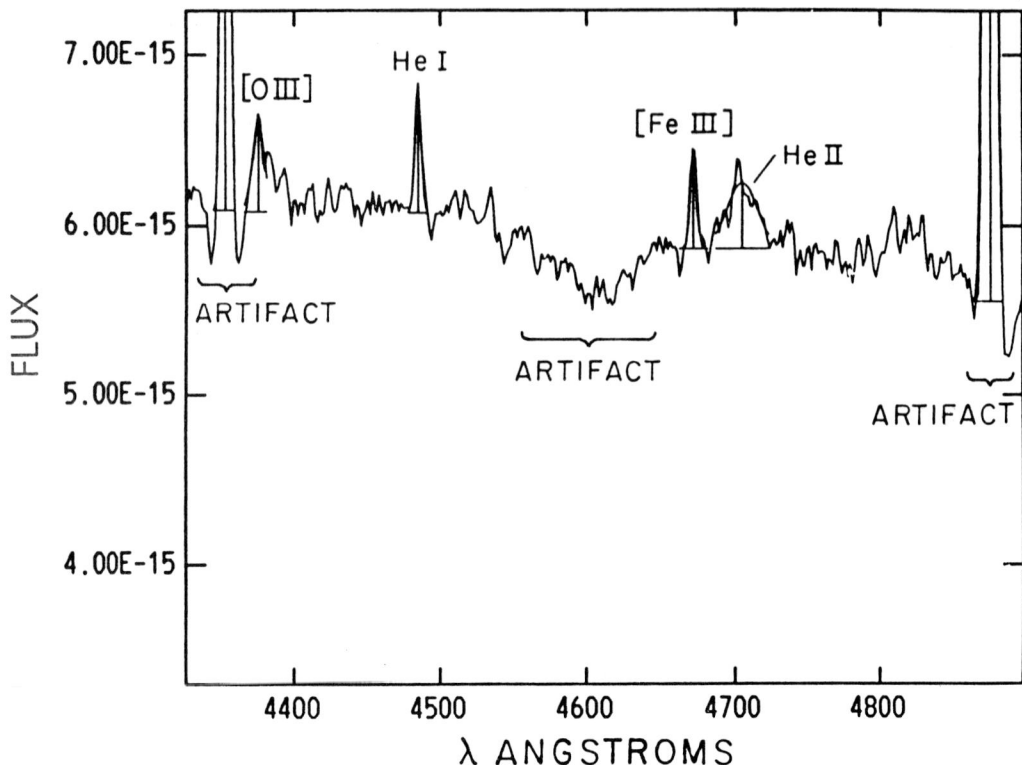

Figure 10. *Large-scale spectroscopy of He2-10, a W-R galaxy, in the vicinity of the He II λ4686 transition. The broadened nature of this line, as compared to the unresolved nebular lines, is apparent.*

tion must await our detailed analysis of the properties of the galaxies themselves. I need to point out that the stellar emission-line features so far identified in W-R galaxies also look very much like those of "transition" Of/WN stars (Bohannan and Walborn 1989). It is possible that such spectra arise in the *most* massive and luminous main-sequence stars whose very strong winds *could* give rise to emission features in the optical. W-R galaxies would be excellent candidates for Hubble Space Telescope observations in order to disentangle the recent star-formation phase—a few million years given the presence of W-R and O-type stars—from the underlying (previous generation?) of stars of type A that one infers are present from the relatively weak Balmer absorption lines and Balmer jump.

5. PRESENT AND FUTURE DIRECTIONS

Hot, luminous, blue stars are a snapshot of the massive-star SFR a few million years ago. They may be used to probe conditions in "starburst" galaxies and other locations where star formation is occurring.

A census of individual stars in various environments may be used to better understand the processes under which massive stars form. For example, O stars may be sampled in the solar vicinity and the Large and Small Magellanic Clouds, each having a different initial composition and a varied past history of star formation. It is likely that a complete census of the SMC can be completed and compared to the solar vicinity

(2.5–3 kpc); such work is now under study by Garmany and Massey and associates. A complete count of the O-star population of the LMC is not completely outside the reach of current telescopes, but will require a long-term, dedicated observational program and cooperation from TACs. Counts of W-R stars within the solar vicinity, and in the Magellanic Clouds, appear nearly complete. W-R stars are also being identified and studied in other galaxies of the Local Group (Armandroff and Massey 1985; Massey et al. 1987; Moffat and Shara 1987) and in more distant sites such as the Sculptor Group (Armandroff and Massey, private communication). Such stars may be used to sample environments differing from the MCs and the solar vicinity.

W-R stars may be used to investigate the spiral structure of our Galaxy in connection with other such extreme Population I indicators as GHII and UCHII regions and GMCs. NIR and IR observations may be able to probe the interstellar extinction which otherwise limits our knowledge of such star-forming regions.

Extragalactic GHII regions may be a paradigm for starburst galaxies, at least as far as their massive-star population is concerned. Analysis of these regions, which are considerably nearer than typical starburst galaxies, may further our understanding of star formation. In particular, W-R galaxies may be the "youngest" examples of the starburst phenomenon and *may* indicate the presence of anomalous IMFs in such regions. H II and GHII regions and W-R galaxies appear to form a *continuum* of their massive-star properties and detailed analysis of *each* will help our understanding of *all*.

I would like to thank Phil Massey and Jean-Marie Vreux for use of their optical and NIR spectrophotometry of Of and W-R stars in our ongoing collaborations. The figures have been made with kind assistance from Ken Brownsberger, Pat Morris, and Bill Vacca and come from various stages in their Ph.D. dissertations. I appreciate continuing support from the NSF under Grant AST88-06594 and from NASA under contract NAG5-1016.

REFERENCES

Abbott, D. C. and Conti, P. S. 1987, *Ann. Rev. Astr. Ap.*, **25**, 113.
Allen, D. A., Wright, A. E., and Goss, W. M. 1976, *M.N.R.A.S.*, **177**, 91.
Armandroff, T. E. and Massey, P. 1985, *Ap. J.*, **291**, 685.
Armus, L., Heckman, T. M., and Miley, G. K. 1988, *Ap. J.*, **326**, L45.
Auer, L. H. and Mihalas D. 1972, *Ap. J. Suppl.*, **24**, 153.
Barlow, M. J. and Hummer, D. G. 1982, in *Wolf-Rayet Stars: Observations, Physics, Evolution*, IAU Symposium 99, eds. C. W. H. de Loore and A. J. Willis (Dordrecht: D. Reidel), p. 387.
Benvenuti, P. 1983, *Highlights of Astronomy*, **6**, 631.
Bohannan, B., Voels, S. A., Hummer, D. G., and Abbott, D. C. 1990, *Ap. J.*, in press.
Bohannan, B. and Walborn, N. R. 1989, *Pub. A.S.P.*, **101**, 520.
Cassinelli, J. P., Mathis, J. S., and Savage, B. D. 1981, *Science*, **212**, 1497.
Conti, P. S. 1973, *Ap. J.*, **179**, 181.
Conti, P. S. and Massey, P. 1981, *Ap. J.*, **249**, 471.
_____. 1989, *Ap. J.*, **337**, 251.
Conti, P. S., Massey, P., and Vreux, J.-M. 1990, *Ap. J.*, **354**, 359.
Conti, P. S. and Vacca, W. D. 1990, *A. J.*, **100**, 431.
D'Odorico, S., Rosa, M., and Wampler, E. J. 1983, *Astr. Ap.*, **53**, 97.

Hillier, D. J. 1989, *Ap. J.*, **347**, 392.
Hillier, D. J., Jones, T. J., and Hyland, A. R. 1983, *Ap. J.*, **271**, 221.
Kennicutt, R. C., Jr. 1984, *Ap. J.*, **287**, 116.
Kunth, D. and Schild, H. 1986, *Astr. Ap.*, **169**, 71.
Lambert, D. L. and Hinkle, K. H. 1984, *Pub. A.S.P.*, **96**, 222.
Little, S. J., Gullixson, C., Dietz, R. D., Hackwell, J. A., Gehrz, R. D., and Grasdalen, G. L. 1989, *A. J.*, **97**, 1716.
Maeder, A. 1990, *Astr. Ap.*, in press.
Massey, P., Conti, P. S., and Armandroff, T. E. 1987, *A. J.*, **94**, 1538.
Massey, P., Garmany, C. D., Silkey, M. and Degioia-Eastwood, K. 1989a, *A. J.*, **97**, 107.
Massey, P., Parker, J. W., and Garmany, C. D. 1989b, *A. J.*, **98**, 1305.
Melnick, J. 1985, *Astr. Ap.*, **153**, 235.
Moffat, A. F. J., Niemela, V. S., Phillips, M. M., Chu, Y.-H., and Seggewiss, W. 1987, *Ap. J.*, **312**, 612.
Moffat, A. F. J., Seggewiss, W., and Shara, M. M. 1985, *Ap. J.*, **295**, 109.
Moffat, A. F. J. and Shara, M. M. 1987, *Ap. J.*, **320**, 266.
Myers, P. C., Dame, T. M., Thaddeus, P., Cohen, R. S., Silverberg, R. F., Dwek, E., and Hauser, M. G. 1986, *Ap. J.*, **301**, 398.
Osterbrock, D. E. and Cohen, R. D. 1982, *Ap. J.*, **261**, 64.
Panagia, N. 1973, *A. J.*, **78**, 929.
Rosa, M. 1980, *Astr. Ap.*, **85**, L21.
Savage, B. D. and Mathis, J. S. 1979, *Ann. Rev. Astr. Ap.*, **17**, 73.
Scalo, J. 1989, in *Windows on Galaxies*, eds. A. Renzini, G. Fabbiano, and J. S. Gallagher (Dordrecht: Kluwer).
Schmutz, W., Hamann, W.-R., and Wessolowski, U. 1989, *Astr. Ap.*, **210**, 236.
Shields, G. A. 1990, *Ann. Rev. Astr. Ap.*, **28**, 525.
Smith, L. F. 1968, *M.N.R.A.S.*, **138**, 109.
Smith, L. F. and Hummer, D. G. 1988, *M.N.R.A.S.*, **230**, 511.
Vacca, W. D. and Torres-Dodgen, A. V. 1990, *Ap. J. Suppl.*, **73**, 685.
Walborn, N. R. 1986, in *Luminous Stars and Associations in Galaxies*, IAU Symposium 116, eds. C. W. H. de Loore, A. J. Willis, and P. Laskarides (Dordrecht: D. Reidel), p. 185.
Walborn, N. R. and Fitzpatrick, E. L. 1990, *Pub. A.S.P.*, **102**, 379.
Weigelt, G. and Baier, G. 1985, *Astr. Ap.*, **150**, L18.
Wood, D. O. S. and Churchwell, E. 1989, *Ap. J. Suppl.*, **69**, 831.

DISCUSSION

N. Devereux: Are the W-R galaxies luminous in the IRAS data?

Conti: Several are strong IRAS sources. Tim Heckman has done some work on them.

T. Heckman: The two galaxies having the most luminous W-R $\lambda 4686$ features (IRAS 01003–2238 and Mrk 309) are both very luminous IRAS sources (few $\times \, 10^{12}$ and few $\times \, 10^{11} \, L_\odot$, respectively). Lee Armus and I have found tentative evidence for W-R stars in a fair number of IRAS-selected galaxies. However, there is as yet no clear statistical link between the "IRAS galaxy" and "W-R galaxy" phenomena.

S. Lamb: I have been looking at interacting starburst galaxies in the UV with IUE. We find that Arp 248b has W-R spectral characteristics in this part of the spectrum. We plan to look at more interacting starburst galaxies in the coming year to see if others of this class also show evidence for large populations of massive stars.

A. Moffat: In the past decade there has been an overzealous trend to *overestimate* the number of W-R stars (*e.g.*, ~ 50 W-R in NGC 604 by D'Odorico and Rosa in 1981 based on slit spectroscopy, versus ~ 10 W-R now based on narrow-band imagery by Drissen, Moffat, and Shara 1990) in giant H II regions and W-R galaxies. I hope that, in future, slit work will be backed up by direct, high-resolution, narrow-band imagery to get a complementary, unbiased estimate of the number of W-R stars.

Conti: I agree. Some of the W-R numbers in my table are *estimates* based upon the integrated fluxes at $\lambda 4686$. These have not all been calculated in a self-consistent manner.

R. Pogge: On W-R galaxies, I point out that Mrk 309 and NGC 6764 are *nuclear* sources whereas the other Kunth and Schild W-R galaxies are blue compact dwarfs or isolated H II regions in small irregulars. I feel uncomfortable lumping the former two in as the environments are quite different: the nuclei of spiral galaxies vs. isolated giant H II complexes in irregulars.

Conti: The current list of W-R galaxies is quite a mixed bag of objects. I was only drawing attention to the similarities in the massive-star (O-type and W-R) content.

F. Bruhweiler: I would like to point out that in galaxies with AGNs as well as starburst activity, you may have extended emission-line regions which can mimic properties of W-R–type features. This can be a problem, especially in the UV. For example, in NGC 1068, there is a very highly ionized, extended emission-line region with N V, C IV, He II, and C III] as revealed in the UV spectra seen with the IUE (see Bruhweiler and Truong—a poster paper in this meeting). To the uninitiated observer this emission superimposed on the bright starbursts may be mistaken for W-R features. Indeed, this has been erroneously suggested in the case of the bright UV knots in NGC 1068.

Conti: In the cases of W-R galaxies I am considering, the *line widths* of He II $\lambda 4686$ are larger than those of the nebular features, which are unresolved (unlike in typical Seyferts). I agree one needs to be careful.

M. Shara: You mentioned the 3.09 μ line as a good place to look for heavily reddened Galactic and extragalactic W-R stars. Why is this line so strongly preferred to the other IR lines you showed, which appear to be as strong, or stronger?

Conti: It is further into the IR and thus has less extinction. Others might also be suitable, but one must avoid the Brackett lines which may have nebular contributions.

I. Gatley: As a practical matter it gets rapidly more difficult to work beyond about 2.5 μ because of thermal background emission. You have plenty of lines to use in the 2 μ region, and I expect they will prove the most useful.

I enjoyed your appreciation of the usefulness of infrared, and think that an important function of this meeting will be to convince the rest of the participants that it's true!

N. Walborn: (1) Another useful IR indicator of W-R stars is the large ratio of He I to Brackett γ in the 2 μ region, as shown by McGregor et al. for the WN9 class in the LMC and by Allen et al. in the Galactic Center. (2) Even finer distinctions among Of stars can be made from wind features in low-resolution UV spectra, e.g., Si IV < C IV and He II present indicate an early-Of spectrum, while in a mid- or late-Of Si IV = C IV.

R. Joseph: How useful is detection of W-R features in galaxies for doing *quantitative* astrophysics? From fluxes in such features, for example, can one infer numbers of W-R stars contributing to the emission?

Conti: In principle one can do that and we are planning to come up with the numbers in some 10 W-R galaxies soon (see Vacca and Conti poster at this workshop).

J. Bland Hawthorn: To underscore Bob Joseph's point, a comparison of the number of W-R stars with the number of, say, O stars may shed light on whether the high-mass star-formation rate is impulsive, continuous, or whatever.

Conti: I concur. Work on this problem is underway (see Vacca and Conti poster). The number of O stars can be estimated from analysis of the nebular lines.

N. Langer: What could be the explanation for the fact that you find only evidence for WN stars in W-R galaxies, and none for WC stars? Could it be that late WNs come from more massive stars than WCs, and the effect is an indication of the starburst being very young? Or is it something completely different?

Conti: I suspect it is the initial composition: in the SMC the WN/WC ratio is large. This has been attributed (e.g., Maeder, this workshop) to its lower abundances relative to the solar vicinity where the WN/WC ratio is near unity. While we do not have the numbers yet I think the W-R galaxies are SMC-like in composition.

N. Walborn: Indeed, the W-R population of giant H II regions is dominated by WN types. My own feeling from the observational morphology is that the most massive (100–200 $M_\odot$) stars may not make it to a WC stage.

L. Drissen: A comment on Langer's question: not only in W-R galaxies is the WC/WN ratio very low; in Local-Group giant H II regions (30 Dor, NGC 604, NGC 595) the WC/WR ratio is below 0.2. Most of these WN stars are of WNL subtype.

Conti: This may have to do with the relatively low abundance of the elements.

A. Campbell: Following the points that there are almost no WC stars in the SMC, and therefore that the presence of W-R stars in star-forming regions may be an abundance effect: we have a sample of 45 H II galaxies, but only the highest-abundance objects show WR features. None with $O/H \lesssim 0.2(O/H)_\odot$ seems to contain W-R stars. Perhaps in metal-poor objects the stars contain too few metals to drive the strong wind necessary to strip an O star down to a W-R star. Also, in H II galaxies, those W-R stars that are seen are of subtype WN—more evidence of insufficient stripping.

A. Moffat: Another important parameter in the W-R/O ratio in starbursts is the age (cf. Drissen 1990, Ph.D. thesis): if the region is too young ($\lesssim 2 \times 10^6$ yr) or too old ($\gtrsim 7 \times 10^6$ yr) there will be no W-R stars (these values apply to solar abundances). For lower metallicity, this "window" gets narrower and the W-R/O ratio falls off, reaching essentially zero at $Z \lesssim Z_\odot/10$ or so. The reason is simple: if the region is too young, W-R stars have not had time to form; if too old, they have evolved away already.

G. Hensler: Are there X-ray fluxes measured from your W-R galaxies? Do the energetics show already the explosions of Type II SN? How does this fit into the evolutionary scenario of starburst galaxies?

Conti: I think several are X-ray sources but I don't recall if they are anomalous. I would expect that W-R galaxies *would not* be environments in which SN have yet played much of a role in the energetics; that will come later when the W-R stars are gone and the SN rate becomes appreciable as the burst ages and stars further down the main sequence evolve.

S. Heap: In a QSO observed by John Hutchings a couple of years ago it was found that in the host galaxy, not in the nucleus, there is He II $\lambda 4686$ in emission. After we deconvolved the images of the galaxy it turned out that it is a barred spiral and that one is simply seeing a young spiral-arm population. It is 5 or 10 arc seconds away from the nucleus and so this brings up a question in my mind of what is the going definition of a starburst. What is the difference between a W-R galaxy and a W-R starburst galaxy? What is your working definition of a starburst?

Conti: My working definition, but I'm not insisting on it, is that you know you have a starburst when you see the intense nebular lines. Because then you know you have O stars there and the argument is you could not have had that going on for the last 10^{10} years because you would have used up all of your gas. Sargent and Searle, I think, made this argument back in 1970 when they first discussed blue compact galaxies. If you see a lot of O stars, you have to have something which is turned on and off. Maybe it is turned on only once, maybe it's turned on several times. But it could not have been going along for the age of the Universe with the number of O stars implied by those nebular lines. It may even be true that if you are making only O stars it still cannot go on for 10^{10} years. So it has to be a burst. That is a minimum definition but I'm not insisting that's the only definition. At a later stage you may just see a lot of stars and the O stars have long since gone.

MASSIVE RED STARS IN GALAXIES

Roberta M. Humphreys
Department of Astronomy
University of Minnesota
Minneapolis, Minnesota 55455
USA

Abstract. The red supergiants are evolved massive stars. Because of their red color and high luminosity they are easily recognized and studied at large distances, including nearby galaxies, with current ground-based techniques. And because of their low temperatures they are readily studied at infrared wavelengths, where there are important advantages for determination of extinction and luminosity. They will be increasingly valuable at large extragalactic distances as distance indicators. However, even the most luminous red supergiants are not among the intrinsically most luminous stars nor are they the visually brightest. Nevertheless, the red supergiants are a critical point in the evolution of massive stars. They define the low-temperature side of the HR diagram for massive stars and are one of the possible endpoints of massive-star evolution. It is important to understand their relationship to other massive stars in the HR diagram.

1. RED SUPERGIANTS IN THE HR DIAGRAM

Red supergiants have evolved from main-sequence O- and B-type stars. But observations of the HR diagram of massive stars in our Galaxy and in the Local Group show that the most massive stars ($> 40\ M_\odot$) do not become red supergiants. The observed upper HR diagram reveals an envelope of declining luminosity with decreasing temperature which becomes essentially constant with luminosity at cooler temperatures ($< 10{,}000$ K; see Fig. 1 in Humphreys 1986). Thus the most luminous hot stars ($M_{Bol} - 10$ to -11) have no cooler counterparts.

Humphreys and Davidson (1979) first drew attention to the physical significance of this empirical luminosity boundary and to the lack of cool, evolved massive stars at the highest luminosities. We suggested then that the observed boundary was due to an instability encountered as the most massive stars ($> 40\text{--}50\ M_\odot$) evolve away from the main sequence, accompanied by high mass loss. The highest mass-loss rates are observed along this luminosity boundary (de Jager 1984, 1988). The temperature-dependent boundary for the hot stars is marked by the presence of some well-known, very luminous unstable stars: η Car, P Cyg, S Dor, and the Hubble-Sandage variables in M31 and M33, now known collectively as the Luminous Blue Variables (LBV's).

This temperature dependence for the hot stars suggests that this boundary defines a critical location in the HR diagram, one that is mass dependent. This, plus the lack of cooler counterparts, indicates that stars above some mass limit probably do not evolve to cooler temperatures, but instead lose mass in ejections as LBV's in transition to the Wolf-Rayet stage.

Obviously, below this mass the massive stars do evolve to the red supergiant region. It is the F, G, K, and M hypergiants in our Galaxy and other Local-Group galaxies that define the temperature-independent luminosity limit for cool stars (see Fig. 1). These cool hypergiants are clearly unstable as evidenced by their light and spectral variations, high mass-loss rates (few $\times\ 10^{-4}\ M_\odot/\text{yr}$), and the presence of circumstellar dust around many.

The most luminous known M supergiants ($M_{Bol} - 9.0$ to -9.5) have probably evolved from stars with initial masses of 30–40 $M_\odot$ and include such well-known stars as μ Cep. However, the majority of normal M supergiants have initial mass between 15 and 30 $M_\odot$ and have evolved from main-sequence stars of similar luminosity. In addition to the relatively normal red supergiants with circumstellar dust, there are also the supergiant OH/IR sources (Jones et al. 1983, 1988). They are likely the most evolved M supergiants, which have lost sufficient mass that their dust shells are now optically thick. The mass-loss rates from the most luminous supergiant OH/IR sources may be very high, up to 10^{-4} to 10^{-3} $M_\odot/\text{yr}$. What will eventually become of these supergiant OH/IR stars? Will they eventually end as supernovae or will they shed sufficient mass at these high rates to evolve back to the blue and become Wolf-Rayet stars?

Most M supergiants probably do not become Wolf-Rayet stars. Conti et al. (1983) and Humphreys, Nichols, and Massey (1985) have shown from observations that most Wolf-Rayet stars have initial masses $> 40\ M_\odot$ and virtually all have evolved from stars $> 30\ M_\odot$. In contrast, the red supergiants are primarily from 15–30 $M_\odot$ progenitors. Only the most luminous red supergiants (M_{init} 30–40 $M_\odot$) have the possibility of becoming Wolf-Rayet stars. IRC+10420 may be a post-M-supergiant OH/IR star blowing off its cocoon of dust and gas as it evolves to warmer temperatures (see discussion in Humphreys 1987, 1990).

Table 1 summarizes various evolutionary scenarios for stars of $> 10\ M_\odot$. For several years the greatest uncertainties in massive-star evolution concerned the uppermost part of the HR diagram and the most massive stars. Although there is still a lot of physics to be learned about the origin of the instability limit, the greatest uncertainties now concern the final stages in the evolution of 10–30 $M_\odot$ stars. Of course, this has been caused by SN 1987A.

2. THE LUMINOSITY/STABILITY LIMIT AND THE MOST LUMINOUS RED SUPERGIANTS—THEIR USE AS DISTANCE INDICATORS

The upper limit to stellar luminosities is usually assumed to be set by the balance between the acceleration due to gravity and the radiation-pressure gradient à la Eddington. However, the observed luminosity boundary is composed of two components—the temperature-dependent boundary for hot stars and its turnover at the cool-star upper limit. The classical Eddington limit due to electron scattering does not show the dependence on temperature observed for the hot stars. However, as the temperature decreases below 30,000 K the opacity increases due to ions of H I, Fe II, and oth-

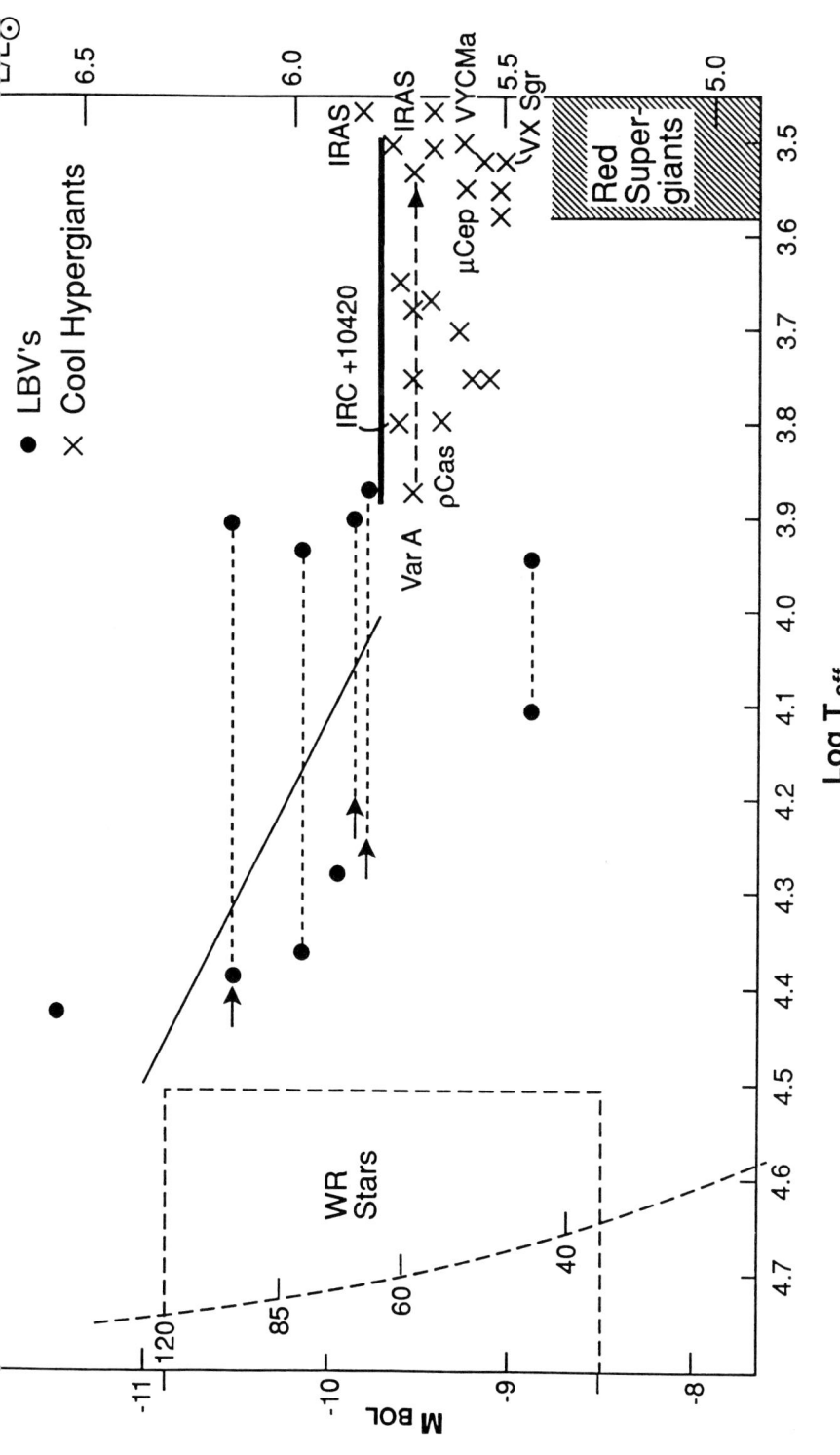

Figure 1. A schematic HR diagram showing the positions of several of the F, G, K, and M hypergiants with respect to the luminosity/stability limit and the positions of the Luminous Blue Variables. The band of the red supergiants and possible location of the progenitors of W-R stars are also shown.

TABLE 1. *Evolutionary scenarios for massive stars*

Initial Mass	
$> 40\ M_\odot$	O star M-S $\to$ Of or BSG $\to$ LBV $\to$ WR $\to$ SN (Type Ib)
$30\text{--}40\ M_\odot$	O star M-S $\to$ BSG $\to$ RSG $\to$ OH/IR $\to$ SN (Type II)
	or OH/IR $\to$ WR $\to$ SN (Type Ib)
$10\text{--}30\ M_\odot^*$	B or O star M-S $\to$ BSG $\to$ RSG $\to$ OH/IR $\to$ SN (Type II)
	or RSG(OH/IR) $\to$ YSG $\to$ BSG $\to$ SN (Type II-SN87A)
	or RSG $\to$ YSG $\to$ BSG $\to$ RSG $\to$ OH/IR $\to$ SN (Type II)

*Different tracks in this mass range may depend critically on initial mass, metallicity, mass loss, and convection.

ers. A modified or opacity-dependent Eddington limit which decreases with decreasing temperature has been proposed and discussed by several investigators (Humphreys and Davidson 1984, Appenzeller 1986, Lamers 1986, Davidson 1987, Lamers and Fitzpatrick 1988). The opacities reach a maximum and the Eddington luminosity a minimum at 10,000 K. Thus the modified Eddington limit will turn up again in the 8000–10,000 K temperature range in agreement with the observed turnover in the luminosity/stability limit. Stars below the critical mass corresponding to the minimum in the Eddington luminosity could then evolve to the red-supergiant region.

But the situation may be more complicated. First attempts to calculate the location of the modified Eddington limit on the HR diagram have not been entirely successful (Lamers and Fitzpatrick 1988). Furthermore, the pressure generated by turbulence in the atmospheres of cool stars may also produce an outward acceleration. Recently, de Jager and collaborators (de Jager 1984, Boer et al. 1988, Piters et al. 1988, de Koter et al. 1988, and Carpay et al. 1989) have shown that the dissipation of mechanical energy in the atmospheres of the cool hypergiants will produce a turbulent-pressure gradient and they have measured supersonic microturbulent motions in these stars' atmospheres.

I think that the observed luminosity/stability limit is a consequence of (1) radiation pressure, *i.e.*, the modified Eddington limit, which dominates in the hot stars, and (2) the turbulent-pressure gradient in the atmospheres of the cool hypergiants which sets an upper boundary to their luminosities independent of radiation pressure in the hot stars.

The observed upper limit to the luminosities of the cool supergiants (F, G, K, and M) near $M_{Bol} - 9.5$ is a consequence of the physics governing the evolution of the most massive stars and the resulting instabilities in the atmospheres of the evolved stars. The luminosities of the brightest M supergiants have been calibrated in several Local-Group galaxies covering 7 magnitudes in galactic luminosity. Figure 2 in Humphreys (1988) shows the luminosity calibrations in M_V, M_K, and M_{Bol} for the confirmed M supergiants in the Local-Group spirals and irregulars plus NGC 2403, M81, and NGC 300. The maximum bolometric luminosity is near -9.5 mag, and $M_{V_{max}}$ and $M_{K_{max}}$ are -8 and -12, respectively, for the most luminous red supergiants. M_{Bol} near -9.5 translates into an M_V near -8, which is nearly constant over five magnitudes of galaxy luminosity. Both M_K and M_{Bol} show some dependence on the luminosity of the galaxy. This is due to the effects of small population samples in the smaller, less luminous galaxies. For those galaxies brighter than $M_V = -18$ we can confidently use this luminosity calibration at $M_V = -8$ and $M_{Bol} = -9.5$ for the brightest red supergiants to determine extragalactic distances out to distance moduli of 33 mag with BVR photometry from WFC on HST, and possibly 35 mag with near-infrared photometry from NICMOS, a second-generation HST instrument.

Given the importance of the red supergiants as endpoints or even turn-around points in massive-star evolution, and their very practical usefulness as distance indicators, how can we identify and measure them, even in distant galaxies?

3. IDENTIFICATION AND MEASUREMENT OF RED SUPERGIANTS VIA THEIR SPECTROSCOPIC AND PHOTOMETRIC SIGNATURES AT OPTICAL AND INFRARED WAVELENGTHS

During the past 20 years the red supergiants have been identified and studied in our Galaxy and in several Local-Group members by spectroscopy and photometry at both optical and infrared wavelengths. In Table 2 I've summarized the major surveys and studies of red supergiants in the different galaxies. It is clear from this list that the red supergiants have been extensively studied in nearby galaxies. The optical and infrared techniques for their identification, confirmation, and measurement have been successfully applied to red supergiants in galaxies as far away as M101 (7 Mpc).

At optical wavelengths, low-dispersion objective-prism spectra have been very successful at identifying candidate red supergiants in our Galaxy and the Magellanic Clouds. At the classical photographic wavelengths the TiO bands are the distinguishing characteristic of the M stars, and the supergiants are easily recognized by the strength of the Fe II lines and weakness of Ca I (4226 Å), the traditional luminosity indicators. However, for supergiants at larger distances beyond the Galaxy and Magellanic Clouds, it is more efficient to use the very luminosity-sensitive Ca II triplet (8500 Å). Wing has developed an 8-color narrow-band system of indices that also yields spectral types and luminosity classification (White and Wing 1978). MacConnell, Wing, and Costa are using it for red supergiants in the Southern Milky Way, and Wing is applying it to the Cloud supergiants.

To identify candidate red supergiants in galaxies at distances greater than the Clouds, multicolor broadband photometry (photographic or CCD) such as $V-R$ vs. $B-V$ or $V-I$ vs. $B-V$ is necessary. A single color such as $V-R$ or $V-I$ is not sufficient because of contamination by foreground red dwarfs. An unpublished figure for the red stars in the field of M101 (Humphreys and Strom 1983) shows the separation of foreground red dwarfs from the probable supergiants. Contamination by foreground red stars is a problem in the fields of Local-Group and other nearby galaxies. In the Clouds, the foreground red stars are all red giants while for the other Local-Group members they are red dwarfs. For this reason, in my studies of the Local Group, I have always supplemented the broadband colors with low-resolution spectroscopy (Ca II triplet) and infrared photometry. However, the contamination by late-type dwarfs will diminish for galaxies at increasing distances; as the magnitude for the brightest red supergiants gets fainter, the number of possible red dwarfs in the field of view will decrease.

Observations and surveys (with IR arrays) in the infrared have several advantages. Broadband JHK photometry plus the CO index can be used to identify red supergiants or as an important supplement to the optical observations. At the infrared wavelengths the interstellar extinction is very low, $A_K = 0.1\ A_V$, and can be measured by the star's

TABLE 2. *Surveys and studies of red supergiants*

Milky Way:	Case low-resolution objective-prism surveys Optical and infrared photometry (Lee 1970) Optical spectroscopy and photometry (Humphreys 1970) Narrow-band photometry (White and Wing 1978) Objective prism, spectroscopy, narrow-band photometry (MacConnell, Wing, and Costa, in progress)
LMC and SMC:	Objective prism (Blanco 1976) Objective prism (LMC, Sanduleak and Philip 1977; SMC, Sanduleak 1989) Objective prism (LMC, Rebeirot et al. 1983; SMC, Prévot et al. 1983) Infrared photometry (Glass 1979) Optical spectra and photometry (LMC, Humphreys 1979a; SMC, Humphreys 1979b; Elias, Frogel, and Humphreys 1985) Infrared photometry (Elias, Frogel, and Humphreys 1985) Spectra and infrared photometry (McGregor and Hyland 1981, 1984; McGregor 1981)
M33:	Photographic survey and photometry (Humphreys and Sandage 1980) Optical spectra and photometry (Humphreys 1980b) Infrared and photometry (Humphreys, Jones, and Sitko 1984)
M31:	Photographic survey and photometry (Berkhuijsen et al. 1988) Spectra and infrared photometry (Humphreys et al. 1988)
NGC 6822 and IC 1613:	Photographic survey (NGC 6822, Kayser 1966) Photographic survey (IC 1613, Sandage and Katem 1976) Optical spectra and photometry (Humphreys 1980a) Infrared photometry (Elias et al. 1981; Elias and Frogel 1985)
NGC 2403, M81, M101:	Photographic surveys (Tammann and Sandage 1968; Sandage 1984a,b, 1983; Humphreys and Strom 1983; Zickgraf and Humphreys, in progress) Spectra and infrared photometry (Humphreys et al. 1986)
NGC 300:	Spectra and infrared photometry (Humphreys and Graham 1986)

position on the $J–H$ vs. $H–K$ two-color diagram. A_K for red supergiants in Local-Group galaxies is typically < 0.1 mag. Furthermore, the bolometric correction to the K magnitude is essentially a constant; it has very little dependence (~ 0.1 mag) on temperature (*i.e.*, spectral type or color), luminosity, or metallicity (Elias, Frogel, and

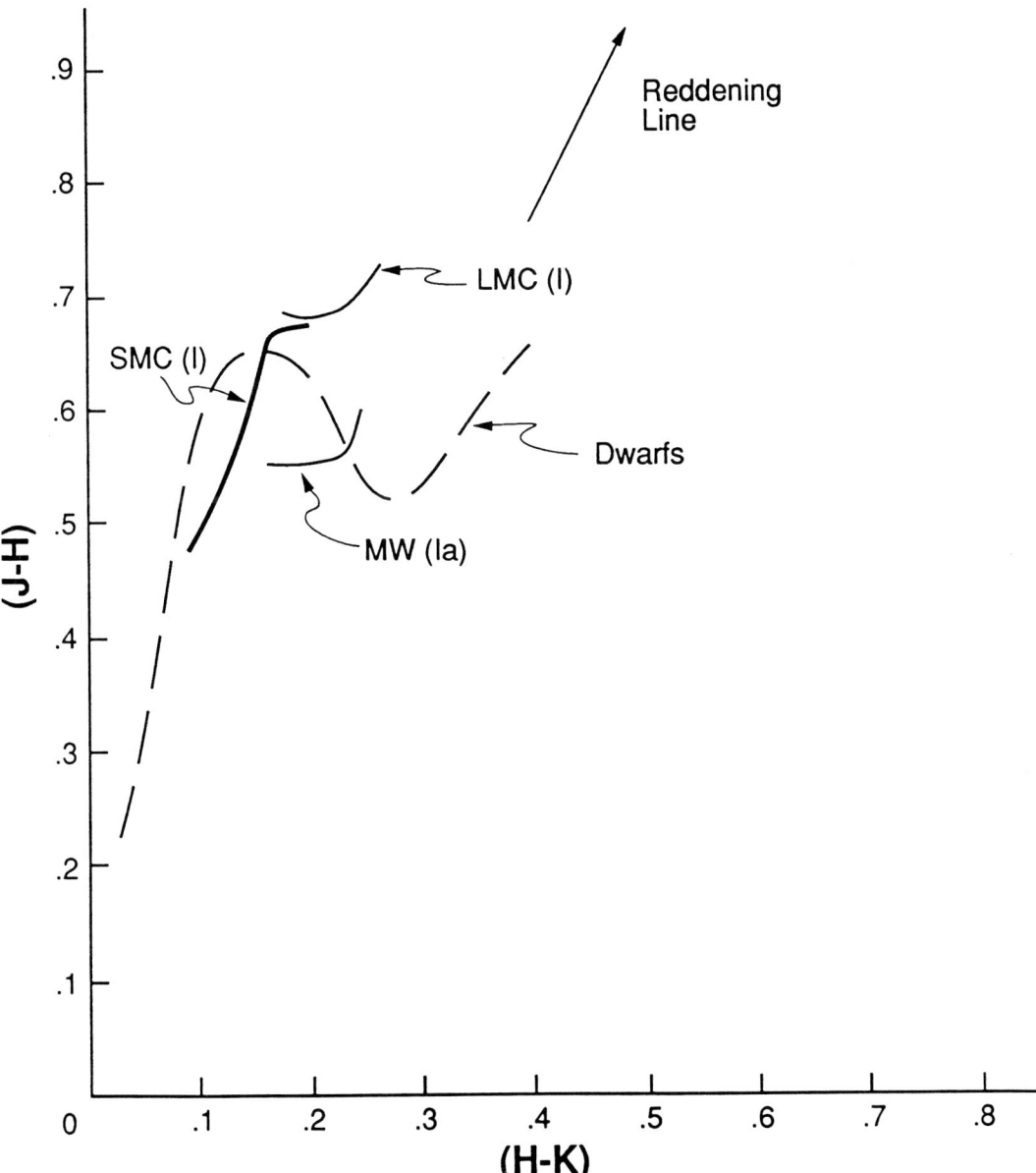

Figure 2. *The J–H vs. H–K intrinsic-color curves for red dwarfs and supergiants in our Galaxy and the Magellanic Clouds.*

Humphreys 1985). Thus, given that a star is a red supergiant, its bolometric magnitude can be determined to within ~ 0.1 mag. Furthermore, supergiants and foreground dwarfs can be separated on the J–H vs. H–K diagram, although the separation is not as unambiguous as we would like. The intrinsic color curves for red supergiants and dwarfs overlap; Figure 2 shows the intrinsic color curves for dwarfs and red supergiants in the Milky Way and the Clouds derived by Elias, Frogel, and Humphreys (1985). However, the red supergiants are usually suffering from some interstellar extinction and can be identified because they lie above the M-dwarf curve (see Figs. 5, 6, 7 in

Humphreys et al. 1988 and Fig. 2 in Humphreys et al. 1986).

But with the addition of the luminosity-sensitive CO index the red supergiants can be unambiguously separated from M dwarfs. The red supergiants have moderate-to-large CO indices but small H_2O indices, while the reverse is true for the red dwarfs. The red giants lie intermediate between the dwarfs and supergiants. Figure 3 shows their behavior on the H_2O/CO plane. The references for these data are Elias et al. 1985, Frogel et al. 1978, and Aaronson et al. 1978. A few supergiant OH/IR sources do show a moderate-to-strong H_2O index with a strong CO index (Jones et al. 1988).

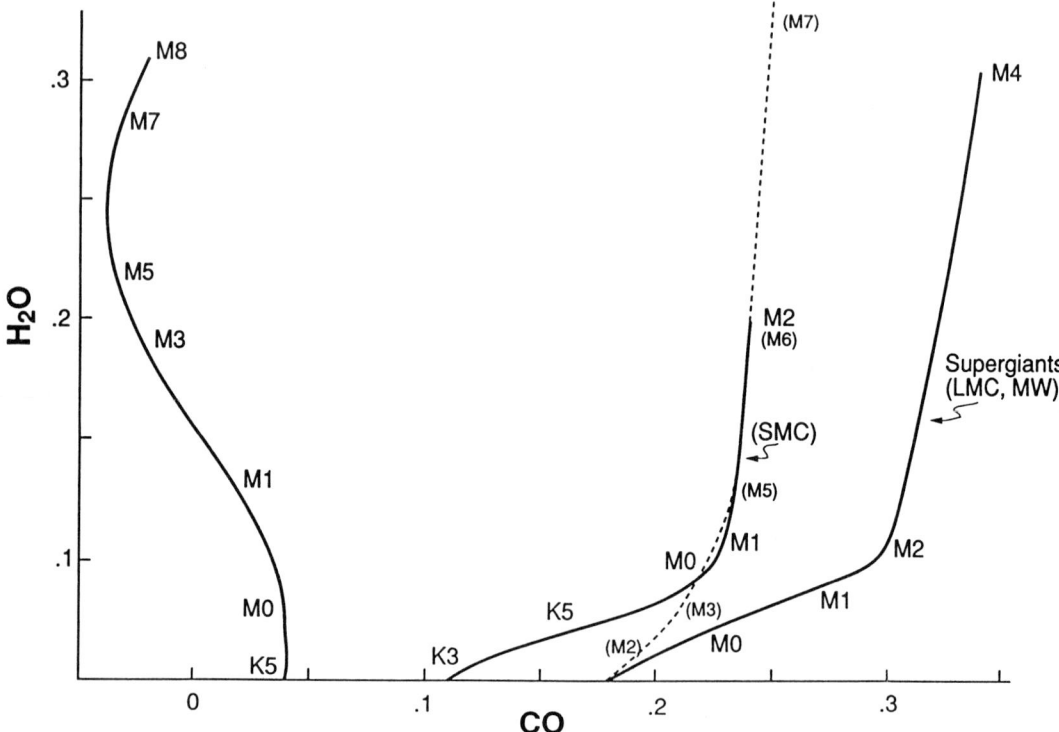

Figure 3. *The loci of the red dwarfs, giants (dashed line), and supergiants in our Galaxy and the Magellanic Clouds in a diagram of the H_2O index vs. CO index.*

There is also an observable metallicity effect on the spectral types and colors of the red supergiants. The comprehensive comparison of the red supergiants in the Clouds and the Milky Way by Elias, Frogel, and Humphreys (1985) clearly showed the shift to earlier spectral types and bluer colors in the Large and Small Clouds (see Fig. 1 in Elias et al. 1985). This shift is most likely due to the well-known shift in the Hayashi track to warmer temperatures with lower metallicity due to the lowered opacities. The behavior of the colors is more complicated but the $V-R$, $V-I$, and $V-K$ colors are bluer. The intrinsic color curves ($J-H$ vs. $H-K$) in Figure 2 for the red supergiants in these three galaxies also show this dependence on metallicity.

Thus JHK photometry plus the CO index will allow the unambiguous identification of the red supergiants, as well as measurement of their extinction and bolometric magnitudes; with the addition of visual photometry V, and R- or B-band measurements, the visual magnitude and information on spectral type and colors can be determined. With the luminosity calibration, the distance to the galaxy can be determined (from M_V or M_{Bol}) and with the resulting colors and luminosities, these stars can be used in stellar-evolution and population studies.

These types of observations can be made with IR arrays and CCD's at the largest ground-based telescopes out to distances of 7–8 Mpc, but at greater distances (> 10 Mpc) the observations are limited by confusion.

4. MASSIVE RED STARS IN STARBURST GALAXIES

In searching for red supergiants or their spectral signatures in starburst galaxies, we must recall that the typical red supergiant is 10 million years old, having evolved from a main-sequence star of 15–40 $M_\odot$. Whether M supergiants are present in a starburst galaxy would depend on the age of the current epoch of star formation. If it is less than 6×10^6 yrs old, then we would not expect red supergiants. But if the star formation has been continuous over at least 10 million years or if there has been a previous burst a few million years before, then the red supergiants should be present. In fact, the presence or lack of red supergiants would be very relevant to mapping the history of star formation and evolution in the starbursts. The red supergiants would also be extremely important for determining the distances to starburst galaxies, especially if observations by HST make it possible to resolve the individual brightest stars in these regions.

The 30 Dor region in the LMC is the closest example of a very active, massive star-forming region in which the brightest stars are resolved. Peter McGregor and his colleagues (McGregor 1981, McGregor and Hyland 1981, McGregor 1984) have studied the blue and red supergiants in the 30 Dor region. They conclude that there are two populations of stars in 30 Dor and that at least two bursts of star formation have occurred in the last 5×10^7 yrs, with the most recent 10^6 yrs ago. My interpretation of their HR diagram for 30 Dor is that more likely there has been continuous star formation over $\sim 10^7$ yrs. (See also Walborn's review in this volume.)

This discussion of the properties of the massive red stars has concentrated on observations of resolved stars; however, the star-forming regions in the starburst galaxies are unresolved. These regions are strong infrared emitters, so it is possible that the signatures of late-type stars, $i.e.$, the H_2O and CO indices, would be detected at long wavelengths as in the infrared studies of the starburst core of M82 by Rieke et al. (1980) and Lester et al. (1990). They report CO band strengths at 2.3μ which, when corrected for dilution by thermal emission, correspond to a population of red supergiants. However, Hyland and Jones (in Harding et al. 1981) report that the integrated near-infrared colors of 30 Dor, in an 11' beam, do not reveal any contribution from late-type supergiants in the 1–2μ region. Alternatively, the CO strength could be due to late-type red giants with higher metallicity than solar. The 2μ and CO measurements at different locations in M82 show the red-star population peaked at the nucleus and spread across the starburst disk, in contrast to the signature of the hot stars (Bγ emission) which is off-center from the nucleus. The red stars must be markers of a previous epoch of star formation 10^7 yrs earlier (or $> 10^8$ yrs if giants) with a homogeneous distribution across the disk.

5. MASSIVE RED STARS IN MORE DISTANT GALAXIES— OBSERVATIONS WITH THE HUBBLE SPACE TELESCOPE

Given the high spatial resolution of the HST and the sensitivity of the WF/PC it will be possible to resolve and measure individual red supergiants in Virgo-Cluster galaxies. For a distance modulus of 31.0 mag, the brightest red supergiants would have V magnitudes of 23. These stars could be measured at V with $S/N = 20$ in integration times of 1000s. Depending on the performance of the WF/PC, we should be able to detect the visually brightest supergiants at distance moduli of 33 mag. This will very likely be the distance limit of the optical wavelength observations.

However, the NICMOS, a near-infrared $(1-2.5\mu)$ camera and spectrometer scheduled for placement in HST in about five years, will greatly extend observations of the luminous red stars. For example, with the NICMOS array the brightest red supergiants in the Virgo Cluster could be detected at H (1.65μ) with integration times of 40s (S/N = 10). Infrared photometry and spectroscopy with NICMOS will permit a detailed look at the regions of massive stars and star formation in galaxies as far away as the Virgo Cluster with a resolution of 10 pc. For the distance scale, the NICMOS observations will permit us to link the well-established distance indicators like Cepheids in the Local Group with the brightest red supergiants which can be readily observed in the Virgo Cluster and in turn with the Type Ia supernovae in more distant galaxies, in reasonable integration times. The brightest red supergiants will be detected in galaxies that significantly overlap with the supernovae. In this way, NICMOS will both validate and calibrate the secondary distance indicators such as supernovae which can be used at the greatest distances.

REFERENCES

Aaronson, M., Frogel, J. A., and Persson, S. E. 1978, *Ap. J.*, **220**, 442.
Appenzeller, I. 1986, IAU Symposium #116, *Luminous Stars and Associations in Galaxies*, p. 139.
Berkhuijsen, E., Humphreys, R. M., Ghigo, F. D., and Zumach, W. 1988, *Astr. Ap. Suppl.*, **76**, 65.
Blanco, V. M. 1976, unpublished.
Boer, B., de Jager, C., and Nieuwenhuijzen, H. 1988, *Astr. Ap.*, **195**, 218.
Carpay, J., de Jager, C., Nieuwenhuijzen, H., and Moffat, A. 1989, *Astr. Ap.*, **216**, 143.
Conti, P. S., Garmany, C. D., de Loore, C., and Vanbeveren, D. 1983, *Ap. J.*, **274**, 302.
Davidson, K. 1987, *Ap. J.*, **317**, 760.
de Jager, C. 1984, *Astr. Ap.*, **138**, 246.
de Jager, C., Nieuwenhuijzen, H., and van der Hucht, K. 1988, *Astr. Ap.*, **193**, 375.
de Koter, A., de Jager, C., and Nieuwenhuijzen, H. 1988, *Astr. Ap.*, **200**, 146.
Elias, J. H., and Frogel, J. A. 1985, *Ap. J.*, **289**, 141.
Elias, J. H., Frogel, J. A., and Humphreys, R. M. 1985, *Ap. J. Suppl.*, **57**, 91.
Elias, J. H., Frogel, J. A., Humphreys, R. M., and Persson, S. E. 1981, *Ap. J. (Letters)*, **249**, L55.
Frogel, J. A., Persson, S. E., Aaronson, M., and Matthews, K. 1978, *Ap. J.*, **220**, 75.
Glass, I. S. 1979, *M.N.R.A.S.*, **186**, 317.
Humphreys, R. M. 1979a, *Ap. J. Suppl.*, **39**, 389.
_____. 1979b, *Ap. J.*, **231**, 384.
_____. 1980a, *Ap. J.*, **238**, 65.
_____. 1980b, *Ap. J.*, **241**, 587.
_____. 1986, in IAU Symposium #116, *Luminous Stars and Associations in Galaxies*, p. 45.

_____. 1987, in *Instabilities in Luminous Early-Type Stars*, ed. Lamers, H. J. G. L. M. and de Loore, C. W. H.), p. 3.

_____. 1988, in *The Extragalactic Distance Scale*, A.S.P. Conf. Series, Vol. 4, p. 103.

_____. 1990, in *Sixth Cambridge Workshop on Cool Stars, Stellar Systems, and the Sun*, A.S.P. Conf. Series, Vol. 9, p. 387.

Humphreys, R. M., Aaronson, M., Lebofsky, M., McAlary, C. W. Strom, and Capps, R. W. 1986, *A. J.*, **91**, 808.

Humphreys, R. M., and Davidson, K. 1979, *Ap. J.*, **232**, 409.

_____. 1984, *Science*, **223**, 243.

Humphreys, R. M., and Graham, J. A. 1986, *A. J.*, **91**, 522.

Humphreys, R. M., Jones, T. J., and Sitko, M. L. 1984, *A. J.*, **89**, 1155.

Humphreys, R. M., Nichols, M., and Massey, P. 1985, *A. J.*, **90**, 101.

Humphreys, R. M., Pennington, R. L., Jones, T. J., and Ghigo, F. D. 1988, *A. J.* 96, 1884.

Humphreys, R. M., and Sandage, A. 1980, *Ap. J. Suppl.*, **44**, 319.

Humphreys, R. M., and Strom, S. E. 1983, *Ap. J.*, **264**, 458.

Jones, T. J., Hyland, A. R., Fix, J. D., and Cobb, M. L. 1988, *A. J.*, **95**, 158.

Jones, T. J., Hyland, A. R., Wood, P. R., and Gatley, I. 1983, *Ap. J.*, **273**, 669.

Kayser, S. E. 1966, Ph.D. Thesis, California Institute of Technology.

Lamers, H. J. G. L. M. 1986, in IAU Symposium #116, *Luminous Stars and Associations in Galaxies*, p. 157.

Lamers, H. J. G. L. M., and Fitzpatrick, E. 1988, *Ap. J.*, **324**, 279.

Lee, T. A. 1970, *Ap. J.*, **162**, 217.

Lester, D. F., Carr, J. S., Joy, M., and Gaffney, N. 1990, *Ap. J.*, **352**, 544.

MacConnell, D. J., Wing, R. F., and Costa, E., 1990, in progress.

McGregor, P. J. 1981, Ph.D. Thesis, Australian National University.

_____. P. J. 1984, *Ap. J.*, **277**, 149.

McGregor, P. J., and Hyland, A. R. 1981, *Ap. J.*, **250**, 116.

Piters, A., de Jager, C., and Nieuwenhuijzen, H. 1988, *Astr. Ap.*, **196**, 115.

Prévot, L., Martin, N., Maurice, E., Rebeirot, E., and Rousseau, J. 1983, *Astr. Ap. Suppl.*, **53**, 255.

Rebeirot, E., Marten, N., Mianes, P., Prévot, L., Robin, A., Rousseau, J., and Peyrin, Y. 1983, *Astr. Ap. Suppl.*, **51**, 277.

Rieke, G. H., Lebofsky, M. J., Thompson, R. I., Low, F. J., and Tokunaga, A. T. 1980, *Ap. J.*, **238**, 24.

Sandage, A. 1983, *A. J.*, **88**, 1569.

_____. 1984a, *A. J.*, **89**, 621.

_____. 1984b, *A. J.*, **89**, 630.

Sandage, A., and Katem, B. 1976, *A. J.*, **81**, 743.

Sanduleak, N. 1989, *A. J.*, **98**, 825.

Sanduleak, N., and Philip, A. G. D. 1977, *Publ. Warner and Swasey Obs.*, 2, No. 5.

Tammann, G. A., and Sandage, A. 1968, *Ap. J.*, **151**, 825.

White, N. M., and Wing, R. F. 1978, *Ap. J.*, . **222**, 209.

Zickgraf, F.-J., and Humphreys, R. M. 1990, in preparation.

DISCUSSION

T. Heckman: You showed that in the near-IR H_2O vs. CO band-strength plot, the M dwarfs and supergiants are well separated. However, for interpreting the integrated near-IR spectra of starburst galaxies, the real issue is discriminating between supergiants and *giants*. Can you comment? Where are the giants in the H_2O vs. CO plot?

Humphreys: The M giants are intermediate between the dwarfs and supergiants in the H_2O–CO diagram. Actually, they are close to the SMC-supergiant locus. Conversely, a more metal-rich giant population could have a CO index comparable to that of supergiants.

D. Weedman: What are the variability characteristics of red supergiants and can they be used as a selection criterion?

Humphreys: Yes. Most red supergiants show some variability. They are typically semi-regular or irregular variables with variations of a few tenths of a magnitude over a few months to a year. A few show much larger variations (like Miras). These are stars like VX Sgr. Variability can be used to identify the supergiants but then we must have observations over several months or years.

N. Scoville: I am a bit confused about the role of dust and the circumstellar envelopes because I would have thought that the very most luminous supergiants would have large mass-loss rates, 10^{-4} to 10^{-5} solar masses per year, which would imply a column density of 10^{22} or even higher and an extinction of one magnitude even at 2 microns. It's hard to understand this upper envelope you see in the visual magnitudes, because they could in principle have 10 magnitudes of visual extinction through the shell and I would have thought that the geometry would be highly influencing the visual fluxes you actually see. I wouldn't expect a clean upper envelope.

Humphreys: Some of them are very dusty, *e.g.,* VY Canis Majoris, and yet they are still optically thin. We are still seeing them in the visual part of the spectrum. For example, μ Cephei, which is probably visually the most luminous red supergiant in the local region of the Milky Way, has a 10-micron silicate feature but the extinction is actually relatively low.

N. Scoville: I suggest that a good technique would be to see if there was a significant IRAS flux and then if so presumably the optical is being degraded into the infrared significantly.

Humphreys: The two examples in the Large Magellanic Cloud show the effect of an optically thick dust shell and show that when we calculate the bolometric luminosity, it's very similar to what we find for the optically thin ones. Similarly for the supergiant OH/IR stars in the Galaxy.

N. Scoville: I think it's not quite so simple to interpret the CO indices in starburst galaxies because there could be hot dust emission, which of course will reduce the equivalent width of the CO band, even if you do have supergiants.

Humphreys: I know that in the Lester *et al.* work on M82 the indices were corrected for that effect.

P. Conti: My impression is that the number of red supergiants in the 30 Dor region is rather small. Thus I would imagine that the star-formation rate a few 10^7 years ago, compared to the more recent "burst" which produced the current O and W-R stars, was relatively low.

Humphreys: There are also evolved A- and F-type supergiants there and evolved B supergiants as well as the red supergiants. There's an evolved population that shows star formation has been going on there for about 10 million years, although there is a more recent event superimposed on it.

I. Gatley: We observed NGC 604 in M33, and found 4 bright ($K < 14$ mag) probable supergiants in a 1' field. They were systematically about $A_V = 3$ mag and redder than your population. This was a 5-minute "snapshot" on a 1-meter telescope. Therefore: (1) The IR technique is useful as you said. (2) NGC 604 (as well as 30 Dor) does have a red-supergiant population. (3) One answer to Scoville's question about very high extinction in the envelope is to select targets in the infrared.

Humphreys: Also in response to Scoville's question, the bolometric luminosities of the red supergiants in these galaxies have been determined from the IR photometry, which minimizes any uncertainties about extinction.

N. Langer: Do your data indicate that the upper luminosity bound in the HR diagram is metallicity independent, and would that mean that an opacity-dependent Eddington Limit (which should strongly depend on Z) cannot solely explain the observed upper luminosity limit?

Humphreys: Actually, we can't tell from the available data sets on massive stars in different galaxies whether the observed upper limit is metallicity dependent. The opacity-dependent Eddington Limit is metallicity dependent and if it is the entire explanation for the upper luminosity of the red supergiants, then their upper luminosity should depend on metal abundance—high luminosity in metal-poor systems, lower in metal-rich ones. We do not observe this effect, possibly because de Jager's turbulent-pressure gradient is operating in the cool hypergiants and because our data sets are dominated by population statistics. For example, in the SMC (metal poor) the red supergiants are actually slightly less luminous. This is because in smaller, less luminous galaxies there are fewer of the most massive stars at any given time. This is a well-known effect in the luminosity calibration of the blue supergiants.

E. Bica: A comment on the occurrence of red supergiants in star clusters. We have observed a sample of 40 LMC and SMC blue clusters (Bica, Alloin, Santos 1990, *A. and A.*, **235**, 103). We use, as red-SG indicators, the TiO bands and Ca II triplet in near-infrared integrated spectra, and ages come from blue-violet colors. We conclude that M spectral-type supergiants occur only in a narrow age range of 7 to 12 million years, characterizing a well-defined spectral phase in the cluster evolution. Particularly, we have identified a cluster in the SMC (NGC 299) with weak TiO and still strong Ca II triplet, showing the spectral behavior of the red-SG phase for low metallicities.

MODEL ATMOSPHERES OF MASSIVE HOT STARS

R. P. Kudritzki, R. Gabler, D. Kunze, A. W. A. Pauldrach, J. Puls
Institut für Astronomie und Astrophysik der Universität München
Scheinerstr. 1
8000 München 80
Germany

1. INTRODUCTION

The spectra of galaxies are in many cases dominated by the signatures of massive hot stars (MHS): either strong recombination lines resulting from the ionization of the interstellar gas due to the prodigious FUV and EUV stellar radiation field are seen as indirect tracers of massive stars or the direct contribution to the continous light can be identified together with the typical spectral characteristics of hot stars like stellar wind P-Cygni and emission lines or photospheric absorption lines. It is obvious that quantitative interpretation of such spectra can only be achieved by means of detailed and deep knowledge about *stellar* spectra. This comprises both, observational stellar spectroscopy and the corresponding quantitative interpretation on the basis of model atmospheres. In this paper the achievements of hot star model atmospheres are summarized from a very personal point of view induced by the shopping list that was provided by the workshop organizers. A more general overview can be obtained from the proceedings of the recent Boulder-Munich Workshop on "Properties of Hot Luminous Stars" (edited by C. D. Garmany, 1990) or the recent ARAA-review by Kudritzki and Hummer (1990). The structure and content of this paper is partially similar to the review given by the first author of this paper a year ago on the occasion of the 50th anniversary of McDonald Observatory (see "Frontiers of Stellar Evolution", ed. D. Lambert, to appear in the *P.A.S.P.* Conference Series). However, in particular sections III, IV and V contain a variety of new aspects and results that have become available only during the past year.

2. MODEL ATMOSPHERES—THE STATE OF THE ART

Two effects complicate the modelling of MHS atmospheres: severe departures from LTE and stellar winds. A detailed NLTE treatment is inevitable throughout the entire atmosphere even in the deepest layers of the continuum formation. The existence of stellar winds—a relatively strong ($\dot{M} \approx 10^{-6}$ to 10^{-5} $M_\odot$/yr) outflow of matter with supersonic velocities up to 3000 km/s—requires a hydrodynamic treatment, additionally. However, in practice, for many purposes it is sufficient to divide the atmospheres of hot stars into two different, but interacting entities; the inner subsonic "photospheres", where hydrostatic equilibrium is a good approximation (see for instance Kudritzki, 1988), and the outer supersonic "winds".

2.1. Photospheres

Photospheres of normal OB stars emit the optical and UV continuum, the weak lines and the wings of the hydrogen and helium lines. It is important to note that the EUV continuum shortward of the He II-edge at 228Å is normally not formed in the photospheres, but emerges from the supersonic wind layers. (For Wolf-Rayet stars—the late stage of MHS evolution characterized by strong and dense winds—photospheres do not exist and the whole continuum is formed in the wind. In view of the enormous recent progress in the quantitative spectroscopy of WR-Stars (Hamann et al. 1988, Hillier, 1984, 1987a, b, 1988, Schmutz et al. 1989) it is unfortunate that we cannot discuss these objects here). Photospheres of MHS are characterized by severe departures from LTE which are caused by the intense radiation field dominating over electron collisions. Although this has been known for 20 years, it is worthwhile to give a significant example. Fig. 1 shows theoretical profiles of two strategic photospheric absorption lines, H_γ ad He II 4542, in LTE and NLTE. LTE leads to profiles that are much too weak and fail completely to match the observations. The same is true for the lines of neutral helium, almost all of the metal lines and even for the shape of the continua. Thus, LTE models are useless for a precise determination of O-star parameters.

It is of course interesting to know where NLTE effects become less important and where the much simpler LTE treatment is accurate enough. This is shown in Fig. 2, where the borderline is defined by the difference of H_γ equivalent widths between LTE and NLTE. In the NLTE domain the difference is larger than 15% whereas in the LTE domain it is smaller than 15%. We see that for stars with $M(ZAMS) \geq 25 M_\odot$ a NLTE treatment is inevitable. It is, however, important to note that even in the LTE domain careful NLTE calculations for complex metal atoms are needed for the determination of accurate metal abundances (for examples, see Kudritzki 1988).

An example for the present state of the art of detailed NLTE calculations is given in Fig. 3 for the case of C II in B stars. Atomic NLTE models of a similar degree of sophistication for O II, N II, Si II, III, IV were developed by Becker and Butler (1988 a,b,c, 1989, 1990 a,b). The absolute need for such very elaborate models is demonstrated in Fig. 4. The older less complete models by Lennon (1983) led to a significant discrepancy with the observed C II λ 4267 line—the strongest C II line in the optical spectra of B stars. The new calculations resolve this discrepancy and lead to reasonable abundances. Further examples for the impressive state of the art in calculating NLTE photospheric models are given by the papers of Anderson (1990), Kunze (1990), Werner et al. (1990) all published in the Boulder-Munich Workshop cited above.

2.2. Winds

As has been proved convincingly by Lucy and Solomon (1970) and Abbott (1979, 1985), the winds of MHS are basically radiation driven. The reason is again, as for the

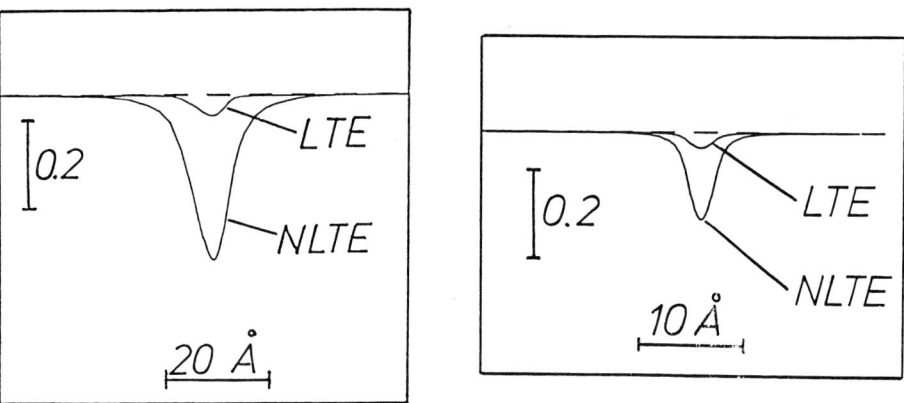

Figure 1. *Photospheric profile of H_γ (left) and He II 4542 (right) calculated in LTE and NLTE for a typical O4f-star.*

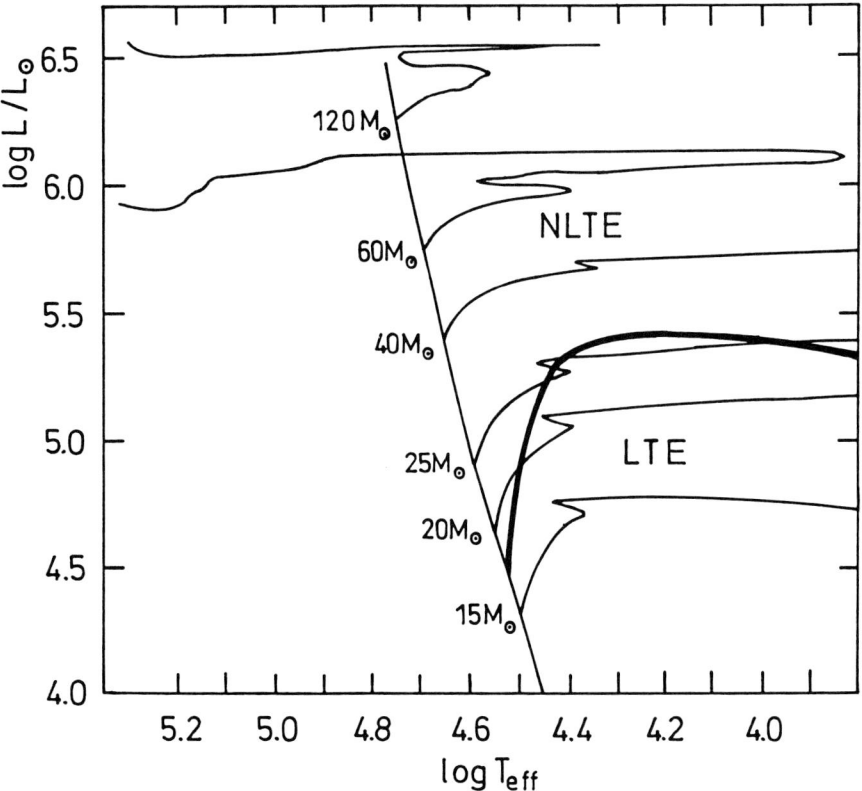

Figure 2. *NLTE and LTE domain in the HR-diagram compared with evolutionary tracks by Maeder and Meynet (1987).*

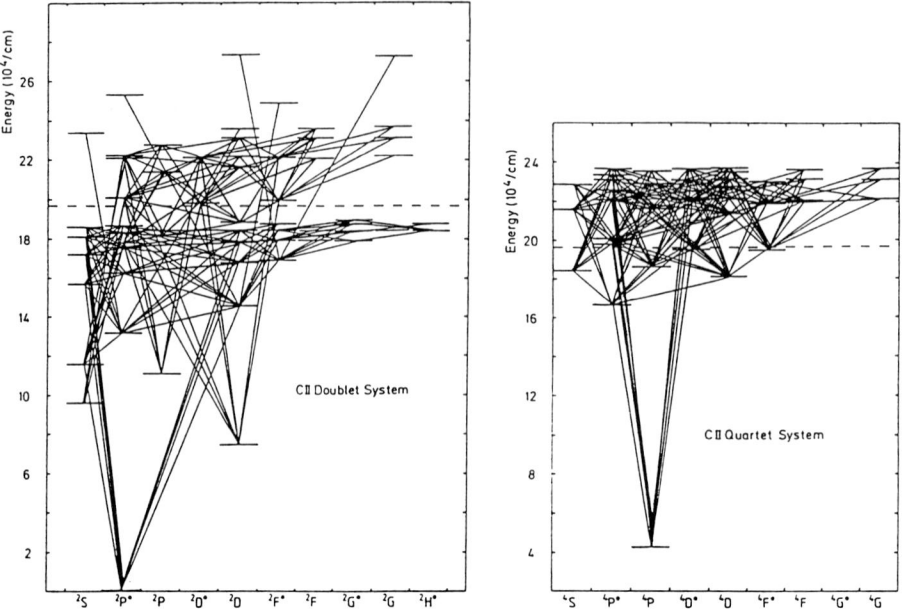

Figure 3. *Grotrian diagram of the atomic model for C II used by Eber and Butler (1988) in their NLTE calculations. Doublets and quartets are treated simultaneously in the rate equations as intersystem collisions are included. Note also the inclusion of individual transitions from autoionization levels.*

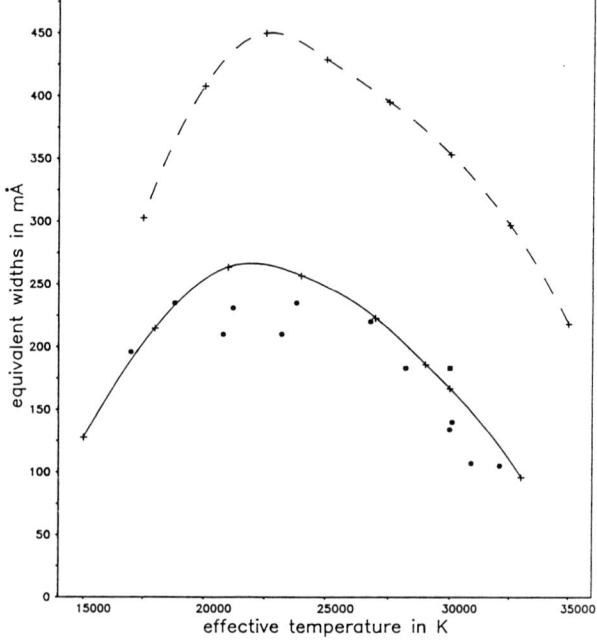

Figure 4. *Equivalent widths of C II λ 4267 as a function of T_{eff} for main sequence models (log g = 4.0). Dashed: older NLTE calculations by Lennon (1983). Solid: new results by Eber and Butler (1988). Filled circles represent the observations.*

departures from LTE, the intense radiation field, which by absorption due to UV metal lines leads to an outward accelerating force strong enough to initialize and to maintain stellar winds.

The basis for the theory of radiation driven winds was originally formulated in two brilliant papers by Castor, Abbott and Klein (1975, "CAK") and Abbott (1982). It was further extendend by Pauldrach, Puls and Kudritzki (1986) and Friend and Abbott (1986) who included the important "finite cone angle effect" of the accelerating continuous radiation field emerging from the photospheric disk. The next important improvements concern the calculation of the occupation numbers in the stellar winds and the velocity induced line overlap. For a long time occupation numbers in the wind were computed in a very approximate way. However, recently the first realistic calculations became available. Pauldrach (1987) adopted a cool wind ($T_W \approx T_{eff}$) and treated the full multilevel NLTE problem of all relevant elements and ions including electron collisions self-consistently with the radiation driven wind hydrodynamics. He included 26 elements, 133 ionization stages, 4000 levels and 10000 radiative bound-bound transitions (see Fig. 5a, as an example) in the rate equations and the correct continuous radiation field for the bound-free rates calculated from the spherical transfer equation. Puls (1987) extended these improvements significantly by including "multiline effects". These arise from the velocity induced line overlap, which causes radiative coupling between different ions and possible multiple momentum transfer from photons to the wind plasma. The first application of this improved theory to the case of the O4f-star ζ Puppis shows excellent agreement with respect to the dynamical quantities $\dot{M}$ and v_∞. In addition—and this is the really important result—an enormous shift towards higher ionization stages is obtained due to the detailed NLTE treatment. Thus, it was possible to reproduce for the first time the observed high ionization features of O VI, N V, C IV etc. by cool wind models. A typical example demonstrating the state of the art is given in Fig. 5b, which shows how perfectly the observed N V profile in the UV-spectrum of ζ Puppis can be reproduced by the consistent radiation driven NLTE wind model atmospheres. Note that this is not a profile fit, where $\dot{M}$ and $v(r)$ have been adjusted as parameters. It is the result of a consistent calculation, that depends only on the choice of the stellar parameters; $\log L/L_\odot$, T_{eff} and $\log g$.

2.3. Unified Model Atmospheres

Hydrostatic photospheric models neglect the emission of the surrounding stellar wind envelope. They are therefore unable to produce the wind-induced strong emission features in the optical and IR spectra of hot stars (H_α, He II 4686, IR excess, IR emission lines) or the distortion of photospheric absorption profiles (H_β, He I 5876, H I, He II 5412) that are observed in many cases. The normal procedure to account for these effects is to put a wind on top of a hydrostatic photosphere and then to study its influence on the spectrum. This method has crucial limitations as it makes an artificial division between photospheres and winds and introduces unnecessary free parameters which strongly influence the calculations at the borderline between the two regions. The alternative is the concept of "Unified Model Atmospheres" introduced recently by Gabler et al. (1989). This new model atmosphere code takes mass-loss rate, density and velocity structure from the stationary wind code (see section 2.2) and calculates temperature structure, hydrogen and helium occupation numbers by a detailed NLTE treatment using radiative equilibrium and correct spherical radiative transfer. These models need only T_{eff}, $\log g$, $R/R_\odot$ at the inner atmospheric boundary

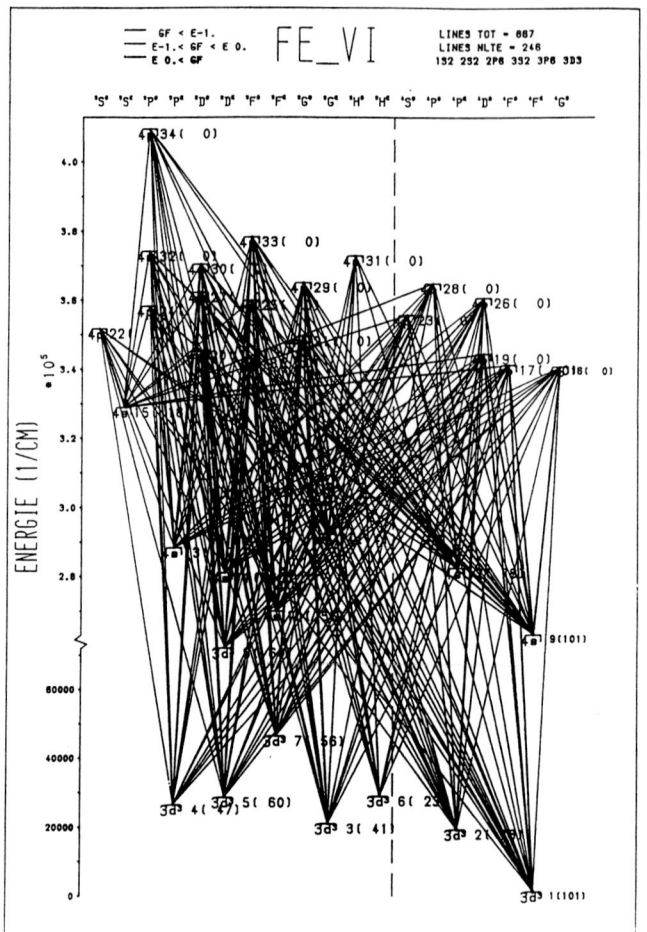

Figure 5a. *Grotrian diagram of FeVI as used in NLTE in the Munich stellar wind code. (From Pauldrach and Puls, 1990).*

as the natural free parameters and are in principle selfconsistent. They are spherically extended and yield the entire sub- and supersonic atmospheric structure, the stellar energy distribution and the hydrogen and helium line spectra including the effects of the outflow velocity field. The weakness of these models in their present stage is that the transfer of photon energy from the continuum into kinetic wind energy due to metal-line absorption is not yet considered in the energy equation. On the other hand, they are as good as the standard hydrostatic photospheric NLTE models as far as opacities and radiative equilibrium are concerned, and they are therefore ideal for studying differentially the influence of sphericity and winds on the observed energy distribution and on hydrogen and helium lines. These effects are quite drastic: The energy distribution is changed dramatically in the far IR and in the EUV. The unified models reproduce the observed IR excess (see Fig. 6a) and yield a factor of a thousand more photons in the EUV for wavelengths shorter than the He II edge at $\lambda = 228$ Å (Fig. 6b). This latter effect is important for the ionization of H II regions (see next section) and planetary nebulae.

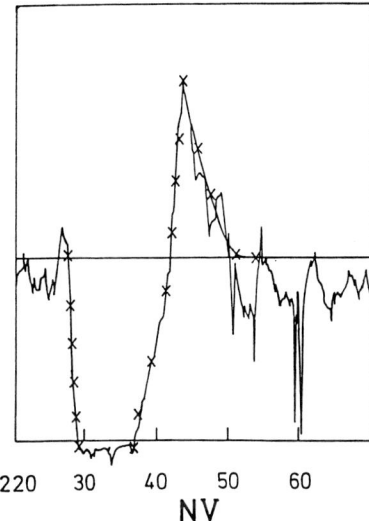

Figure 5b. *The IUE high resolution stellar wind profile of N V λ 1240 of the O4f-star ζ Puppis compared with the result of selfconsistant radiation driven wind model (from Puls, 1987).*

The observed IR emission lines (P_α, B_α, He II 10124, He I 10830) and as well as the emission lines in the optical (H_α, He II 4686) are also reproduced (see Fig. 7a,b). Particulary in view of the forthcoming advances in ground-based and space-IR spectroscopy this is an extremely promising result, which will allow one to use the IR-emission-line characteristics as a tool for quantitative analyses.

3. IONIZING FLUXES

Thanks to the new NLTE radiative transfer techniques developed during the past years—in particular the Accelerated Lambda Iteration method by Werner and Husfeld (1985)—the present photospheric NLTE models are able to take into account the metal opacity in a more realistic way than before. Fig. 8 shows a purely photospheric NLTE calculation by D. Kunze carried out at Munich Observatory for a late-O main-sequence star. Evidently, the emergent EUV radiation field is significantly modified by metal absorption. We have, therefore, calculated a new grid of O-star model photospheres with different metal content to investigate the effects of metallicity. The results are summarized in Fig. 9, which displays the number of emerging ionizing photons as a function of effective temperature for main-sequence stars and supergiants.

Several comments are necessary: effects of metallicity appear to be negligible in the hydrogen continuum but become important towards shorter wavelengths, in particular the O II continuum. The effects are less important for supergiants. As the comparison between Fig. 9a and b shows, both the number of photons and the "hardness" of the ionizing spectra (*i.e.*, for instance, the ratio of H to He I or O II ionizing photons) depends strongly on luminosity class, as soon as O stars become cooler than 42000 K. For population synthesis/analysis, which uses the strength of observed recombination lines as a tool to determine the upper mass cutoff of the IMF this might be of importance.

The calculations presented in Fig. 9 neglect an important effect that was first inves-

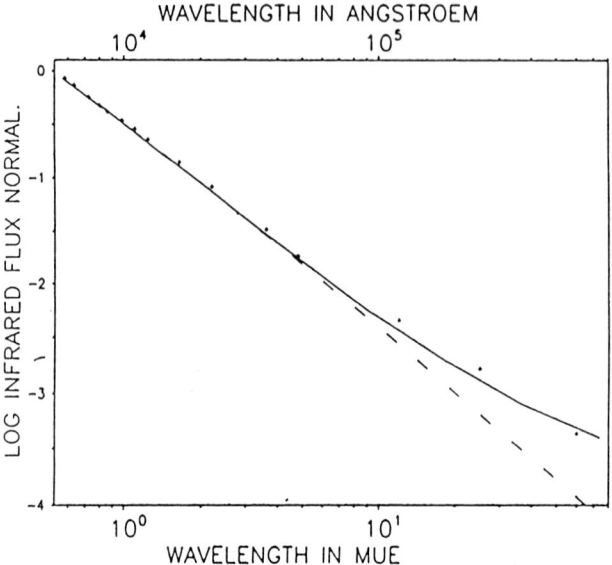

Figure 6a. *IR energy distribution for ζ Puppis calculated with the unified model atmosphere code (solid curve) and with a standard hydrostatic NLTE code (dashed-dotted). The crosses represent observations. (From Gabler et al. 1989).*

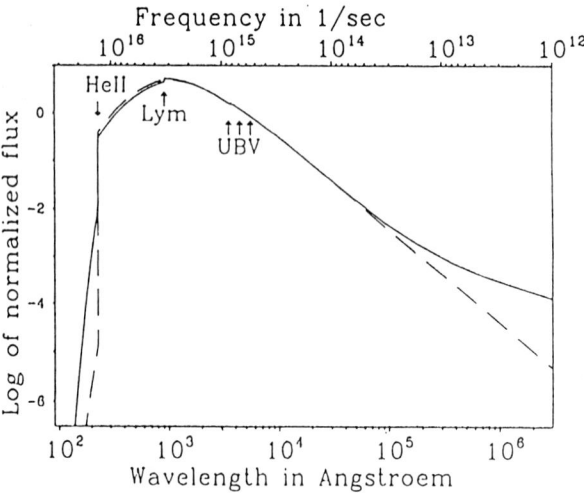

Figure 6b. *The difference in emergent EUV flux shortward of 228Å between unified models and standard hydrostatic NLTE models. (From Gabler et al. 1989)*

tigated by Hummer (1982) and later studied by Abbott and Lucy (1985) and Abbott and Hummer (1985): photons streaming from the photosphere must be partially scattered back by the surrounding winds, due to absorption and spontaneous reemission in metal lines. This "wind blanketing" causes a heating of the photosphere, which means that the density of the radiatively driven wind influences the thermodynamics (and, therefore, spectral appearance) of the deeper atmospheric layers.

We have used the Munich stellar-wind code (as described in the foregoing section) to include the effects of wind blanketing in the grid of models of Fig. 9. Fig. 10

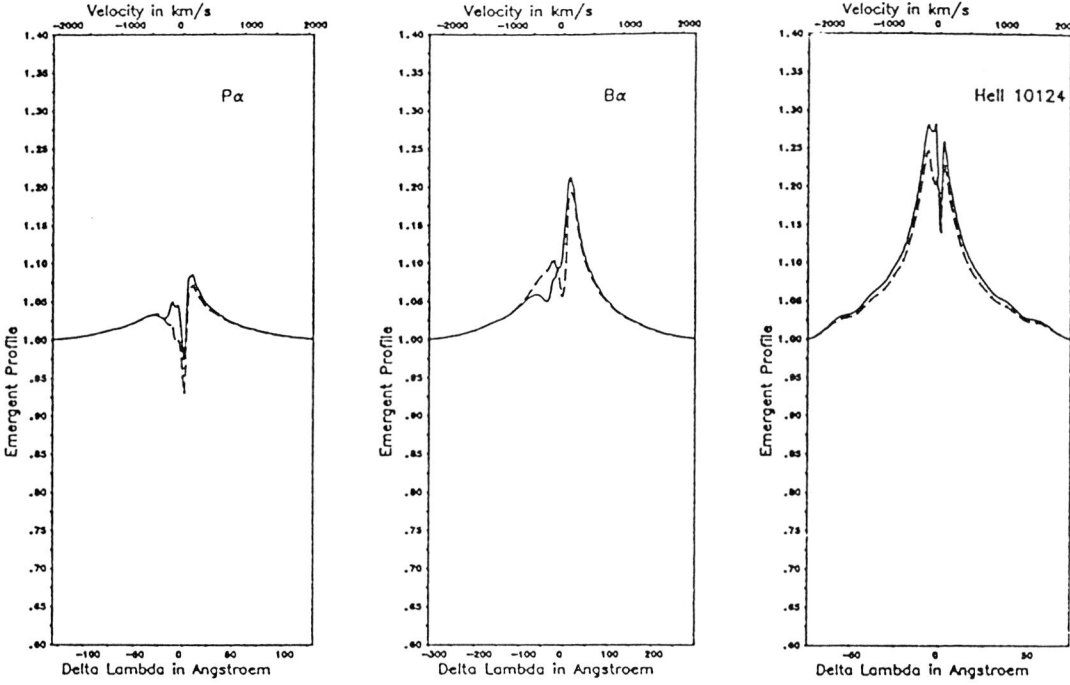

Figure 7a. *IR O-star emission lines as computed from unified models. (From Gabler et al. 1989)*

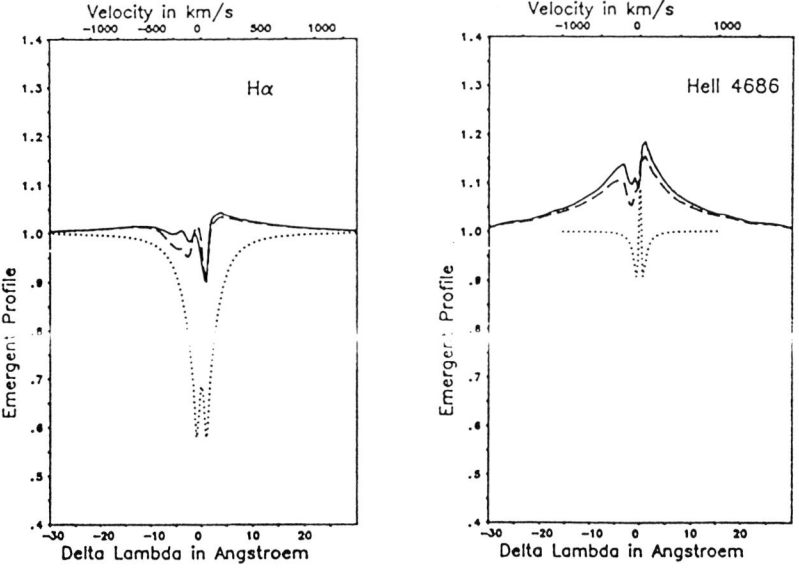

Figure 7b. *H_a and HeII 4686 from unified models compared with hydrostatic NLTE models (dotted). (From Gabler et al. 1989)*

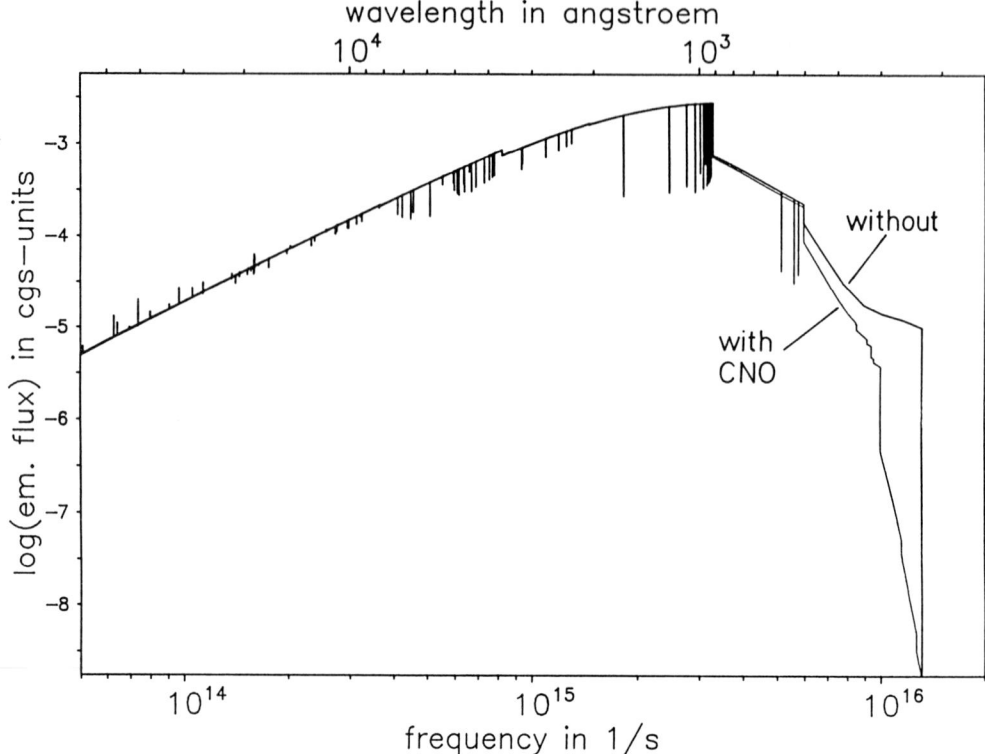

Figure 8. *Logarithm of emergent flux as a function of (logarithmic) frequency (or wavelength) for a late main-sequence O star (T_{eff} = 35000 K, $\log g$ = 4) using the hydrostatic NLTE code developed by D. Kunze (Munich Observatory). The influence of CNO-opacity is clearly seen.*

demonstrates the metallicity dependence of emergent line spectrum, wind albedo and photospheric temperature structure for a typical O4f supergiant. Fig. 11 is similar to Fig. 9 but includes now the wind blanketing effects. Surprisingly, the differences are very small. The amount of photospheric photons blocked and scattered back by the stellar wind is obviously compensated by the additional emission due to the photospheric heating. Thus, the conclusions with regard to luminosity and metallicity drawn from Fig. 9 remain unchanged.

Some extragalactic H II regions show the recombination line of He II 4686, which means that the ionizing flux shortward of 228Å of the central objects must be extraordinarily strong as for the case of planetary nebulae. Hydrostatic NLTE models in a parameter range reasonable for O stars fail to predict such fluxes because of their enormous He II absorption edge. On the other hand, unified models including the He II emission of the surrounding wind produce much higher fluxes for $\lambda < 228$Å, as was demonstrated in section II.

We have therefore also calculated a grid of unified models for main-sequence stars and supergiants to compare the number of predicted He II photons. The result is shown in Fig. 12, which reveals a difference of 2.5 dex in $\log N_{phot}$ (He II) due to wind effects. It would be interesting to investigate whether in this way very hot O-stars (spectral type O3) could be made responsible for the high-excitation class of these peculiar H II regions.

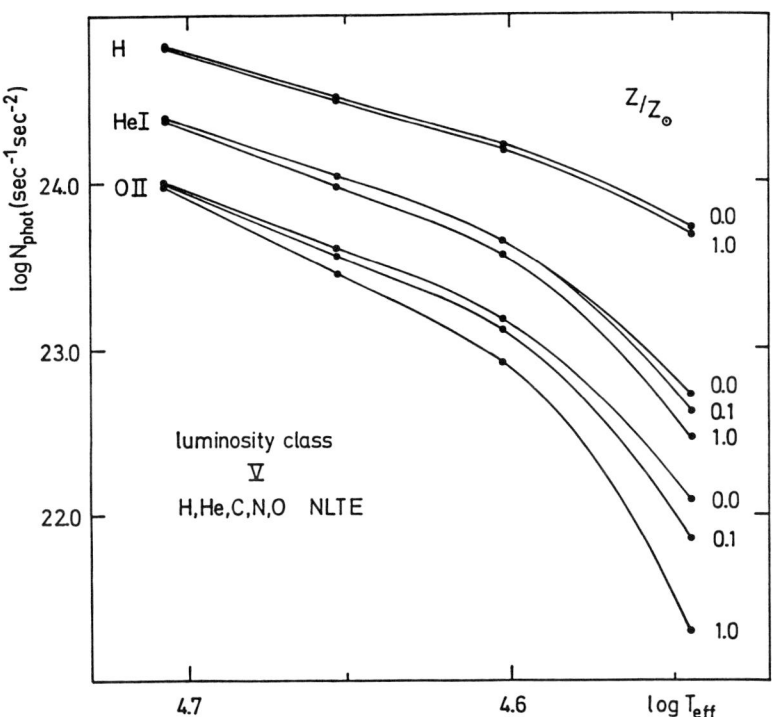

Figure 9a. Number of ionizing photons emerging per second and per cm^2 from the stellar surface as a function of T_{eff} and metallicity. Curve H means photons with $\lambda < 911 Å$, He I $\lambda < 504 Å$, O II $\lambda < 353 Å$. The calculations are carried out for main-sequence stars.

4. WINDS AND METALLICITY

One of the most frequent questions with regard to the evolution of stars and galaxies concerns the metallicity dependence of stellar winds: the evolution of massive stars depends crucially on the rates of mass loss. The yields of processed matter provided by evolved stars due to mass-loss are extremely important for the chemical evolution of galaxies. The amount of mechanical energy of the stellar winds provided continuously during the evolution from the main sequence to pre- and post-RGB phases is an important source for heating the ISM, which might be crucial for the energetics of sequential star formation processes.

The ideal laboratory to investigate this problem are the Magellanic Clouds in comparison with the Galaxy. If it is true that LMC and SMC have a significantly lower metallicity (Dufour, 1984) and that the winds of MHS are driven by metal line absorption, then one would expect the winds in the Clouds to be weaker than in the Galaxy.

The observations of terminal wind velocities fulfill this expectation. Garmany and Conti (1985) and Garmany and Fitzpatrick (1988) used low resolution (!) IUE spectra to demonstrate that (statistically) the inequality

$$v_{\infty}^{SMC} < v_{\infty}^{LMC} < v_{\infty}^{Galaxy}$$

holds. The situation is not that clear with regard to mass-loss rates. Adopting ad

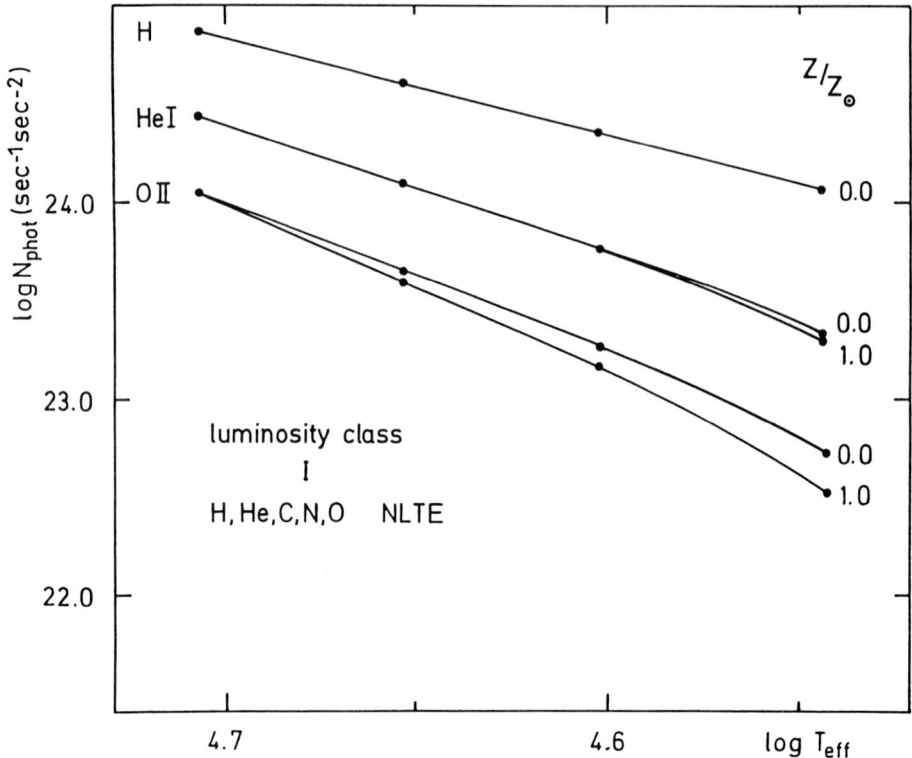

Figure 9b. *Same as 9a, but for supergiants.*

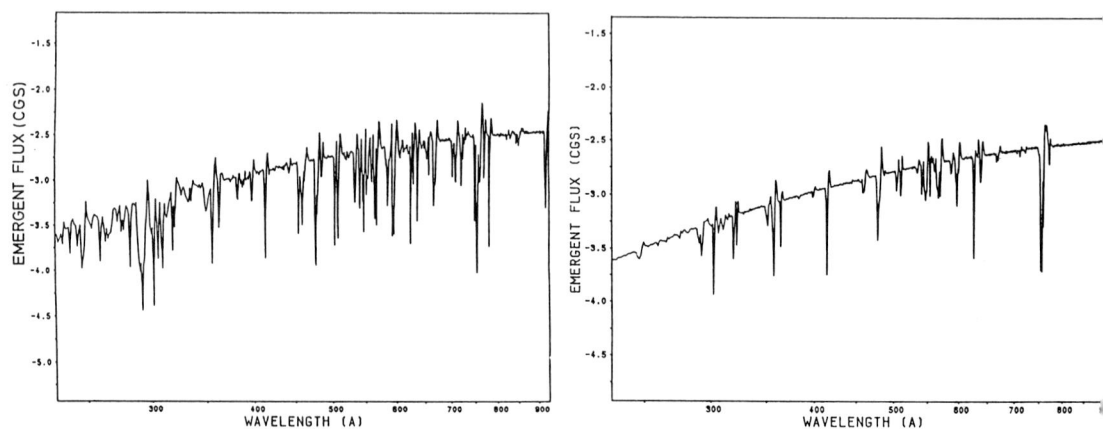

Figure 10a. *Emergent EUV-flux of an O4f-supergiant with $Z = Z_\odot$ (left) and $Z = Z_\odot/10$ (right).*

hoc 1/10 solar nitrogen and carbon abundance for their SMC O-stars Garmany and Fitzpatrick deduce from the strengths of the N V and C IV wind lines that the mass-loss rates in the SMC and the Galaxy are comparable. However, as they point out in their paper, this might be a marginal conclusion because of the low resolution quality of the data. On the other hand Leitherer (1988a,b) came to a similar conclusion using a completely different technique, the observed strengths of stellar H_α-lines. But again,

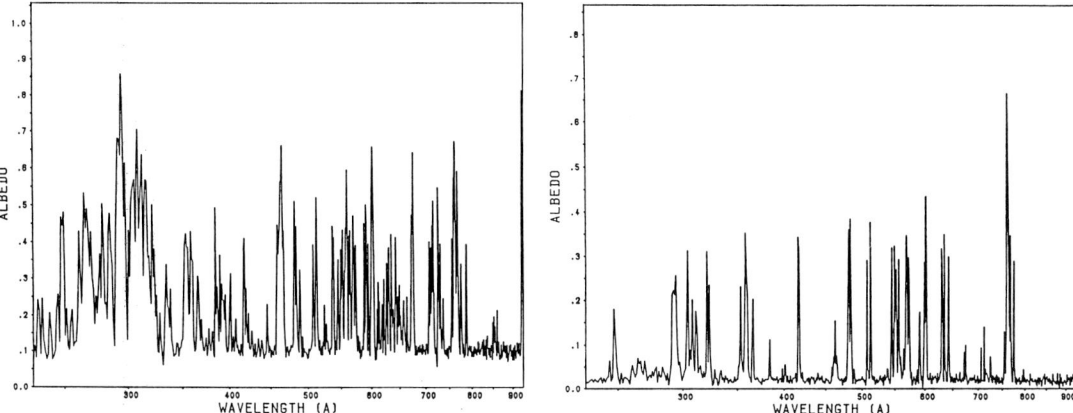

Figure 10b. *Wind albedo corresponding to Fig. 10a.*

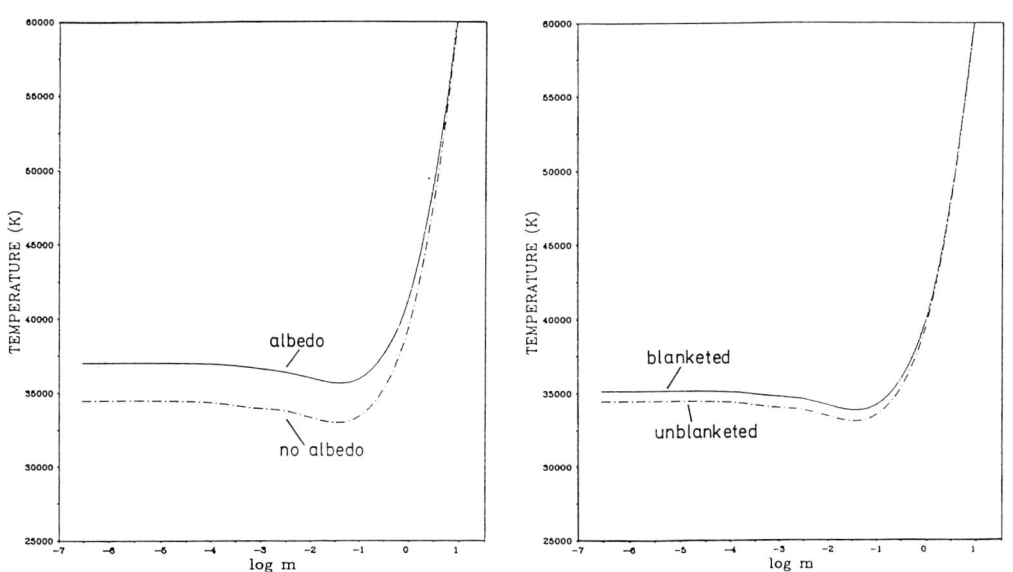

Figure 10c. *Temperature structure corresponding to Fig. 10b.*

his LMC/SMC H_α observations were only of intermediate resolution and S/N so that in particular the contamination with nebular H_α recombination (although accounted for in the reduction) might have affected the measurement of equivalent widths.

From the viewpoint of theory the situation seemed satisfactory until very recently. Kudritzki *et al.* (1987) had calculated radiation driven wind models with different metallicity along evolutionary tracks by Pylyser *et al.* (1985) of corresponding metal content. These calculations could reproduce the v_∞ observations in the SMC, LMC and the

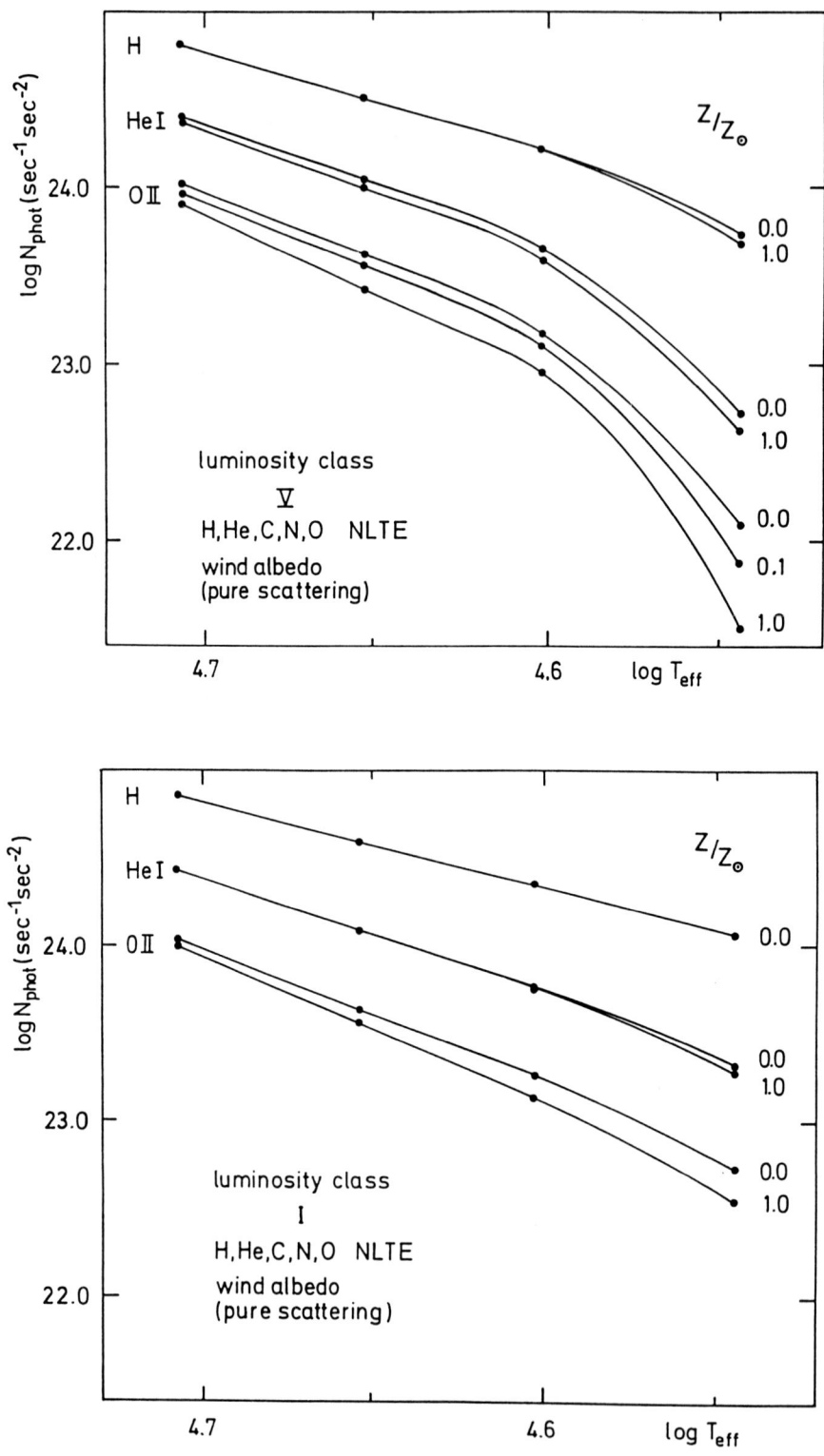

Figure 11a,b. *Same as Fig. 9a,b, but including effects of wind blanketing.*

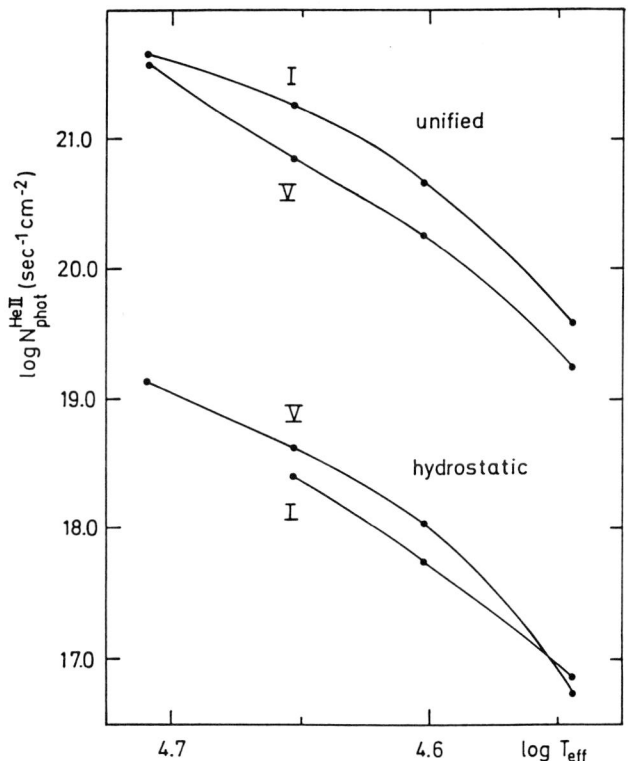

Figure 12. *Number of He II ionizing photons ($\lambda < 228$) calculated from hydrostatic NLTE models and Unified Models for a sequence of Main Sequence Stars and Supergiants.*

Galaxy and provided at least an explanation for the small differences between the mass-loss rates in each galaxy. Fig. 13 summarizes the results of these old calculations with regard to mass-loss rates: Although for tracks with same initial mass but different metallicity $\dot{M}$ is smaller in the SMC than in the LMC or the Galaxy, it would be difficult to disentangle this effect observationally, because the evolutionary tracks cross each other in the ($\log \dot{M}$, $\log L$)-plane.

Thus, the conclusion was that, although $\dot{M}$ scales roughly with $(Z/Z_\odot)^{1/2}$ this cannot be observed, unless very detailed and precise spectral diagnostics of individual objects are carried out.

However, the old computations by Kudritzki et al. (1987) had two basic disadvantages. First, they were done, before Pauldrach (1987) developed his full NLTE treatment of radiation driven winds. Therefore, they applied only a very approximate recombination type NLTE-treatment for the occupation numbers in the wind, as it was suggested by Abbott (1982) and also used by Pauldrach et al. (1986). However, due to the work of Pauldrach (1987) it became evident that this approximate treatment is rather poor and yields misleading results for the ionization and the wind dynamics. Second, the evolutionary tracks by Pylyser et al. (1985) used the Roxbourgh-criterion for the treatment of overshooting. As it became clear later, the amount of overshooting and convective mixing is strongly overestimated in this treatment. Thus, the evolu-

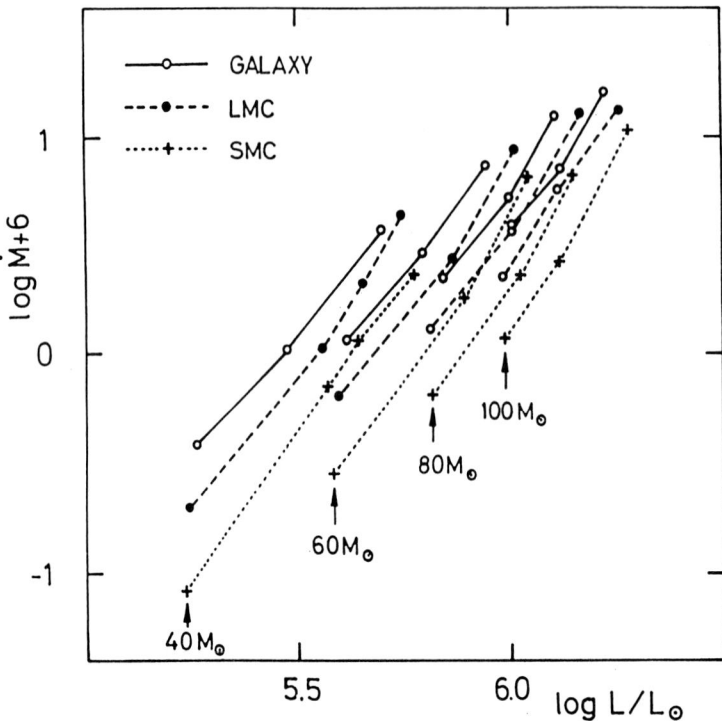

Figure 13. *Logarithm of mass-loss rates as function of logarithm of luminosity calculated along evolutionary tracks of different mass. The calculations are carried out for galactic metallicity $Z_\odot$ (Galaxy), $0.3Z_\odot$ (LMC), $0.1Z_\odot$ (SMC). These are the old calculations by Kudritzki et al. (1987) along tracks by Pylyser et al. (1985).*

tionary tracks used by Kudritzki et al., although the only tracks available including mass-loss and overshooting at lower metallicity at that time, were not very reliable.

Very recently, we have therefore decided to perform new calculations. We have used the improved wind theory as described in section II and calculated wind models along evolutionary tracks. For galactic metallicity we used the tracks by Maeder and Meynet (1987). (This causes, of course, an inconsistency, as these tracks used mass-loss rates for evolution different from those computed by us afterwards. This will be iterated in the near future). For the LMC ($0.25Z_\odot$) and SMC ($0.1Z_\odot$) metallicities we collaborated with M. El Eid and N. Langer (both Göttingen) to couple our wind code with their stellar evolution code (details of this collaboration will be published in *Astron. Astrophys.*) so that mass-loss rates and evolution are consistent.

These new calculations differ remarkably from the old ones described above. Let us first discuss the terminal velocities. Fig. 14 shows v_∞ as calculated along the tracks and compared with observations. Although the topology of the $(v_\infty, \log T_{eff})$ diagram is now completely different from Kudritzki et al. (1987)—a consequence of the different evolutionary tracks—the agreement between theory and observations is satisfactory. The lower terminal velocities in the SMC are nicely reproduced by these calculations.

Fig. 15 shows the mass-loss rates resulting from these new computations. The comparison with the old computations reveals two striking differences. First, the new rates are lower by 0.3 to 0.5 dex. Second, no crossings between tracks of different metallicity does occur, any more. The consequences are seen in Fig. 16, which overlays the H_α mass-loss rates by Leitherer (1988a,b). (Note that the log $\dot{M}$-values determined by Leitherer have been reduced ad hoc by 0.2 dex to account for the He II-blend at H_α, which was neglected but is important, as shown very recently by Gabler et al. 1990). For the Galaxy the observed and calculated rates do almost agree (the offset of ~ 0.1 dex is no reason to worry). However, there appears no way, how the calculations for 1/10 solar metallicity can match Leitherer's H_α mass-loss rates.

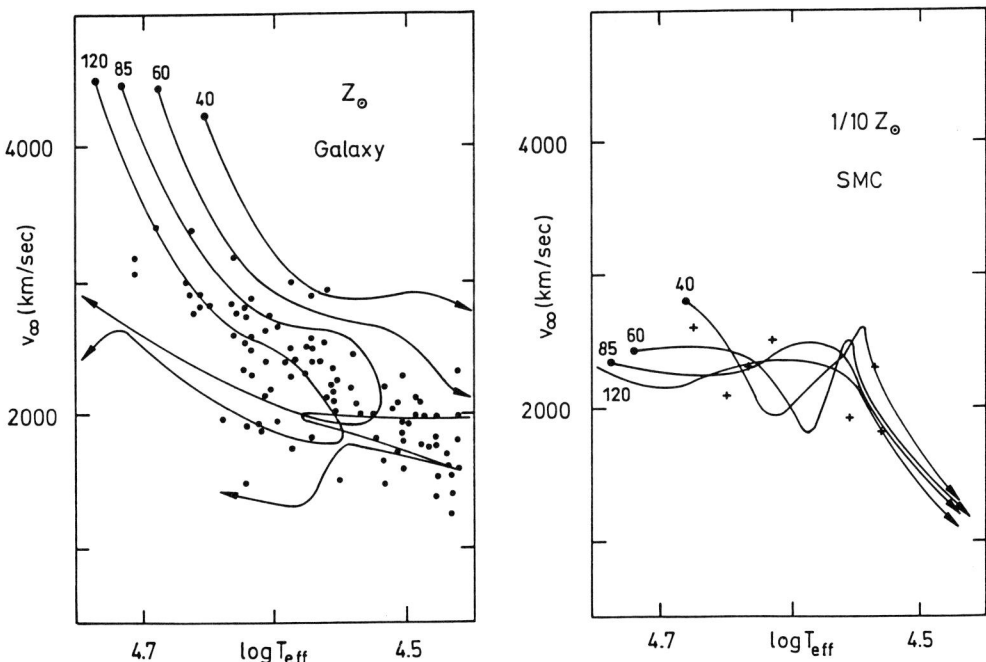

Figure 14. *Terminal velocity versus $\log T_{eff}$ along evolutionary tracks of different mass for solar metallicity (Galaxy) and 1/10 solar (SMC). Details of these new computations are described in the text. The observations for the Galaxy (dots) and SMC (crosses) refer to O stars with $\log L/L_\odot \geq 5.4$ corresponding to the luminosities of tracks with $M/M_\odot \geq 40$. The observed data are from Howarth and Prinja (1989) for the Galaxy and from Garmany and Fitzpatrick (1988) for the SMC. Tracks by Langer/El Eid (priv. comm.).*

To summarize, the new calculations, which make use of much better physics than the old ones, match the terminal velocities in SMC and Galaxy and the mass-loss rates in the Galaxy. However, they fail to reproduce the observed rates in the SMC, if 1/10 solar metallicity is adopted. On the other hand the comparison with Fig. 15 indicates that no problem would arise, if the metallicity would be 1/4 solar. It is important to note, in this connection, that the crucial elements driving the wind are the iron group elements. Thus, if iron would not be as strongly depleted in the SMC as, for instance, CNO, than we would perhaps have a solution, although we then would have to test again, whether the terminal velocities still agree. On the other hand one should not forget that, as discussed above, the quality of the observational data collected so far to investigate the wind properties in LMC and SMC is far from being excellent. This

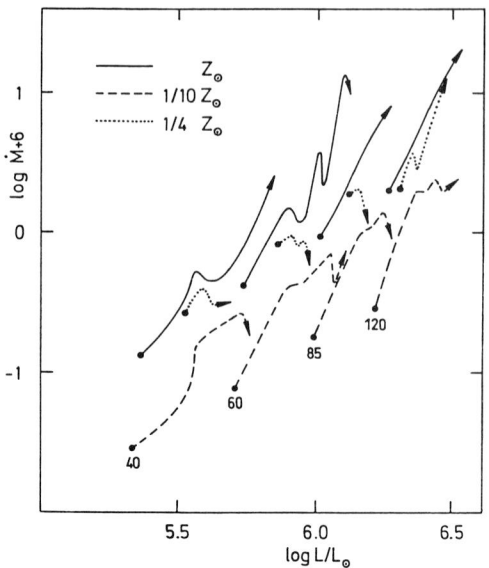

Figure 15. *Logarithm of mass-loss rates resulting from the new calculations (compare with Fig. 13). Note that now $Z(LMC) = 0.25 Z_\odot$.*

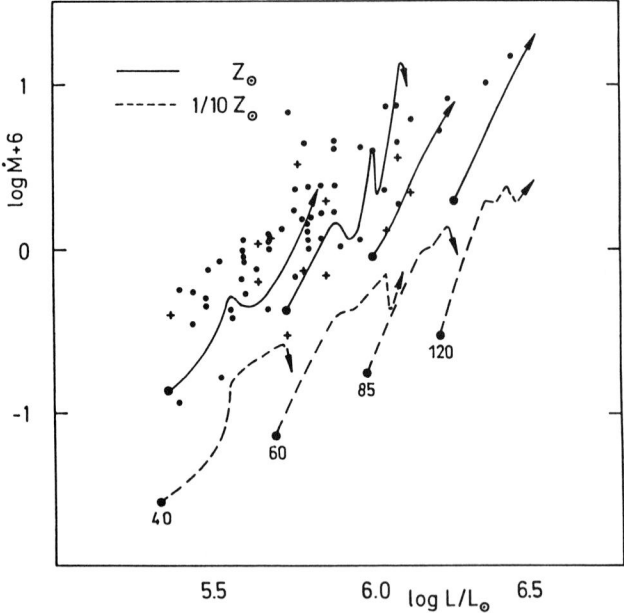

Figure 16. *Observed mass-loss rates for the Galaxy (dots, Leitherer, 1988a) and the SMC (crosses, Leitherer, 1988b) compared with the theory. For discussion see text.*

might have introduced a bias in the determination of mass-loss rates. It is interesting that in the only case, were we have a reasonable high resolution IUE spectrum available, namely Sk 80 = AV 232, our wind theory with $1/10\ Z_\odot$ leads to a perfect fit of the observed N V and C IV profiles (see Fig. 17). v_∞ of our model is 1200 km/s and

$\log \dot{M} = -5.7$. This mass-loss rate is 0.5 dex smaller than the one quoted by Leitherer (1988b).

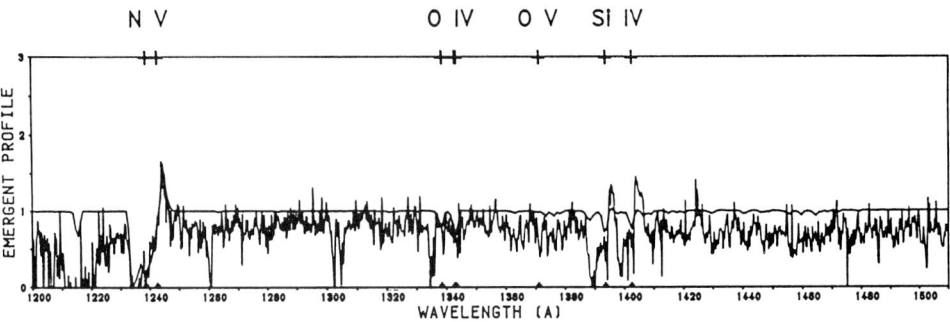

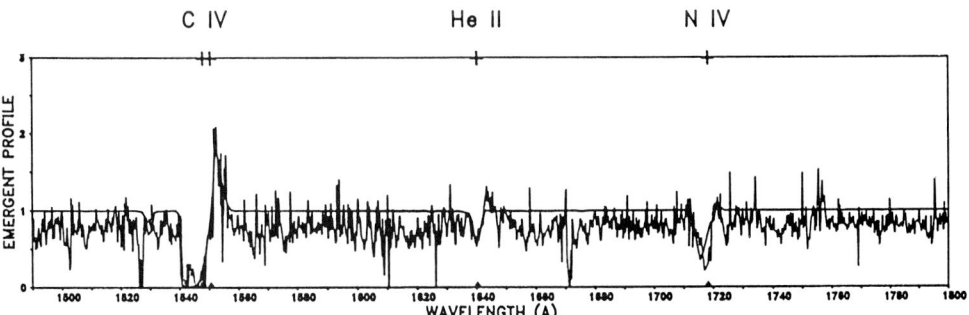

Figure 17. *Observed high resolution IUE spectrum of N V and C IV of Sk80 = AV232 in the SMC compared with a selfconsistent radiation wind model of 1/10 solar metallicity.*

The lack of adequate observational data will be overcome very soon, when HST will provide us with high quality UV spectra of LMC and SMC stars. This will allow us to determine precisely abundances, mass-loss rates and velocities for individual objects in all evolutionary stages and to compare with the predictions of the theory. We are confident that this will settle the question of winds and metallicity.

Finally, to give an impression, how the winds of MHS can influence the energetics of the ISM in galaxies of different metallicity, we have plotted in Fig. 18 the kinetic power and total energy provided by radiation driven winds. A simple fit formula for the energy as function of ZAMS-mass and metallicity is

$$E_{wind} = 3.7 \times 10^{46} (M/M_\odot)^{2.25} (Z/Z_\odot)^{0.86} erg.$$

5. DETAILED DIAGNOSTICS

In this section we discuss the quantitative diagnostic tools that are provided by model atmosphere techniques applied to spectral observations of MHS. We start with the analysis of photospheric spectra, continue with the use of particular stellar wind lines as luminosity indicators and present finally a purely spectroscopic method to determine stellar masses and radii (*i.e.*, distances !) from v_∞ and $\dot{M}$.

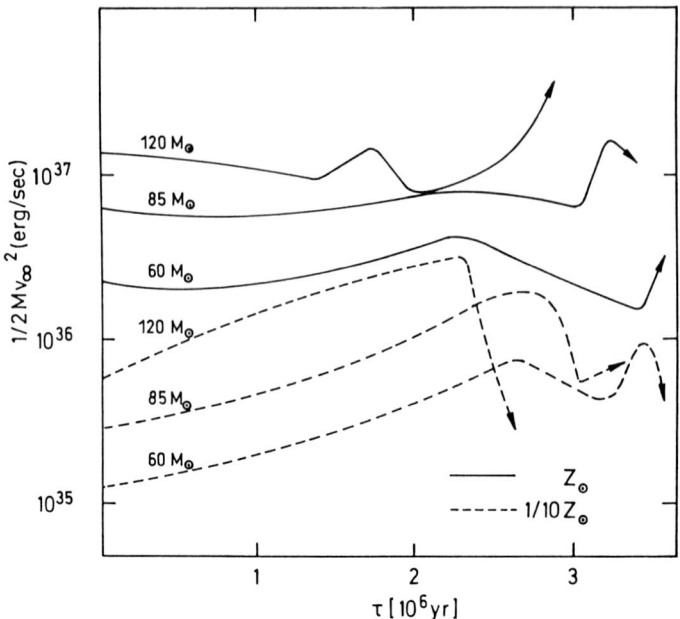

Figure 18a. *Kinetic power provided by radiation driven winds as a function of stellar age for stars of different mass and metallicity.*

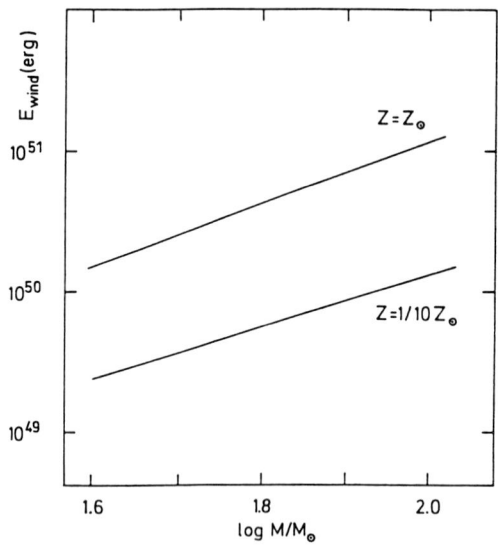

Figure 18b. *Kinetic energy provided by radiation driven winds (time integral of curves in Fig. 18a) as function of mass and metallicity.*

5.1. Quantitative spectroscopy of photospheric lines

Photospheric lines by application of NLTE model atmosphere and line formation techniques allow the determination of stellar parameters like T_{eff}, $\log g$ and abundances. The basic analysis method is sketched in Fig. 19: The profiles of the hydrogen lines H_γ, H_δ contain information about the gravity ($\log g$), whereas the profiles of the neutral helium lines depend mainly on T_{eff} and helium abundance. The He II lines of the Pickering series, which depend on $\log g$ and helium abundance, can then be used as the independent information fixing the helium abundance.

This technique has been applied to more than 50 galactic and Magellanic Cloud OB stars with excellent results. The intrinsic accuracy of the method is 3% for T_{eff}, 0.1 dex for $\log g$, 0.1 dex for $\log(N_{He}/N_H+N_{He})$. (For a list of objects analyzed so far and for a discussion of the systematic errors see Kudritzki and Hummer, 1990). Note that this is the only way to obtain reliable stellar parameters. The simpler method to measure the ratio of equivalent widths of He I λ 4471 and He II λ 4542, which defines the spectral type of O stars, and then to assign an effective temperature to each spectral type using an "average" calibration of main sequence stars leads to large systematic errors, since the degree of helium ionization depends on both T_{eff} and $\log g$. This is demonstrated in Fig. 20, which indicates that low gravity supergiants of same spectral type are significantly cooler than main sequence stars (see also Kudritzki et al. 1983, Voels et al. 1989).

In the following subsections we discuss recent results obtained by means of this technique in more detail.

5.1.1. The most massive stars in the Galaxy and the SMC

The most massive stars in our Galaxy are found in the Carina Nebula. They have been analyzed quantitatively by Kudritzki (1980) and Simon et al. (1983), who used photographic IIIa-J-emulsion spectra. Very recently, Imhoff (1990) and Imhoff et al. (1990), were able to secure ESO 3.6m CASPEC/CCD spectra and to reanalyze these objects. The results are summarized in Table 1 and Fig. 21. Fig. 22 gives an example for the quality of profile fits that can be achieved.

TABLE 1. *Spectroscopic parameters of the most massive stars in the Galaxy*

No.	Name	Sp. type	T_{eff} ($10^3 K$)	$\log g$ (cgs)	$\dfrac{N_{He}}{N_H + N_{He}}$
1	HD 93250	O3 V((f))	51.0 ± 1.5	3.9 ± 0.1	0.09 ± 0.02
2	HD 93129 A	O3 If*	50.5 ± 1.0	$3.75^{+0.1}_{-0.05}$	0.09 ± 0.02
3	HD 93128	O3 V((f))	52.0 ± 1.5	3.9 ± 0.1	0.09 ± 0.02
4	HD 303308	O3 V((f))	48.0 ± 1.5	3.9 ± 0.1	0.09 ± 0.02
5	ζ Puppis	O4 f	42.0 ± 2.0	3.50 ± 0.1	0.17 ± 0.03

(results for ζ Puppis from Bohannan et al. 1986)

Evidently, HD 93250 and HD 93129A, according to their luminosity, are objects with ZAMS-masses larger than $120 M_\odot$ (the determination of their present masses will be discussed in section V), which confirms that very massive stars can be formed in the galactic disk (see Nakano, 1989). In addition, the enhanced helium abundance of ζ Puppis, (the most evolved object in Fig. 21), indicates that the process of mixing nuclear burned matter towards the photospheric surface by either mass loss, convection and/or rotationally induced mixing has obviously been effective during the evolution.

The most massive stars in the SMC are located in NGC 346, an extremely young compact star cluster which excites N 66, the largest and brightest H II region of this

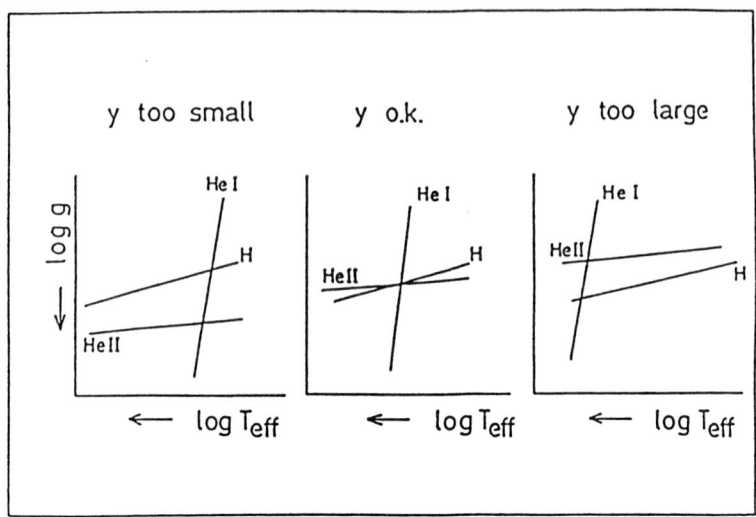

Figure 19. *Sketch of the method to determine T_{eff}, $\log g$ and helium abundance $y = N_{He}/(N_H + N_{He})$ from the fit of the photospheric profiles of H, He I and He II: Fit diagrams in the $(\log g, \log T_{eff})$-plane are constructed along which observed and calculated profiles agree. For the correct helium abundance a well defined intersection area is obtained, which allows one to read off T_{eff} and $\log g$.*

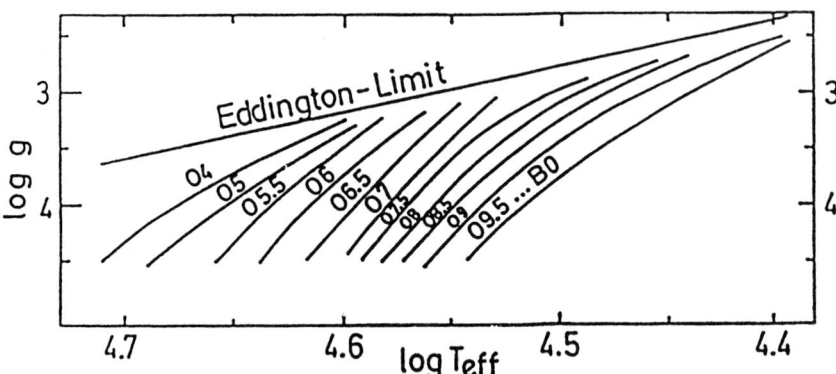

Figure 20. *The domains of O-star spectral types in the $(\log g, \log T_{eff})$-diagram. The spectral types are defined by the ratio of He I to He II equivalent widths. Note that the spectral type depends on both T_{eff} and $\log g$.*

galaxy. Recent CCD wide-band photometry and spectral classification (Niemela et al. 1986, Walborn and Blades, 1986, Massey et al. 1990) have revealed the presence of very early O-type stars within this cluster. The Munich group (Kudritzki et al. 1987 a,b) has now completed the analysis of the brightest O-stars in NGC 346 and of the O7 Iaf supergiant SK 80 which is located very close to this cluster. The results based on ESO 3.6m CASPEC/CCD spectra are shown in Table 2 and Fig. 23 and 24.

The profile fits demonstrate that for these rather faint objects (12 mag $\leq m_V \leq$

TABLE 2. *Spectroscopic parameters of the most massive stars in the young open cluster NGC 346 in the SMC*

Star	No.	Sp. type	T_{eff} ($10^3 K$)	$\log g$ (cgs)	$\frac{N_{He}}{N_H + N_{He}}$
NGC 346	1	O4 III(f)	43.5 ± 1.5	3.6 ± 0.10	0.09 ± 0.02
	3	O3 III(f)	$55.0^{+5.0}_{-2.5}$	3.9 ± 0.10	0.09 ± 0.02
	4	O5-6 V	42.0 ± 2.0	3.8 ± 0.10	0.09 ± 0.02
	6	O4-5 V	40.0 ± 2.0	3.7 ± 0.15	0.09 ± 0.02
Sk 80		O7 Iaf	34.0 ± 1.0	2.9 ± 0.1	0.33 ± 0.05

14 mag) quantitative spectroscopy is feasible with the same precision as for the much brighter galactic O stars. Clearly, very massive stars with ZAMS-masses larger around $100 M_\odot$ also exist in the SMC. The helium enrichment found for Sk 80 is a hint that mixing processes are also important for the evolution of massive stars with lower metallicity.

5.1.2. B Supergiants in Galaxy, LMC and SMC

As a part of an extensive spectroscopic project on the formation and evolution of massive stars in parent galaxies of different metallicity, B supergiants in the Galaxy, LMC and SMC are presently studied by the Munich group (R. P. Kudritzki, D. L. Lennon, H. G. Groth, K. Butler, D. Husfeld, S. Becker, F. Eber) together with E. Fitzpatrick (Princeton). The results concerning the atmospheric parameters obtained so far are summarized in Table 3. Fig. 25 gives an example for the profile fits.

TABLE 3. *Spectroscopic parameters of B supergiants in Galaxy, LMC, SMC*

star	Sp. type	T_{eff} ($10^3 K$)	$\log g$ (cgs)	$\frac{N_{He}}{N_H + N_{He}}$	galaxy
Sk 21-65°	B0 Ia	27	2.70	0.35	LMC
41-68°	B0.5 Ia	25	2.60	0.23	LMC
HDE 268685	B1 Ia	22	2.35	0.23	LMC
Sk 13-69°	B1 Ia	22	2.40	0.15	LMC
HDE 269504	B2 Ia	20	2.25	0.23	LMC
Sk 159	B0.5 Ia	25	2.55	0.35	SMC
119	B2 Ia	20	2.30	0.15	SMC
ϵ Ori	B0 Ia	26	2.75	0.20	GALAXY
χ Ori	B0.5 Ia	25	2.70	0.20	GALAXY

Very strikingly, *all* objects are helium enriched with a large scatter in helium abundance. Fitzpatrick and Garmany (1990) point out that this is consistent with the interpretation of their HR-diagram "ledge" that most of the observed blue supergiants are in a post-red-supergiant phase of evolution, where CNO-burnt matter is expected to become visible at the photospheric surface.

The optical spectra of the B supergiants also allow the determination of NLTE metal abundances, at least differentially between the objects of same effective temperature in

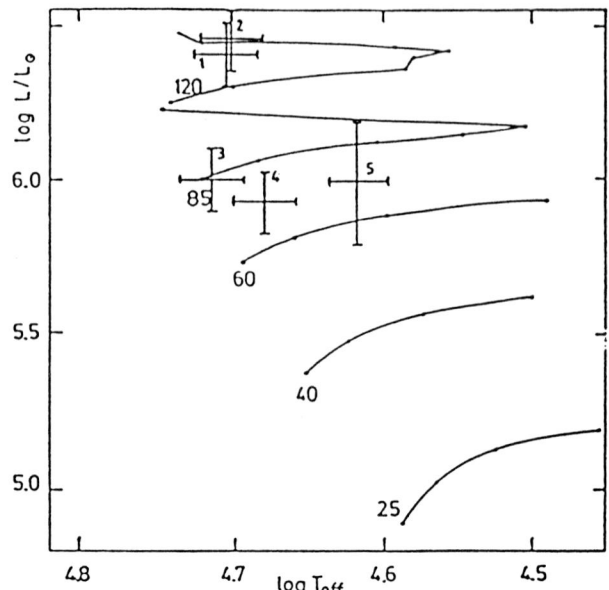

Figure 21. *HR-diagram of the most massive stars in our Galaxy compared with evolutionary tracks from Maeder and Meynet (1987). (See also Table 1).*

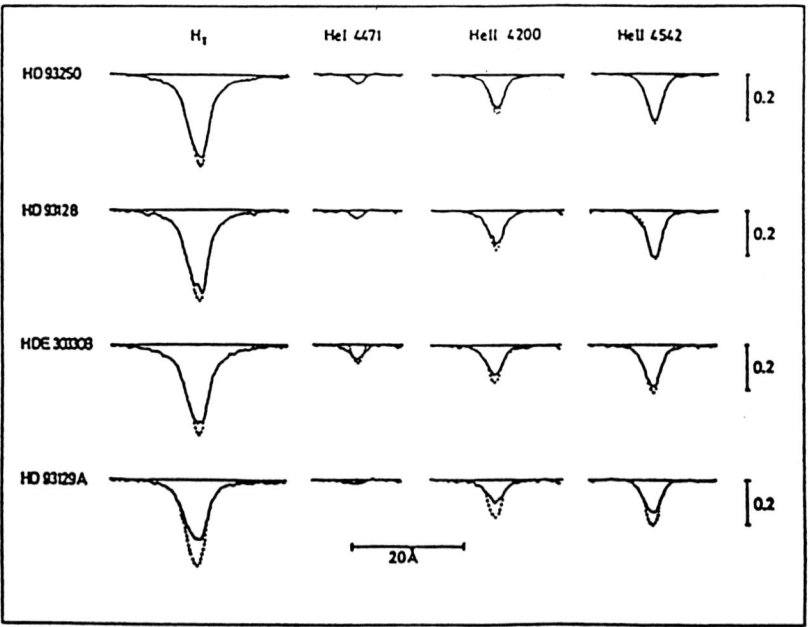

Figure 22. *Profile fits of the strategic hydrogen and helium lines.*

different galaxies. This project has been started by D. Lennon and R. P. Kudritzki (1990) and collaborators in Munich. Fig. 26 demonstrates qualitatively the presence of significant abundance differences between the Galaxy and the SMC. Quantitatively, the results are summarized in Table 4.

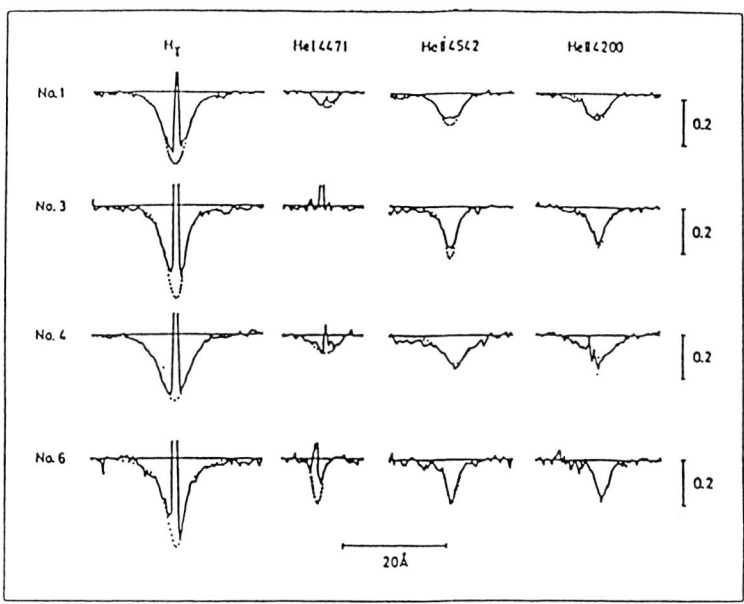

Figure 23. O stars in the SMC cluster NGC 346: The observed line profiles used for the analysis are compared with those of the final models.

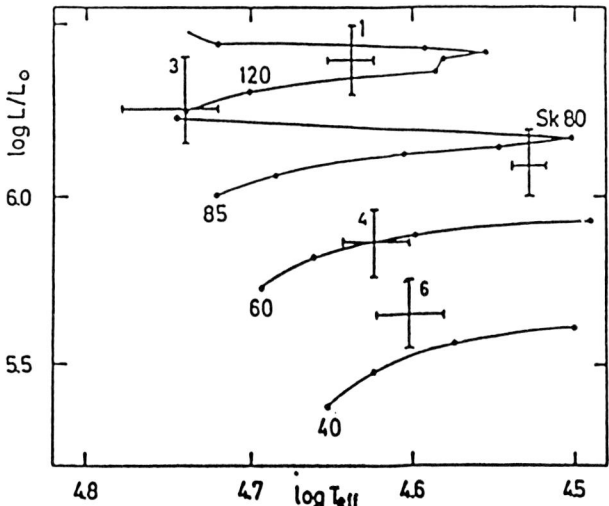

Figure 24. HR-diagram of the four O stars in NGC 346 plus the O7Iaf-supergiant Sk 80 compared with evolutionary tracks by Maeder and Meynet (1987).

The metal deficiency of Sk 159 is mostly striking for oxygen which is underabundant by a factor of 40. But also all the other elements are significantly underabundant. (Note that the numbers for C and Mg are upper limits.) Both LMC supergiants do also have real deficiencies in O, Mg and Si. The situation for C and N is not clear, as both LMC objects show important differences between themselves. (The abundances for these

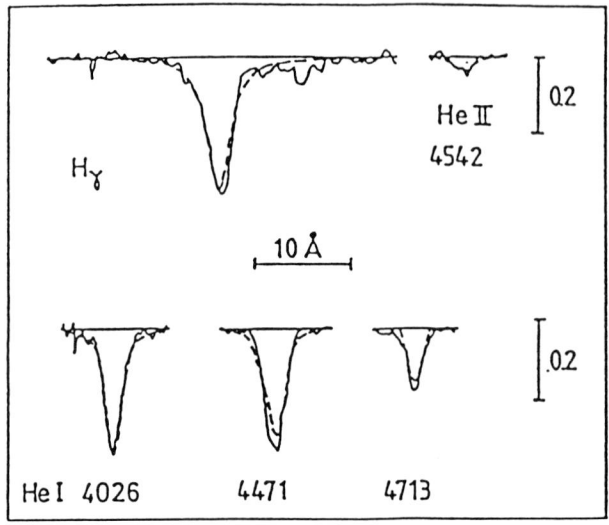

Figure 25. *The fits of the hydrogen and helium lines of the SMC supergiant Sk 159.*

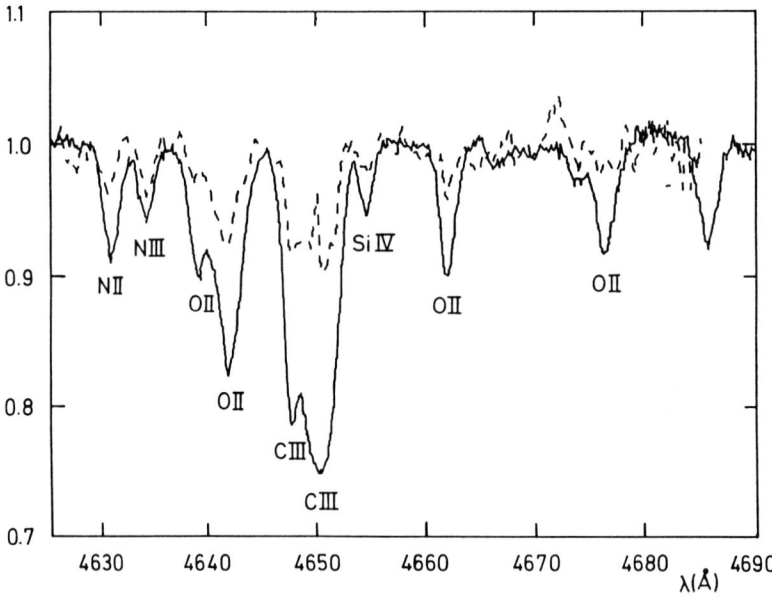

Figure 26. *Metal lines in the SMC supergiant Sk 159 (dashed) and in χ Ori (solid).*

two elements in Sk 21-65° are however based upon one line for each species only and are therefore uncertain). The extension of this work to a larger sample of objects will clarify the abundance situation in the Magellanic Clouds further. The future use of "unified model atmospheres" will allow for the effects of velocity fields and atmospheric extension in the abundance determination so that not only relative abundances but also reliable absolute abundances will become available. In this way, the question of photospheric mixing will be discussed again by means of the absolute abundances of C, N and O.

TABLE 4. *Relative abundances* ($\log \epsilon_{MC}/\epsilon_{Galaxy}$). *The galactic comparison star for Sk 41-68° and Sk 159 is χ Ori and the comparison for Sk 21-65° is ϵ Ori. H II region data are from Dufour (1984).*

Element	LMC			SMC	
	Sk 41-68°	Sk 21-65°	H II	Sk 159	H II
C	+0.06	−1.12	−0.77	≤ −1.09	−1.51
N	+0.03	−0.45	−1.02	−0.70	−1.53
O	−0.42	−0.82	−0.49	−1.59	−0.90
Si	−0.14	−0.87	—	−0.75	—
Mg	−0.78	−0.62	—	≤ −1.0	—

5.2. Optical and IR emission lines as luminosity indicators

The calculation of HeII 4686-, H_α- and IR line profiles by means of the Unified Model Atmosphere technique shows that the strengths of the emission of these profiles is a strong function of luminosity. Fig. 27 gives an example for this effect.

Having in mind that present day intermediate resolution spectrographs allow to measure such profiles for objects as faint as 18th to 19th magnitude we regard a systematic investigation of this effect as extremely worthwhile.

5.3. UV P-Cygni lines as luminosity indicators

From the work of Walborn and collaborators (Walborn and Panek, 1984a,b, 1985, Walborn and Nichols-Bohlin, 1987, Walborn et al. 1985) it became evident that UV stellar wind spectra show a systematic dependence on luminosity, temperature and abundance. The most important feature is the strong luminosity dependence of the Si IV resonance doublet. Pauldrach et al. (1990) have used the Munich radiation driven wind code to test whether the theory can reproduce the change of spectral morphology in different evolutionary phases. They calculated wind models along evolutionary tracks and found that the theory is basically able to reproduce the observations (see Fig. 28). The quantitative use of, for instance Si IV, for the determination of luminosity therefore appears to be feasible. In view of the UV-capacity of the HST this is an important aspect for future quantitative spectroscopy of MHS.

5.4. The direct spectroscopic determination of stellar masses and radii

One of the most important goals of stellar spectroscopy is to obtain precise information about masses and radii (distances) for stars that are not eclipsing binaries. The winds of hot stars provide a unique possibility to obtain such information directly from the stellar spectra.

5.4.1. Direct mass determinations for O-stars

First, we treat the simpler case, where the distance is known from other methods and only the information about the masses is needed. Quantitative spectroscopy of photospheric lines then yields T_{eff} and $\log g$. Using the distance, the observed and calculated model flux one can determine the radius R and the luminosity L. These

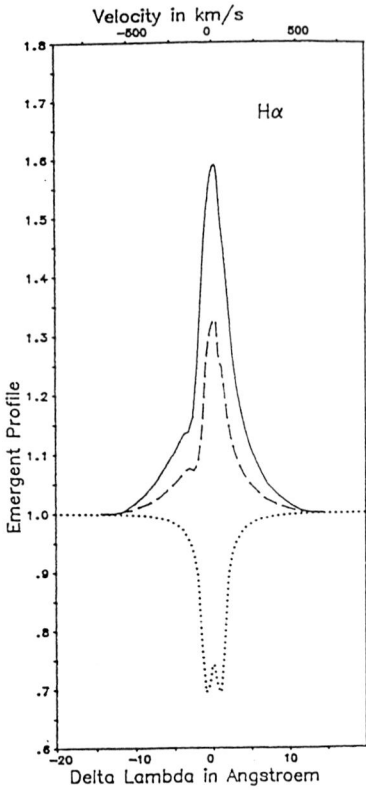

Figure 27. H_α *line profiles for two unified models with same* $T_{eff} = 32000$ K *and* $\log g = 2.9$ *but different* $\log L/L_\odot = 5.8$ *and* 6.1. *The dotted absorption profile results from a hydrostatic model.*

quantities are normally well defined as long as the distance is not too uncertain. Direct information about the mass is normally obtained from the gravity using the radius. However, because of the uncertainties in $\log g$ and R masses derived in this way are rather inaccurate. As an alternative, we propose, therefore, to use the easily observable terminal velocity v_∞ as an indicator for the mass. The basis for this procedure is the theory of radiation driven winds which predicts

$$v_\infty = v_{esc} * f(T_{eff}, L/L_\odot, M/M_\odot) \tag{1}$$

Since $v_{esc} = (2GM(1 - L/L_E)/R)^{1/2}$ is mainly a function of mass and radius (L_E is the Eddington luminosity) and since the function $f(T_{eff}, L/L_\odot, M/M_\odot)$ is well studied (see Kudritzki et al. 1989), eq. (1) can be used for a direct mass determination. Fig. 29 gives an example for this method using stars discussed in Table 1 of section III, 1. For different luminosities (the range is given by the uncertainties of distance, R and T_{eff}) v_∞ is plotted as a function of mass and the observed v_∞ yields directly the stellar mass. Table 6 compares masses obtained in this way with derived from $\log g$.

The agreement between the two methods is striking, however, the v_∞-masses have much smaller uncertainties. We are convinced that the v_∞-method is ideal to investigate the actual masses of O stars in different phases of their evolution with strong mass loss.

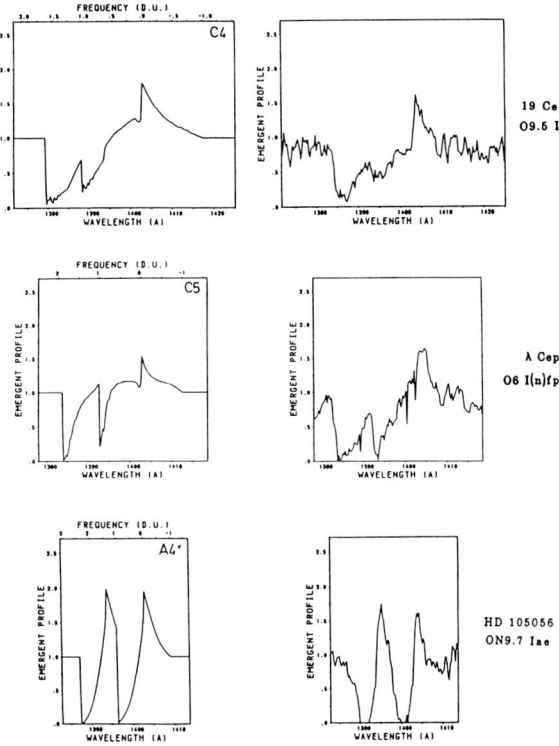

Figure 28. *The strong Si IV stellar wind resonance line for Of supergiants: theoretical profiles versus observations. (From Pauldrach et al. 1990)*

5.4.2. Direct determination of mass and radius

In view of the use of luminous O stars as standard candles it is important to develop a method to determine directly radii and masses without any a priori knowledge about the distance. This is not possible with photospheric spectroscopy alone. The winds of hot stars, on the other hand, provide this possibility. The theory of radiation driven winds yields analytical expressions for both mass-loss rate $\dot{M}$ and v_∞ as function of T_{eff}, $\log g$ and R (see Kudritzki et al. 1989):

$$\dot{M} = f_1(T_{eff}, \log g, R/R_\odot)$$
$$v_\infty = f_2(T_{eff}, \log g, R/R_\odot) \quad (2)$$

This means that if $\dot{M}$ and v_∞ are determined from the stellar wind lines and T_{eff} has been derived from photospheric spectroscopy, then $\log g$ and $R/R_\odot$ can be obtained from eqs. (2). Since $\log g$ can simultaneously be determined from the photospheric spectroscopy, one has an additional possibility to test the procedure.

Fig. 30 and 31 demonstrate how this method works in reality. The examples are the O4 f-star ζ Puppis and the extreme B supergiant and Luminous Blue Variable P Cygni. For ζ Puppis we obtain $\log g = 3.62 \pm 0.05$ and $R/R_\odot = 17 \pm 5$, which yields $\log M/M_\odot = 1.64 \pm^{0.18}_{0.25}$ in agreement with Tables 1 and 6. While the uncertainties in mass and radius appear to be rather large, it is important to realize that the error in radius transforms to only ± 0.5mag in distance modulus. An extension to much fainter

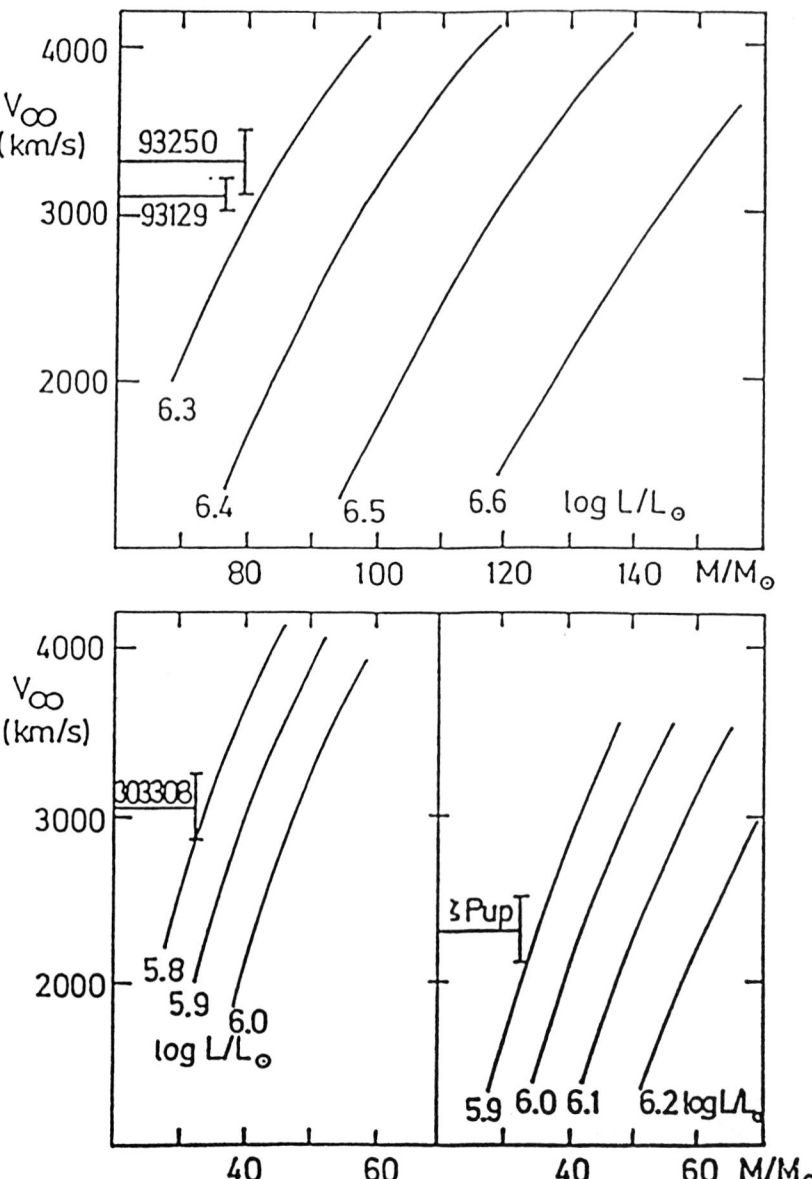

Figure 29. *Terminal velocity as a function of stellar mass calculated for four O stars.* $\log L/L_\odot$ *is used as parameter within the range its incertainties. The observed* v_∞ *including its uncertainty is shown as the horizontal bar.*

objects, using HST in the UV for v_∞ and optical spectroscopy for H_α and He II 4686-emission, to obtain information on $\dot{M}$ without losing accuracy in the distance modulus appears to be straightforward.

For P Cygni Fig. 31 leads to $\log g = 2.05^{+0.05}_{-0.02}$, $R/R_\odot = 70 \pm 15$ and $\log \dot{M}/\dot{M}_\odot = 1.30 \pm 0.15$. (Note that the discontinuities in Fig. 31 are caused by the bistability of radiation driven winds in this case, see Pauldrach and Puls, 1990). Again, this procedure leads only to a very small error in distance modulus of ± 0.4 mag.

Because of the high resolution IUE spectrum available (see Fig. 17) we were also

TABLE 6. *Stellar masses from $\log g$ and v_∞ for objects with known distances (and radii)*

Star	Log $M/M_\odot$ from v_∞	from $\log g$	Source
HD 93129	2.08 ± 0.10	2.07 ± 0.17	Kudritzki, 1990
HD 93250	2.01 ± 0.11	2.06 ± 0.18	Kudritzki, 1990
HD 303308	1.16 ± 0.12	1.75 ± 0.21	Kudritzki, 1990
ζ Puppis	1.68 ± 0.10	1.62 ± 0.20	Kudritzki, 1990
HD 15629	1.40 ± 0.16	1.36 ± 0.26	Herrero, Kudritzki, Vilchez, 1990
HD 15558	1.70 ± 0.15	1.70 ± 0.26	Herrero, Kudritzki, Vilchez, 1990
HD 34656	1.42 ± 0.28	1.42 ± 0.40	Herrero, Kudritzki, Vilchez, 1990
HD 193514	1.46 ± 0.16	1.44 ± 0.26	Herrero, Kudritzki, Vilchez, 1990
HD 192639	1.37 ± 0.15	1.26 ± 0.26	Herrero, Kudritzki, Vilchez, 1990
Sk 80	1.54 ± 0.09	1.50 ± 0.20	Kudritzki, Pauldrach, Puls, Hummer, in prep.

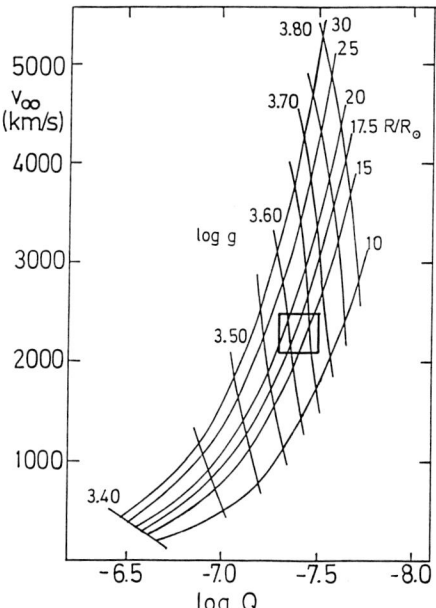

Figure 30. *Terminal velocity v_∞ and $\log Q = \log \dot{M}(M_\odot/yr) - 3/2 \log R/R_\odot$ of the O4f-star ζ Puppis calculated for wind models with different $\log g$ and $R/R_\odot$. The observed values are indicated by the rectangular box. (Note that $\log Q$ is obtained directly from radio observations or H_α—or He II 4686-emission). The diagram allows one to read off directly the stellar gravity and radius.*

able to apply this method to Sk 80 = AV 232 in the SMC. The result is shown in Fig. 32, where $\log \dot{M} - \log R/R_\odot$ was obtained from the fit of the UV lines. Again the gravity of $\log g = 2.95 \pm 0.05$ is in excellent agreement with optical spectroscopy. For the radius we obtain $R/R_\odot = 34^{+11}_{-9}$, which transforms into ±0.6 mag in distance modulus.

We are very optimistic that the HST will allow us not only to obtain much higher precision but also to determine the distances of a much larger sample of objects in the

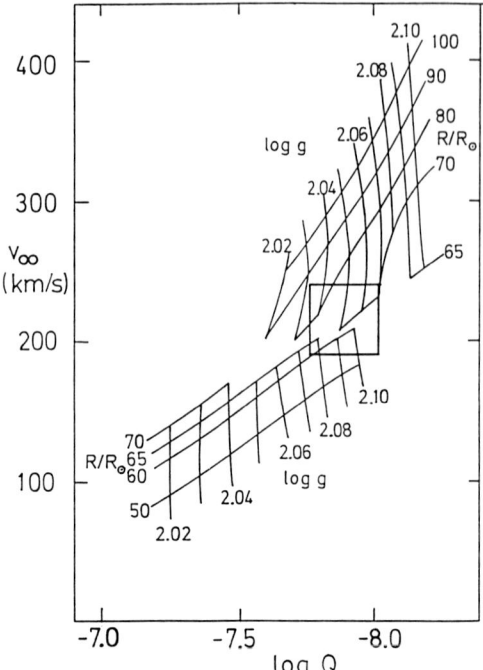

Figure 31. *Same as Fig. 30 for the case of P Cygni. Note the precision of the log g determination in this case.*

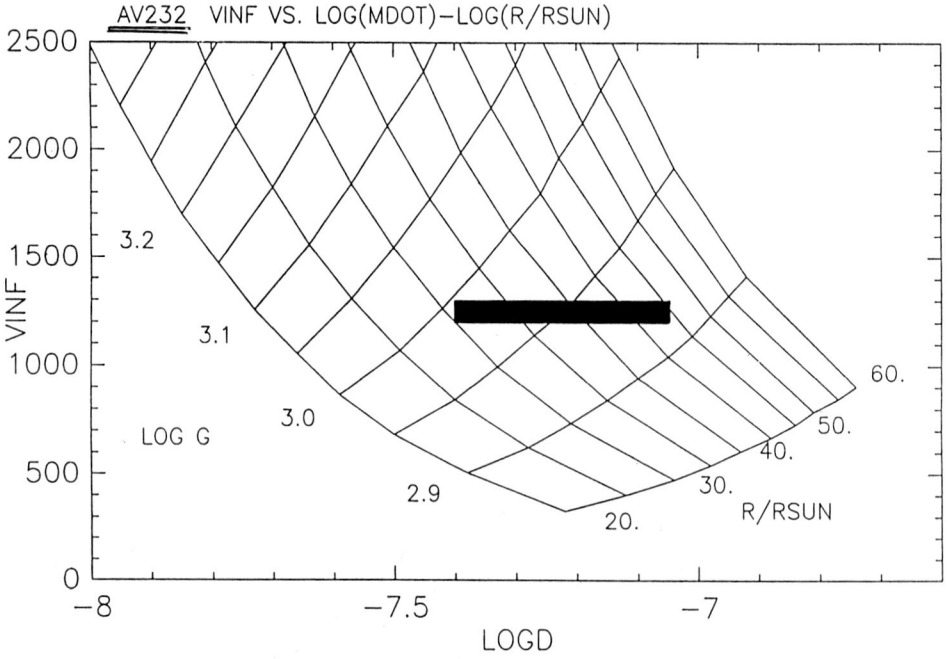

Figure 32. *Same as Fig. 30, 31 but for Sk 80 = AV 232 in the SMC.*

LMC and SMC and to compare the results of this independent method with those of other methods.

This work was supported by the DFG under grants KU 474/11 and 13. We also acknowledge grant 010R9008 by the Bundesminister für Forschung und Technologie.

REFERENCES

Abbott, D. C., 1979, *IAU Symp. 83*, p. 237.
———. 1982, *Ap. J.*, **259**, 282.
———. 1985, In "*Relations Between Chromospheric-Coronal Heating and Mass Loss in Stars*, 1984 Workshop, Sacramento Peak Observatory, eds. R. Stalio and J.B. Zirker.
Abbott, D. C., and Lucy, L. B., 1985, *Ap. J.*, **288**, 679.
Abbott, D. C., and Hummer. D. G., 1985, *Ap. J.*, **294**, 286.
Becker, S., and Butler, K., 1988a, *Astr. Ap.*, **201**, 232.
———. 1988b, *Astr. Ap. Suppl.*, **74**, 211.
———. 1988c, *Astr. Ap.*, **76**, 331.
———. 1989, *Astr. Ap.*, **209**, 244.
———. 1990a, *Astr. Ap.*, in press.
———. 1990b, *Astr. Ap. Suppl.*, in press.
Bohannan, B., Abbott, D. C., Voels, S. A., and Hummer, D. G., 1986, *Ap. J.*, **308**, 728.
Castor, J., Abbott, D. C., and Klein, R., 1975, *Ap. J.*, **195**, 157.
Dufour, R. J., 1984, *IAU Symp. 108*, 353.
Eber, F., and Butler, K., 1988, *Astr. Ap.*, **202**, 153.
Fitzpatrick, E. L., and Garmany, C. D., 1990, *Ap. J.*, in press.
Friend, D., and Abbott, D. C., 1986, *Ap. J.*, **202**, 153.
Gabler, R., Gabler, A., Kudritzki, R. P., Puls, J., and Pauldrach, A., 1989, *Astr. Ap.*, **226**, 162.
Gabler, A., Gabler, R., Pauldrach, A., Puls, J., and Kudritzki, R. P., 1990, in "*Hot Star Workshop: Intrinsic Properties of Hot Luminous Stars*", Astron. Soc. of the Pacific Conference Series, ed. C.D.Garmany, p. 218.
Garmany, C. D., and Conti, P. S., 1985, *Ap. J.*, **293**, 409.
Garmany, C. D., and Fitzpatrick, E. L., 1988, *Ap. J.*, **332**, 711.
Hamann, W. R., Schmutz, W., and Wessolowski, 1988, *Astr. Ap.*, **194**, 190.
Herrero, A., Kudritzki, R. P., and Vilchez, I. M., 1990, in Proc. European IAU Conf., in press.
Hillier, D. J., 1984, *Ap. J.*, **280**, 744.
———. 1987a, *Ap. J. Suppl.*, **63**, 947.
———. 1987b, *Ap. J. Suppl.*, **63**, 965.
———. 1988, *Ap. J.*, **327**, 822.
Howarth, I., and Prinja, R., 1989, *Ap. J. Suppl.*, **69**, 527.
Hummer, D. G., 1982, *Ap. J.*, **257**, 724
Humphreys, R. M., and Davidson, K., 1979, *Ap. J.*, **232**, 409.
Imhoff, J., 1989, Diplomarbeit, Universität München
Imhoff, J., Kudritzki, R. P., Husfeld, D., and Groth, H. G., 1990, *Astr. Ap.*, in prep.
Kudritzki, R. P., 1980, *Astr. Ap.*, **85**, 174.

―――――. 1988, "*Radiation in Moving Gaseous Media*", 18th Advanced Course, Swiss Society for Astrophysics and Astronomy, Saas-Fee 1988, eds. Y. Chmielewski, and T. Lanz.

―――――. 1990, in "*Frontiers of Stellar Evolution*", ed. D.L. Lambert.

Kudritzki, R. P., Cabanne, M. L., Husfeld, D., Niemela, V. S., Groth, H. G., Puls, J., and Herrero, A., 1989a, *Astr. Ap.*, **226**, 235.

Kudritzki, R. P., Gabler, A., Gabler, R., Groth, H. G., Pauldrach, A., and Puls, J., 1989, in *Proc. IAU Coll. 113*, "*Physics of Luminous Blue Variables*", ed. K. Davidson et al. p.67.

Kudritzki, R. P., and Hummer, D. G., 1990, *Ann. Rev. Astr. Ap.*, **28**, 303–45.

Kudritzki, R. P., Husfeld, D., Groth, H. G., Hummer, D. G., and Conti, P. S., 1990b, *Astron. Astrophys.*, in prep.

Kudritzki, R. P., Pauldrach, A., and Puls, J., 1987, *Astr. Ap.*, **173**, 293.

Kudritzki, R. P., Pauldrach, A., Puls, J., and Abbott, D. C., 1989, *Astr. Ap.*, **219**, 205.

Kudritzki, R. P., Simon, K. P., and Hamann, W. R., 1983, *Astr. Ap.*, **118**, 245.

Leitherer, C., 1988a, *Ap. J.*, **326**, 356.

―――――. 1988b, *Ap. J.*, **334**, 626.

Lennon, D. J., 1983, *M.N.R.A.S.*, **205**, 829.

Lennon, D. J., Kudritzki, R. P., Becker, S. R., Eber, F., Butler, K., and Groth, H. G., 1990, in "*Hot Star Workshop: Intrinsic Properties of Hot Luminous Stars*", ed. C. D. Garmany, Astron. Soc. of the Pacific Conference Series, p. 315.

Lucy, L. B., and Solomon, P., 1970, *Ap. J.*, **159**, 879.

Maeder, A., 1983, *Astr. Ap.*, **120**, 113.

Maeder, A., and Meynet, G., 1987, *Astr. Ap.*, **182**, 243.

Massey, P., Parker, J., and Garmany, C. D., 1990, *A. J.*, **98**, 1305.

Nakano, T., 1989, *Ap. J.*, **345**, 464.

Niemela, V. S., Marraco, H. G., and Cabanne, M. L., 1986, *Pub. A.S.P.*, **98**, 1133.

Pauldrach, A. W. A., 1987, *Astr. Ap.*, **183**, 295.

Pauldrach, A. W. A., Puls, J., and Kudritzki, R. P., 1986, *Astr. Ap.*, **164**, 86.

Pauldrach, A. W. A., Puls, J., Kudritzki, R. P., and Butler, K., 1990, *Astr. Ap.*, **228**, 125.

Pauldrach, A. W. A., and Puls, J., 1990, *Astr. Ap.*, in press.

―――――. 1990, Reviews in Modern Astronomy, Vol. 3.

Puls, J., 1987, *Astr. Ap.*, **184**, 227.

Schmutz, W., Hamann, W. R., and Wessolowski, U., 1989, *Astr. Ap.*, **210**, 236.

Schönberner, D., Herrero, A., Becker, S., Eber, F., Butler, K., Kudritzki, R. P., and Simon, K. P., 1987, *Astr. Ap.*, **197**, 209.

Simon, K. P., Jonas, G., Kudritzki, R. P., and Rahe, J. 1983, *Astr. Ap.*, **125**, 34.

Voels, S. A., Bohannan, B., Abbott, D. C., and Hummer, D. G., 1989. *Ap. J.*, **340**, 1073.

Walborn, N. R., 1970, *Ap. J.*, **161**, L 149.

―――――. 1971, *Ap. J.*, **164**, L 67.

―――――. 1976, *Ap. J.*, **205**, 419.

Walborn, N. R., and Blades, J. C., 1986, *Ap. J.*, **304**, L 17.

Walborn, N. R., and Nichols-Bohlin, J., 1987, *Pub. A.S.P.*, **99**, 40.

Walborn, N. R., and Nichols-Bohlin, J., and Panek, R. J., 1985, NASA Reference Publication 1155 (IUE High Res. O Star Atlas).

Walborn, N. R., and Panek, R. J., 1984a, *Ap. J.,* **280**, L27.
_____. 1984b, *Ap. J.,* **286**, 718.
_____. 1985, *Ap. J.,* **291**, 806.
Werner, K., and Husfeld, D., 1985, *Astr. Ap.,* **148**, 417.

DISCUSSION

C. Leitherer: Did you check the internal consistency of your wind models and the evolutionary tracks you used? Maeder and Meynet's evolutionary models are input for your models in the sense that you use their stellar parameters, especially the M-L relation. On the other hand you predict a mass-loss rate, which should be consistent with the one adopted in the evolutionary models.

Kudritzki: No, in this case not.

C. Leitherer: So one would like to see if your mass loss rate and the evolutionary mass-loss rates agree.

Kudritzki: In this case there's still an inconsistency in these calculations. The treatment for the Small and Large Magellanic Cloud is more consistent because that was our original idea what we wanted to do in collaboration with N. Langer. Unfortunately they have not done similar calculations for solar metallicity, so of course that could change the picture again. I agree.

G. Koenigsberger: I've got a comment on iron. We have been observing Wolf-Rayet binaries for a few years now, and we have done practically all the binaries that have moderate to large orbital inclinations. It turns out that we see very strong atmospheric eclipses in the Fe V and VI pseudo continuum, which lies between 1280 and 1470 Å. This occurs even for binaries which have 45° orbital inclination. One of the sources we did observe is HD 5980 in the Small Magellanic Cloud. Its spectral type is very similar to HD 90657 in the Galaxy so that we can estimate the iron deficiency in the Small Magellanic Cloud with respect to the Galaxy assuming that both Wolf-Rayet stars are the same. We get an upper limit of 1/10 in the relative iron abundance, assuming that both stars have the same nitrogen abundance. Of course, if we go a little step further and say that there's less nitrogen in the Small Cloud than there is in the Galaxy, then you get even smaller iron abundances. That's one observational point. The other point that I also want to make is that we see very strong scattering as a function of orbital phase in the Fe V and VI pseudo continuum. This means we see the scattering occurring out as far as several stellar radii. From our very intuitive point of view I say, well, we have the scattering way out there, therefore, we have a driving force that's way out there. Thus we have a very extended acceleration zone and we can easily get higher terminal velocities.

B. Rocca-Volmerange: Did you try to compare the number of Lyman continuum photons produced by your O star models with the classical estimates from NLTE models (Mihalas models for example)?

Kudritzki: As far as hydrogen photons are concerned, my curve is almost identical with the old Panagia (1973) curve. When I checked this, I was of course disappointed that we have no differences here, but I have to admit this. I was also disappointed that the effect of metallicity is so small. If we do these calculations with wind blanketing, there is still one approximation, that is the treatment of photons reflected back to the photosphere and the pure scattering approximation, which might be poor. We are just at the moment calculating models which take into account the effect of collisions for this problem. This could change the thing again, and we might obtain different numbers of emergent photons as function of metallicity, but it's too early to give a definitive answer. This was the reason why I mentioned this pure scattering approximation on my transparencies. But up to now, as far as the hydrogen photons are concerned, our curves do not differ from Panagia.

R. Humphreys: I've noticed that in your atmospheric analysis of the group of B supergiants, you found that all of them had enhanced helium abundances. I think that we are not going to assume that all those had been red supergiants prior to becoming B supergiants. But, would you care to comment on what might be other causes of enhanced helium in these B stars?

Kudritzki: I was very pleased by the very recent paper by Fitzpatrick and Garmany (1990) on the HR diagram in the Large Magellanic Cloud and their careful discussion of evolutionary tracks. What I learned from this paper is that there's a very high probability that you find objects coming back from the RGB in this luminosity range. If this is the case, then I would be pleased because then for me it's pretty clear that objects coming back from the RGB have a reasonable chance to be helium enriched in the atmosphere. And then I would be able to understand this.

R. Humphreys: So you do think that *all* objects you analyzed are post-red-supergiants? Do you have an explanation for this?

Kudritzki: Perhaps I was unlucky. I don't know! That is worrying me. I know that this problem is there. I can't say anything except what we got.

P. Conti: According to recent analyses by Spite and Spite, and independently by Bessell and associates, the A supergiants in the SMC have iron abundances about 1/4 that of the solar vicinity. If this is correct in general, it might help explain the current discrepancy in the mass loss rates for SMC.

Kudritzki: I had no time to speak about A supergiants. We have a joint project on these objects. It's too early to say something about the abundances but you are right, that would be the simplest way out although I don't know then what to do with your observations of Wolf-Rayet stars. But Wolf-Rayet's are always different.

P. Conti: I noted that there was a metal abundance effect on the O II ionization balance. This might also be the case for higher ionization levels. Could you say whether or not this might affect the oxygen equilibrium calculations for H II regions? For example, the increase in oxygen line strength with decreasing abundance is well known. Beyond this, the equilibrium calculations seem to indicate an *increase* in the temperature of the exciting star, and an implied *increase* in its mass with *diminished* abundance. Could this inference now be suspect?

Kudritzki: I can't give you the answer, but I hope that if some people see these graphs, they start to think about this and find the answer.

S. Heap: About five years ago you and David Hummer made a prediction that you would find that you systematically overestimated the masses because you systematically underestimated the gravities from photospheric analysis, because your models didn't take into account radiation pressure in the lines. But I don't see any big change here, so did that not turn out to be an effect?

Kudritzki: Yes, the effect is there. But, for a star like ζ Puppis, this effect is on the order of 0.1 in log g, and this is just about the error we normally quote when we have excellent spectra. This is one of the reasons why masses inferred from gravities are a little bit uncertain. I would say for F stars, if you just simply compare with hydrostatic line profiles, you make an error of, let's say, 0.1 in gravity. We're doing the same type of calculations for neutral helium as well to see whether the neutral helium lines are affected. But it is too early to say something concerning the ionization balance as well.

A. Maeder: You have given this nice list of very massive stars, do you have the C/O ratio for those stars? The point is that I would very much like to know at which limiting mass there are non-solar C/O ratios. This is very indicative of any additional mixing processes.

Kudritzki: I still do not have good CN abundances for these very luminous objects. The reason is again, first of all, that those lines you see in the optical are clearly affected by stellar wind effects. Typical lines observed in the spectra are sometimes in emission and sometimes in absorption. The effects of stellar winds are very strong so you need the approach of unified model atmospheres to get the metal abundances out of these lines. And that is just something which we're doing with several Ph.D. students at the moment. Years ago I was very optimistic that we would be able to get metal abundances out of the optical and ultraviolet lines by simple photospheric non-LTE calculations for C III, N III and C IV for these luminous stars. But I learned, meanwhile, that this is not the case. You have to take into account the effects of stellar winds. However, we have done this kind of work for stars of lower luminosity, let's say corresponding to 25–30 solar masses. We were able to determine the C and O abundances in OBN stars and apparently normal stars, and the normal stars have normal abundances as our work has shown, whereas the OBN stars show the signature of the C and O process in the spectra.

S. Owocki: You showed a spectral model from Joachim Puls' thesis, I think it was nitrogen from ζ Puppis. The profile looked very, very black towards v_∞ or towards the edge of the profile. The implication is that you can't get such a black profile without putting in multiple resonance layers associated with normal velocity fields as was pointed out by Leon Lucy, 5 or 6 years ago. Is it now possible to get this just from pure non-LTE effects you're describing, or how is it possible that your spectrum has this blackness?

Kudritzki: Joachim Puls still had a contribution of what he called "microturbulence" of 100 km/s in his calculation. But only 100 km/s. The rest was just simply the difference from the normal velocity field. He has a real density distribution which becomes hydrostatic in a smooth way. That's the major reason why, for instance, he doesn't need such an amount of microturbulence to fit the profile around low velocities. So what essentially comes out from his kind of work is that the microturbulent velocities you need for fitting the profiles are much smaller than the 200–300 km/s they had before.

MASSIVE STAR EVOLUTION

André Maeder
Geneva Observatory
CH-1290 Sauverny
Switzerland

1. INTRODUCTION: ARE MASSIVE STARS THE SAME THROUGHOUT THE UNIVERSE?

The properties of massive stars are essential for our understanding of starbursts in the Universe. Due to high luminosities and short lifetimes massive stars offer striking signatures of starbursts in galaxies. For a long time, metallicity was not considered to play a major role in massive star evolution. There were two main reasons. Firstly, in our Galaxy massive stars observed in the solar neighborhood always belong to Population I and all have similar metallicity Z. Secondly, the dominant opacity source in massive stars is electron scattering, which is essentially independent of Z. Thus, the structural effects of Z were considered to be small.

Massive stars, however, copiously evaporate during their evolution. Recent mass-loss rates for O stars, supergiants and W-R stars indicate that all stars with initial masses larger than 30 $M_\odot$ in Population I finish their life with final masses between 5 and 10 $M_\odot$ (cf. Maeder, 1990a; see Fig. 6 below). Even a star with initially 100 $M_\odot$ is likely to have only about 10 $M_\odot$ left at the time of its supernova explosion. Thus, it is not exaggerated to say that evaporation by stellar winds is the dominant factor in massive star evolution.

In this context we immediately realize that a key point about massive star evolution in different galaxies is the relation between mass loss rates $\dot{M}$ and metallicity Z. If there were no such relation, massive star evolution would be about the same everywhere. However, recent models of stellar winds (cf. Abbott, 1982; Kudritzki et al. 1987; Leitherer and Langer, 1990) have suggested the existence of such a relation. In this way Z-effects enter massive star evolution. In the SMC, in blue compact galaxies or in elliptical galaxies at high redshift, Z is low and so are the mass-loss rates, while in the solar neighborhood, Z is high and mass loss effects are substantial.

This mass loss versus metallicity relation influences all the outputs of massive star evolution (cf. Maeder, 1990a):
- the lifetimes in the various phases and thus the distribution of populations in galaxies,

- the nucleosynthetic yields,
- the supernova progenitors.

The object of this study is to review the main evolutionary effects depending on mass and Z, as available today and to show possible tests which can presently be performed.

2. MASS AND Z EFFECTS IN THE COURSE OF EVOLUTION

In order to explain the observed differences of the Wolf-Rayet populations in galaxies, Maeder, Lequeux and Azzopardi (1980) proposed that the observed differences result from a relation between mass-loss rates and metallicity Z. The sensitivity of evolutionary properties on $\dot{M}$ was found so strong that even a mild dependence of $\dot{M}$ on Z would be able to account for the observations. Abbott (1982) proposed a relation of the form $\dot{M} \sim Z^\alpha$ with $\alpha = 1$ at a given luminosity and T_{eff}. From new stellar wind models, Kudritzki et al. (1987) derived an exponent of about 0.5, which is also supported by more recent work by Leitherer and Langer (1990).

The evolutionary consequences of the changes of mass-loss rates have been extensively studied for years by many authors, and they are reviewed by Chiosi and Maeder (1986). The main effects on the lifetimes, and consequently on the stellar populations, can be summarized as follows.

For the currently observed mass-loss rates, the consequences of stellar winds on main sequence (MS) evolution are small. The MS lifetime is $t(MS) \sim qM/L$. The mass of the convective core qM decreases with increasing mass loss, while the luminosity L does the same. Thus, the net result for $t(MS)$ is small: at most an increase by 5 to 10%, associated with moderately small differences in the tracks.

On the other hand, mass loss effects in the He-burning phase are enormous. The lifetime $t(He)$ in the He-burning phase is usually shared, for massive stars, between blue supergiants, red supergiants and W-R stars:

$$t(He) = t(BSG) + t(RSG) + t(WR)$$

The sharing and balance between these 3 stages depend very sensitively on $\dot{M}$. Typically, for a star with initially 30 $M_\odot$, in case of no mass loss $t(He) = t(BSG)$; the whole He-phase is spent as a blue supergiant. This is mainly due to the large intermediate convective zone (cf. Stothers and Chin, 1979; Maeder, 1981) which keeps a large zone more homogeneous both in composition and temperature, thus the blue location. For increasing mass-loss rates, the time fraction spent in the blue declines, and thus the star moves to the red supergiant stage earlier and the time fraction spent in this phase is thus increased. However, the larger the mass-loss rates in the red supergiant phase, the earlier the star will be peeled off and evolve to the stage of a bare core, generally identified with the W-R stage. Henceforth, there is a surprising behavior of $t(RSG)$, which first goes up as $\dot{M}$ increases and then goes down as $\dot{M}$ further increases. Of course, $t(WR)$ always increases with the growth of the mass-loss rates in previous stages.

The above effects lead to different evolutionary sequences according to the ranges of initial stellar masses considered:

$M > M_1$: O - Of - BSG - LBV - WR - SN
$M_1 > M > M_2$: O - BSG - YSG - RSG - WR - SN
$M_2 > M$: O - RSG (with or without Cepheid loop) - SN

BSG, YSG, RSG mean blue, yellow and red supergiant respectively. LBV stands for luminous blue variables, W-R for Wolf-Rayet stars and SN for supernovae. The limiting masses between these sequences very much depend on the exact value of the mass-loss rates. For the currently observed rates in Population I stars, $M_1 = 50\ M_\odot \pm 10\ M_\odot$ and $M_2 = 35\ M_\odot \pm 5 M_\odot$.

New grids of models of massive star evolution at $Z = 0.002, 0.005, 0.020$ and 0.040 have been computed in the range of 120 $M_\odot$ to 15 $M_\odot$ (cf. Maeder, 1990a). Figs. 1 and 2 illustrate the results for the two extreme values of Z considered. The lifetimes, luminosities, T_{eff} and chemical compositions are given at well corresponding ages, allowing easy population synthesis and modeling of starbursts in galaxies of various metallicities. The extension of these grids to lower masses is now in progress. These models use up-to-date nuclear cross sections and new opacity tables made by G. Schaller at Geneva Observatory from the Los Alamos Opacity Program. Proper account is given in the opacity tables to the fact that the O/Fe and α-nuclei/Fe abundance ratios are larger than solar at low Z. The initial model abundances, including helium, have also been modified accordingly.

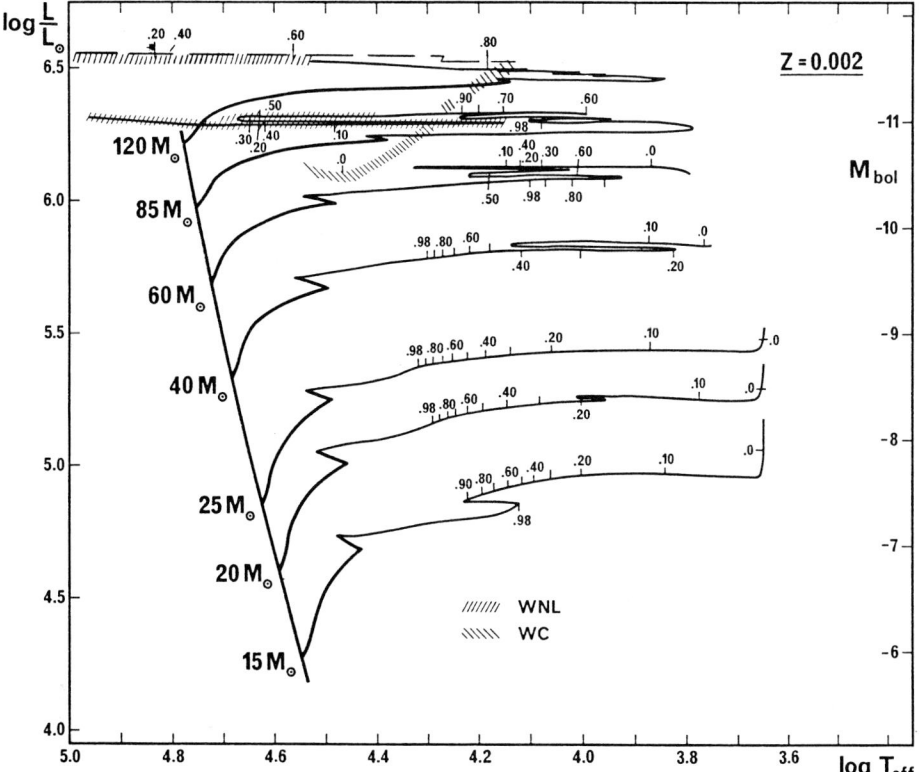

Figure 1. *Evolutionary tracks in the log L vs. log T_{eff} diagram for models with Z = 0.002. The data along the tracks indicate the values of the central He-contents Y_c during the He-burning phase. The W-R phases are marked with hatched lines.*

The mass-loss rates for Pop. I stars with $Z = 0.020$ are based on the recent data analyses by de Jager et al. (1988). For other Z-values the mass-loss rates of non-W-R stages have been scaled according to a law $Z^{0.5}$, as discussed above.

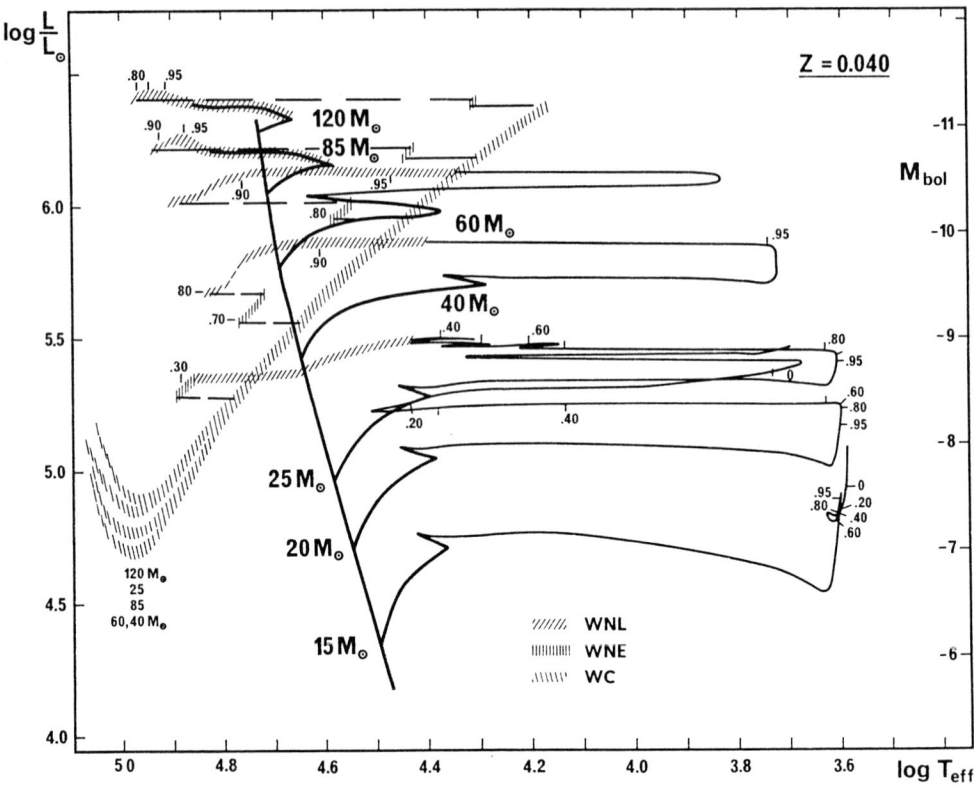

Figure 2. *Evolutionary tracks for $Z = 0.040$. Same remarks as for Fig. 1.*

For W-R stars, we no longer use the average observed mass loss rates, as we and others generally did in the past. As shown by Schmutz et al. (1989), this was leading to theoretical W-R luminosities much higher than those observed. There are several indications in favor of mass-loss rates for WNE and WC stars depending on the actual masses of W-R stars, (cf. Abbott et al. 1986; St-Louis et al. 1988; Langer, 1989; Smith and Maeder, 1989).

It has repeatedly been suggested (cf. Maeder and Meynet, 1989) that stellar models should include the core-overshooting effect. Also, the T_{eff} values of W-R stars have been corrected for the optical thickness of the wind, supposed to be due mainly to electron scattering. Due to the T_{eff} correction, the WNE and WC stars follow unique well defined tracks in the HR diagram, almost independently of their initial stellar masses. This can be seen for example in Figs. 2. These tracks are explained by the fact that as their mass and luminosity decline, the mass loss rates and the amplitudes of the T_{eff} correction are gradually reduced. Thus W-R stars move downwards and bluewards in the HR diagram.

3. LIFETIMES AS SUPERGIANTS AND QUESTIONS RELATED TO SN1987A

Presently many questions arise regarding the evolutionary status of blue supergiants and the evolutionary connections between the various types of supergiants and W-R stars. Let us first consider the general results of evolution with standard hypotheses, including overshooting. We examine here the case of a 25 $M_\odot$, since it also evolves in some cases to W-R stars. Fig. 3 illustrates the T_{eff} evolution for various Z. We recognize the long blue phases of low Z models. The duration of the red supergiant phase is seen to increase from $Z = 0.002$ to $Z = 0.020$ and then decrease again for higher Z. For the considered stellar mass, low Z models do not enter the W-R phase, while high Z models do. We emphasize that this behavior is extremely dependent on the mass-loss rates and that a change of these rates by a factor of 2 may shift the whole picture, and that the supergiant phases are particularly sensitive, (cf. also Leitherer, 1990)

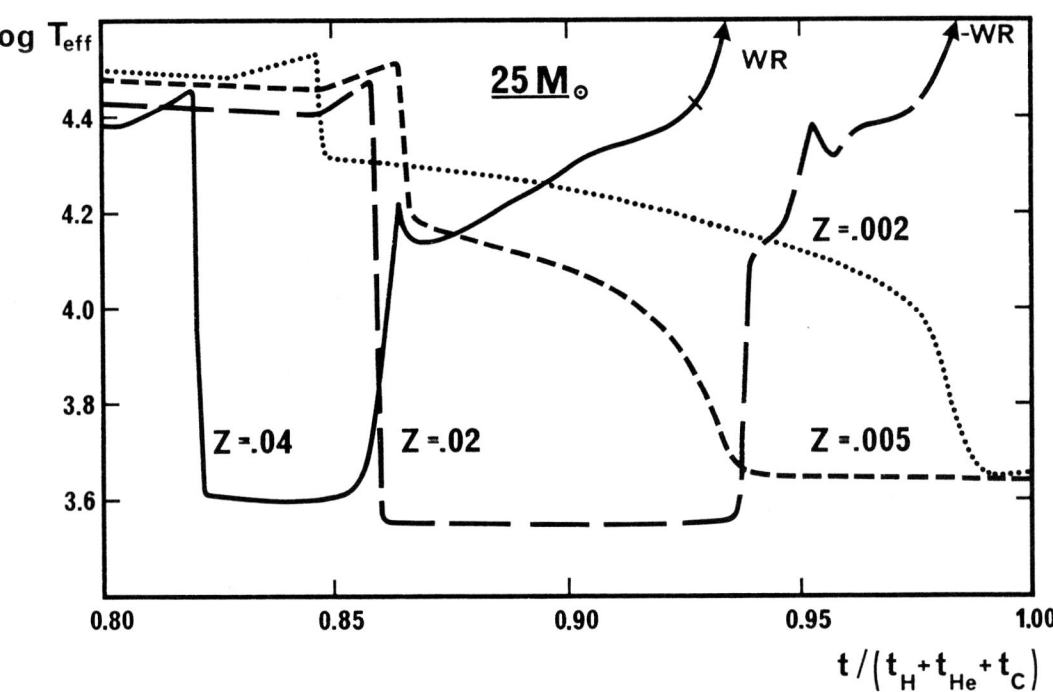

Figure 3. *Evolution of log T_{eff} for star models with an initial mass of 25 $M_\odot$ and various metallicities. The abscissa is the age t divided by the total stellar lifetime from the zero-age main sequence to the end of the C-burning phase.*

The occurrence of SN1987A from a blue progenitor has provoked many questions about massive star evolution. Various physical effects may intervene, the importance of which is often still uncertain and may greatly influence the final location of supernova progenitors. Let us examine those effects.

1. Mass loss and overshooting are included in the above models. These two effects influence the occurrence of the blue loops and modify the internal structure. Thus, mass loss plays a crucial role in the final T_{eff} of supernova precursors, (see point

3 below). Also, semiconvection in the H-burning phase is substantially reduced by mass loss and overshooting or it may even disappear (cf. Chiosi and Maeder, 1986).

2. Insufficiencies in the atomic opacities are very likely. Around $\log T_{eff} = 5.5$, Rosseland opacities may be larger by a factor of 2.5, and this would considerably affect the lifetimes and distribution of stars in the upper HR diagram, since the evolutionary tracks are modified by opacity changes, as shown by Nasi and Forieri (1990).

3. The use of the Ledoux criterion for convection has been proposed by Woosley (1988) to account for the blue progenitor, (cf. also Weiss, 1989). However, when a moderate mass loss is taken into account, the final blue location occurs too early, $i.e.$, during central He-burning. And this is not in agreement with the suggestion that the precursor of SN1987A was a red supergiant a few thousand years before the explosion.

4. A treatment of semiconvection intermediate between the Ledoux and Schwarzschild criterion has been proposed by Langer et al. (1989) to account for the blue SN progenitor. For some fine tuning of the convection parameter, this treatment well produces the blue progenitor. However, it does not simultaneously account for the observed He and N/C enrichment in SN1987A. An additional mixing or diffusion process has then to be called for to explain these surface enrichments. Langer et al. (1989) is using there rotational diffusion, as proposed by Maeder (1987) to explain some of the ON blue stragglers in young associations. If this applies also to SN1987A, this means that two fine tunings are required to explain simultaneously the blue location and the N/C enrichment. This is not impossible, but this also means that the generality of that solution can be questioned.

5. The possibility of mixing of the external layers of a red supergiant at the end of the He-burning phase has been envisaged by Saio et al. (1988). Indeed, our models also confirm that at this stage, the external and intermediate convective zones become extremely close to each other, so that any tiny convective extension is likely to bring the two zones into contact. The resulting mixing produces simultaneously the He- and N/C enrichments as well as the blue location of the SN progenitor and this makes the suggestion by Saio et al. (1988) an attractive one.

In conclusion of this discussion, we see that presently there is not one unique interpretation possible for the blue progenitor of SN1987A. My own view is that the sensitivity of the supergiant phases to opacity is so great (cf. point 2 above) that the right answer may only come when the new expected opacity tables will be available and lead to a revival of our conceptions in that very sensitive part of the HR diagram.

Let us also emphasize that observations of chemical abundances and HR diagrams of young clusters and associations may also greatly contribute to constrain the evolutionary models—perhaps in a more general way than merely considerations on the blue progenitor of SN1987A.

4. W-R STARS: PROPERTIES AND NUMBERS IN GALAXIES

W-R stars are most important for the study of starbursts. They are very luminous and also the most easily identifiable stars. Originating from the most massive stars, their age is small although they represent an advanced evolutionary stage. Thus, they are excellent tracers of starbursts in galaxies. Also W-R stars emit a substantial fraction of their energy ($i.e.$, about 5%) in the form of mechanical energy, which copiously feeds up giant HII regions, to which they are generally associated.

The uncertainties in evolution we have mentioned in the previous section do not affect W-R stars very much. As the stars become bare cores (*i.e.*, W-R stars), the different modifications affecting their outer layers have been washed out in the stellar winds.

Surface abundances play a key role in the identification of models and W-R subtypes. As is known, the observed abundances of WNL, WNE, WC and WO stars are consistent with the progressive exposition of nuclearly processed materials (cf. Chiosi and Maeder, 1986).

- **WNL (WN6-9) stars** are generally characterized by N/He values compatible with CNO equilibrium values, with generally H still present, (cf. Conti *et al.* 1983)
- **WNE (WN2-6) stars** have the same N/He as WNL stars, but with generally very little or no H left. Many exceptions to the rule seem to occur.
- **WC stars** have no H left, high abundance of He, C and probably O, as compatible with evidence of the products of partial He-burning. Smith and Hummer (1988) have determined C/He ratios in WC stars and shown that these ratios increase for early WC subtypes. Smith and Hummer and Smith and Maeder (1990) give the following (in numbers):

WC9: (C+O)/He=	0.03 to 0.06	WC5:	0.55
WC8:	0.1	WC4:	0.7 to 1.0
WC7:	0.2	WO :	> 1.0
WC6:	0.3		

Fig. 4 gives an overview of the lifetimes $t(WR)$ in the W-R stage as a function of mass and Z. The clear and easily understandable trend is an increase of $t(WR)$ with initial mass and Z. Also we notice that the minimum mass for W-R formation is lower at higher Z. Similar graphs can be established for the lifetimes in the WNL, WNE, WCL and WCE phases.

The lifetimes in the various phases can be used to derive the relative number frequencies WR/O, WC/WR, WC/WN. Assumptions on the star formation rate (SFR) and the initial mass function (IMF) have of course to be made. For a galaxy or a large galactic ring, the assumptions of a constant SFR over the last ten million years is reasonable. (For a single HII region, this is not necessarily acceptable and ageing effects are likely to intervene). For the IMF, two forms $dN/dM = AM^{(-1-x)}$ with $x = 1.35$ and 1.7 are considered. Table 1 shows the results for the number ratios WR/O, WC/WR and WC/WN at various metallicities. Two cases for the lower T_{eff} limit of O stars are considered, one at $\log T_{eff} = 4.52$ and the other at 4.53.

The results of Table 1 can be compared to the observed number ratios in galaxies and galactic rings in the Milky Way (cf. Table 2). The main source references for the observations are Smith (1988) and Arnault *et al.* (1989).

The comparison between the observed and theoretical WR/O number ratios is shown in Fig. 5. The general increase of WR/O is remarkably well reproduced by the models and the progress is considerable with respect to previous analyses. Only the observed point at $Z = 0.029$ corresponding to inner galactic regions shows some noticeable difference. The reason may be that the detection of W-R stars is certainly more complete than that of O-stars in the interior ring. On the whole, Fig. 5 gives powerful support for the evolutionary models and the idea by Maeder *et al.* (1980) that metallicity Z, through its effects on the mass-loss rates of O stars and supergiants, is responsible for the enormous differences of the supergiants and W-R distributions in galaxies with active star formation. The comparisons for the other number ratios in galaxies are discussed by Maeder (1990b).

TABLE 1. *Theoretical number ratios for Wolf-Rayet stars as a function of metallicity Z, for 2 slopes x of the IMF, and for 2 choices of the minimum T_{eff} of O stars.*

Z	WR/O $\log T_{eff} > 4.53$	WR/O $\log T_{eff} > 4.52$	WC/WR	WC/WN
$x = 1.35$				
0.002	0.0034	0.0032	0.057	0.061
0.005	0.0185	0.0182	0.192	0.237
0.020	0.0790	0.0752	0.640	1.784
0.040	0.1628	0.1557	0.736	2.784
$x = 1.70$				
0.002	0.0021	0.0019	0.058	0.061
0.005	0.0135	0.0133	0.170	0.204
0.020	0.0654	0.0616	0.634	1.732
0.040	0.1438	0.1363	0.744	2.908

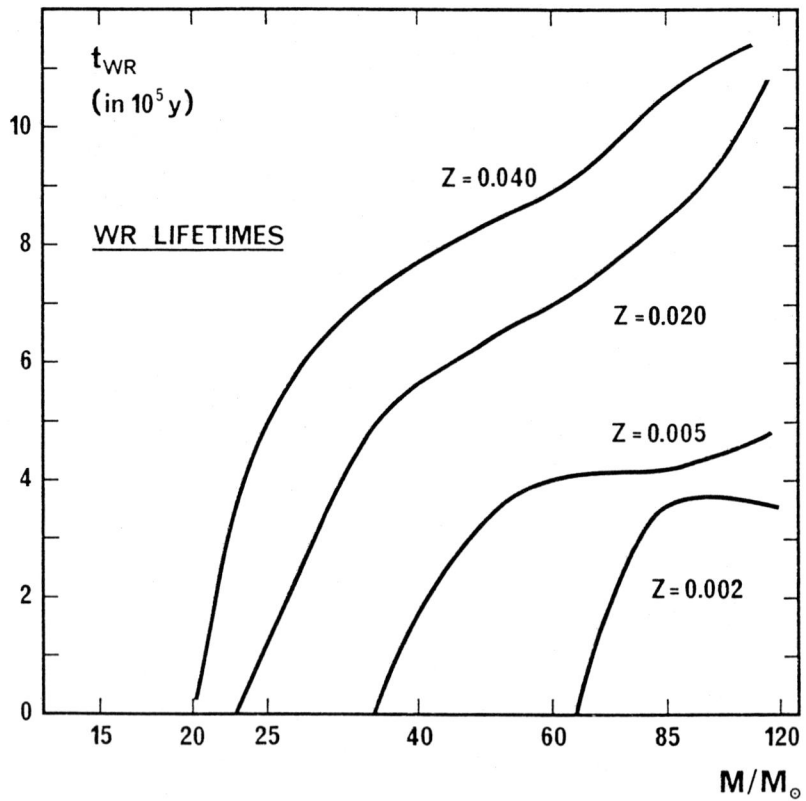

Figure 4. *Lifetimes as W-R stars as a function of initial masses and metallicities.*

The relative number of WN stars with and without hydrogen, identified with WNL and WNE stars respectively, are also derived. The WNL phases generally last much

TABLE 2. *Observed number ratios of W-R and O stars in different galaxies and galactic sites as a function of Z.*

GALAXY	Z	WR/O	WC/WR	WC/WN
M 31	0.035	—	0.74	2.82
sol. n. 6–7.5 kpc	0.029	0.205	0.74	2.85
	0.025–0.035			
sol. n. $d < 2.5$ kpc	0.020	0.116	0.65	1.88
	0.011–0.035			
sol. n. 7.5–9.5 kpc	0.020	0.104	0.55	1.22
	0.016–0.025			
sol. n. 9.5–11 kpc	0.013	0.033	0.5 :	1 :
	0.011–0.016			
M 33	0.013	0.06	0.42	0.71
	0.007–0.020			
LMC	0.0057	0.04	0.20	0.26
NGC 6822	0.0045	0.015–0.03	—	—
SMC	0.0021	0.015	0.12	0.14
IC 1613	0.002	0.02	—	—

Ref: Arnault et al. 1988; Azzopardi and Breysacher, 1979, 1985; Azzopardi et al. 1988; Garmany and Conti, 1982; Meylan and Maeder, 1983; Smith, 1988; van der Hucht et al. 1988 (sol. n. stands for solar neighborhood).

longer for large initial masses, with little dependence on Z. On the other hand, the WNE phases are longer for lower initial masses, also being shorter at $Z = 0.04$ than at $Z = 0.02$. The comparisons with observations of WNL and WNE stars in galaxies show agreement for Z equal to or larger than 0.02. At lower Z, a large fraction of the existing W-R stars probably results from binary evolution and they are mostly WNE stars.

The relative numbers of WC stars with various subtypes are also analysed quantitatively. WC models are classified according to the (C+O)/He number ratios as shown above. The models show that the entry points and lifetimes in the sequence WC9 to WC4 and WO are extremely dependent on Z and mass. The entry into the WC9 subtype only occurs for high initial Z and mass, i.e., Z larger or equal to 0.02 and $M > 50$ $M_\odot$. Lower initial M and Z lead to entry at earlier WC subtypes, with a shorter overall WC phase. After integration of the lifetimes over the mass spectrum, we notice that for high initial Z, late and intermediate WC subtypes are favored, while at lower Z, only early WC and WO stars are found and in much lower frequencies. The theoretical number ratios of late, early WC and WO stars compare quite well with observed number ratios in galaxies of the Local Group.

5. FINAL MASSES, CHEMICAL YIELDS AND MASS LIMITS

Let us now examine what are the final masses at the time of supernova explosion. Figure 6 shows the main results as a function of initial stellar masses and for various metallicities according to recent models (Maeder, 1990b). For no mass loss, the relation in Fig. 6 would of course be linear with a slope of 45 degrees. Fig. 6 also shows the relation betwen final and initial masses in the case of constant mass-loss rates in the

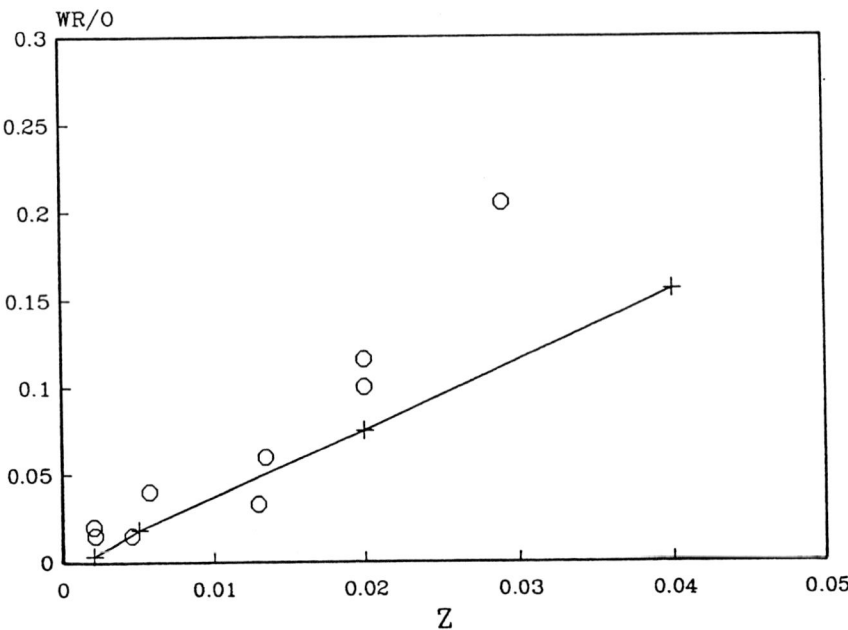

Figure 5. *Comparison of observed (circles, cf. Table 2) and theoretical (continuous line, cf. Table 1) number ratios WR/O of W-R to O stars as a function of metallicity. The theoretical values are derived for an IMF slope of $x = 1.35$ and a T_{eff} limit of O stars at 33000 K.*

W-R stage; in this case, the final masses were found to be rather large in the case of the initially most massive stars. In the more realistic case where the mass-loss rates of W-R stars depend on the W-R masses, the situation is different. For $Z = 0.02$ or more, stars with initial masses larger than about 30 $M_\odot$ all finish their life with small final masses in the range of 5 to 10 $M_\odot$. For $Z = 0.005$, small final masses result only from stars with initial masses larger than about 85 $M_\odot$. At $Z = 0.002$ the final masses are found to be rather large as the mass-loss rates in the pre-W-R phases are much smaller. Let us emphasize again that these specific results very much depend on the adopted parameterization for the mass-loss rates.

The change of the final stellar masses with Z has great consequences for several of the major astrophysical problems, namely:
- the chemical yields,
- the nature of supernova progenitors,
- the existence of black holes.

Up to now, detailed chemical yields have been obtained only in the case of Population I stars. As an example, Fig. 7 shows the yields obtained from presupernova models of Pop. I stars (cf. Meynet, 1990); in these models the $\dot{M}$ for W-R stars has been taken as constant. However, from the curves in Fig. 6, we see that in the mass range of 9 $M_\odot$

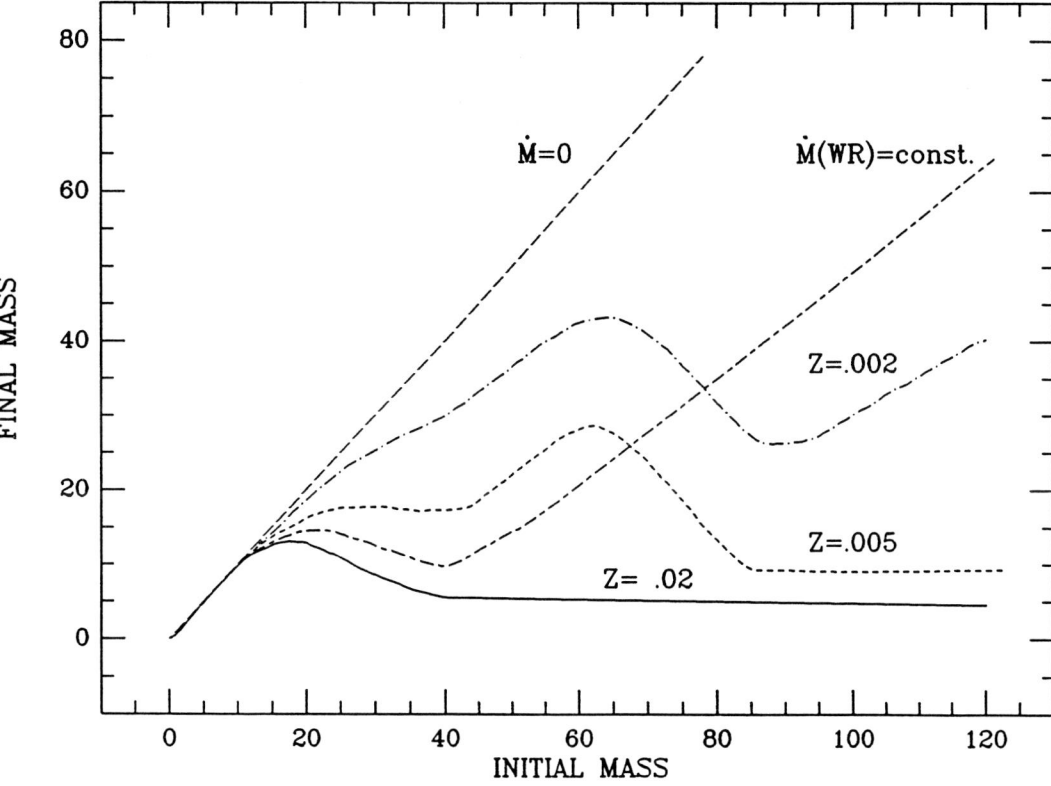

Figure 6. *Relations between the final and initial stellar masses for various metallicities.*

to 40 $M_\odot$, these models have final masses rather close to those of the new models with $Z = 0.005$ and $= 0.020$. We notice in Figure 7 the large wind contributions in ^{4}He and heavy elements, as well as the very large amount of oxygen ejected in supernovae. It was emphasized by Meynet that for a star with an initial mass of 20 $M_\odot$ these yields remarkably correspond to the observed abundances of the elements in SN1987A.

It is evident from the existence of very different models at various Z that the chemical yields will also be different for various metallicities. The amount of elements ejected in the winds (typically ^{4}He, ^{12}C, ^{22}Ne, ^{25}Mg, ^{26}Mg and s-elements) is much larger at higher metallicity. This means, for example, that the relative helium to metal enrichment dY/dZ is much larger at high Z than at low Z. Superficially, it could be thought that the fact that helium is ejected a bit earlier in the wind or a bit later in the SN explosion does not make any difference. This would not be true: if a lot of helium is ejected in the wind, it is not further processed into heavy elements and it thus leads to a higher ratio of helium to metals. The same is true, to a smaller extent, for the C and O ejected in the winds, which are therefore not further transformed: their ejection in the wind contributes to making their yields larger. Another significant consequence of the high wind of W-R stars is that the left-over CO cores may be significantly reduced with respect to the case of no mass loss, and this also influences the yields. All these effects clearly show that the yields have to be computed carefully and in detail for each metallicity and associated mass-loss rate. This work is now currently in progress.

Fig. 6 also shows that the range of stellar masses leading to various supernova types

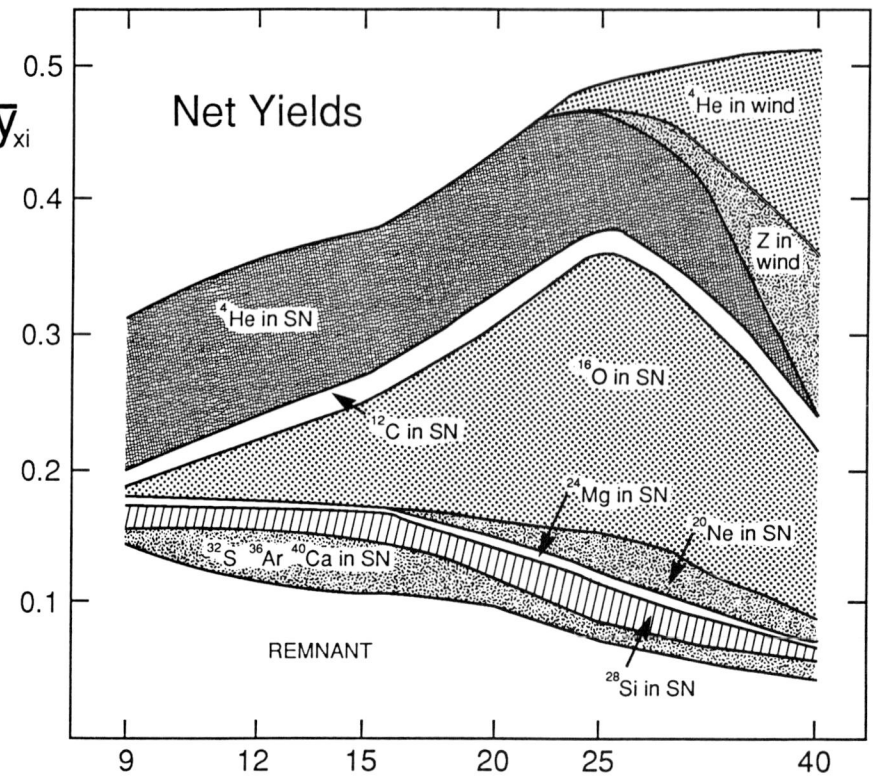

Figure 7. *Stellar yields from pre-supernovae models by Meynet (1990).*

will depend on the metallicities, since the final masses are different. For example, the range of initial masses leading to SN from W-R stars (often identified with SNIb, cf. Ensman and Woosley, 1988; Langer, 1989, 1990) is likely to be much shorter at low Z than at high Z. Thus, SNIb should be rare at low Z. Conversely, the range of initial masses leading to SNII is expected to be larger at low Z. For the estimates of SNII numbers, it is also particularly necessary to account for the modification due to overshooting of the mass limits for electron capture and C-detonation (cf. Maeder and Meynet, 1989). Finally, it is also clear that the occurrence of black holes supposed to result from the core collapse of massive stars should be more likely in low Z populations, since, as shown by Fig. 6 the final masses are much larger.

On the whole, I hope to have shown here that the evolution of massive stars must be entirely reconsidered in terms of metallicity effects. This has many consequences not only for the distribution of massive stars and W-R stars in galaxies as shown above, but also for all the astrophysical outputs of massive star evolution.

REFERENCES

Abbott, D. 1982, *Ap. J.*, **259**, 282.
Abbott, D., Bieging, J. H., Churchwell, E., and Torres, A. V. 1986, *Ap. J.*, **303**, 239.
Arnault, P., Kunth, D., and Schild, H. 1989, *Astr. Ap.*, **224**, 73.

Azzopardi, M., and Breysacher, J. 1979, *Astr. Ap.*, **75**, 120.
———. 1985, *Astr. Ap.*, **149**, 213.
Chiosi, C., and Maeder, A. 1986, *Ann. Rev. Astr. Ap.*, **24**, 329.
Conti, P. S., Leep, E. M., and Perry, D. N. 1983, *Ap. J.*, **268**, 228.
Ensman, L. M., and Woosley, S. E. 1988, *Ap. J.*, **333**, 754.
Garmany, C. D., and Conti, P. S. 1982, *Catalogue of O-type Stars*, Astron. Data Center, Goddard Space Flight Center.
van der Hucht, K. A., Hidayat, B., Admiranto, A. G., Supelli, K. R., and Doom, C. 1988, *Astr. Ap.*, **199**, 217.
de Jager, C., Nieuwenhuijzen, H., and van der Hucht, K. A. 1988, *Astr. Ap. Suppl.*, **72**, 259.
Kudritzki, R. P., Pauldrach, A., and Puls, J. 1987, *Astr. Ap.*, **173**, 293.
Langer, N. 1989, *Astr. Ap.*, **220**, 135.
———. 1990, in *Supernovae, 10th Santa Cruz Summer Workshop*, ed. S. E. Woosley, in press.
Langer, N., El Eid, M. F., and Baraffe, I. 1989, *Astr. Ap.*, **224**, L17.
Leitherer, C. 1990, *Ap. J. Suppl.*, **73**, 1.
Leitherer, C., and Langer, N. 1990, in *IAU Symposium 148, The Magellanic Clouds*, ed. R. F. Haynes, and D. K. Milne (Dordrecht: Kluwer), in press.
Maeder, A. 1981, *Astr. Ap.*, **102**, 401.
———. 1987, *Astr. Ap.*, **178**, 159.
———. 1990a, *Astr. Ap. Suppl.*, **84**, 139.
———. 1990b, *Astr. Ap.*, in press.
Maeder, A., Lequeux, J., and Azzopardi, M. 1980, *Astr. Ap.*, **90**, L17.
Maeder, A., and Meynet, G. 1989, *Astr. Ap.*, **210**, 155.
Meylan, G., and Maeder, A. 1983, *Astr. Ap.*, **124**, 84.
Meynet, G. 1990, Thesis, Univ. Geneva.
Nasi, E., and Forieri, C. 1990, *Astrophysics Space Sci.*, **166**, 229.
Saio, H., Nomoto, K., and Kato, M. 1988, *Nature*, **334**, 508.
Schmutz, W., Hamann, W.-R., and Wessolowski, U. 1989, *Astr. Ap.*, **210**, 236.
Smith, L. F. 1988, *Ap. J.*, **327**, 128.
Smith, L. F., and Hummer, D. G. 1988, *M.N.R.A.S.*, **230**, 511.
Smith, L. F., and Maeder, A. 1989, *Astr. Ap.*, **211**, 71.
———. 1990, *Astr. Ap.*, in press.
St-Louis, N., Moffat, A. F. J., Drissen, L., Bastien, P., and Robert, C. 1988, *Ap. J.*, **330**, 286.
Stothers, R., and Chin, C. W. 1979, *Ap. J.*, **233**, 267.
Weiss, A. 1989, *Ap. J.*, **339**, 365.
Woosley, S. E. 1988, *Ap. J.*, **330**, 218.

DISCUSSION

T. Heckman: If the importance of W-R stars relative to O stars increases with metal abundance *and* the WC/WN ratio *also* increases with metal abundance then I would think that the so-called "Wolf-Rayet Galaxies" discussed by Peter Conti would be dominated by WC stars. Yet they show no evidence for WC's at all! How can one get a population in which the W-R's are pronounced relative to O stars, yet WC stars are deficient?

Maeder: I do not see how you can say that the W-R population is pronounced with respect to the O-star population in W-R galaxies, since I have not seen any reliable estimate of the W-R/O number ratio in such galaxies.

Tell me the age of the burst, the metallicity, the WN and WC numbers and thus we may see whether W-R galaxies fit or not in the scheme I have presented.

T. Heckman: I am asking more of an observer's question. Assume you have a whole population of objects having wide ranges in metal abundances. You will preferentially observe those with a pronounced Wolf-Rayet population. Those ought to be the ones where the WC stars dominate just because you showed in the cases where the WN's dominate over the WC's at the same time the relative number of Wolf-Rayet to O stars is extremely small. And so just empirically it doesn't seem to work.

P. Conti: That's a good question and I do not have a definitive answer. We will try to come up with some numbers once the analysis of W-R galaxies is complete. In the meantime, some possible answers could include: (1) W-R galaxies are relatively young enabling one to detect the integrated stellar emission at $\lambda 4686$ due to WN stars. This emission is at a maximum since few stars formed in previous bursts. (2) There is a relatively *flat* IMF in W-R galaxies compared to the norm, thus there are relatively more massive progenitors. (3) The W-R stars detected are not classical WN types but might more properly be thought of as hydrogen-burning Of/WN stars; these are relatively bright and relatively longer lived than helium-burning WN objects.

N. Panagia: Although it is fashionable to claim that type Ib supernovae come from W-R stars, there is no evidence that this is true. In fact, there are good arguments against that claim (see *e.g.,* Panagia and Laidler, 1989, Lecture Notes in Physics, No. 316, pp. 187–191). Among other aspects, the rate of SNIb events is too high (about 1/3 of that of SNII; Branch, 1986, Ap. J. Lett. 300, L51; van den Bergh et al. 1987, Ap. J. 323, 44) to be accounted for by W-R stars. Let me stress that my point is not that W-R stars do not make SN explosions, but that they do not make SNIb events. For example, I am wondering if the core collapse of a W-R star may account for "dim" explosions such as that believed to have given rise to Cas A.

Maeder: I don't know whether somebody has considered carefully the statistics of Ib and Wolf-Rayet numbers. Do you have an idea about the statistical arguments?

N. Langer: I know the arguments and I don't know how reliable these numbers are. I would like to answer to such questions that, as has been shown by Ensman and Woosley, if a low-mass Wolf-Rayet star explodes, it looks like a Ib supernova. That does not mean that all type Ib supernovae come from Wolf-Rayet stars. So, the question of statistics may be solved that way.

N. Panagia: No doubt that one needs a relatively low mass at explosion because otherwise the light curve would never be explained. Whether it's Wolf-Rayet or not, that is a different story.

B. Rocca-Volmerange: In chemical evolution models, including SNI and SNII explosions the iron in our Galaxy can only be fitted if we assume that a part of SNI is unclassified: SNIb could be a low mass SNII with W-R stars as progenitors. The production of iron and other elements (C, N, O, ...) is well fitted with the absolute solar abundances.

R. Kudritzki: My question is not related to statistics, but I have these recent results by Hamann and collaborators on Wolf-Rayet stars in the Large Magellanic Cloud. The interesting thing they find is that the luminosity of these objects in the Large Magellanic Clouds is in many cases significantly smaller than in the Galaxy. Although I find the results you have extremely convincing, I doubt that your tracks can explain these observational results. In a HR diagram of the Large Magellanic Cloud you find W-R stars at luminosities already below $\log L = 5.00$. I'm wondering if you can explain this?

Maeder: Yes, we have looked at this point but you have to be careful on the subtype of the Wolf-Rayet star. At the given subtype they would not have the same luminosity whether they are formed in high Z or low Z. For a significant comparison, one has to compare a given subtype in the LMC with the corresponding subtype in the Milky Way. I did this together with Lindsay Smith, and we found acceptable agreement.

G. Koenigsberger: I'm afraid, now I'm thoroughly confused about how you get a Wolf-Rayet star. We have three mass-loss episodes: (i) the mass-loss rate of the progenitor, (ii) some sort of abrupt mass loss where a large fraction of mass is lost, and (iii) the Wolf-Rayet phase where you have yet another mass-loss rate. If the evolution is so sensitive to your mass-loss rates, how do you take into account the abrupt mass loss, which nobody really understands?

Maeder: It is clear that evolution is extremely sensitive to mass loss. We adopted the $\dot{M}(L, T_{eff})$ relation published by de Jager *et al.* (1988) including a Z dependence for all the mass-loss rates before the Wolf-Rayet stage. For the Wolf-Rayet phase, we took the dependence on the assumed mass but not on Z, because it is not expected that Wolf-Rayet stars formed in a low-Z or high-Z medium would have a very different mass-loss rate. For now, we don't know anything except their composition. For example, a WN star is essentially close to a pure helium star, and whether it is formed in a low-Z or high-Z medium, it would probably have the same mass-loss rate at a given mass—at least to the first order. And the same is true for the WC. Therefore we don't include any Z-dependence in the Wolf-Rayet stage. Maybe in the future it could be possible.

G. Koenigsberger: But is there not a large difference in the mass loss of an O star and a Luminous Blue Variable?

Maeder: Yes. The parameterization by de Jager *et al.*, which I have used for the mass-loss rates in the HR diagram, reproduces more or less correctly the mass-loss rate of about 10^{-3} in the LBV phase. Therefore, I did nothing more than taking their mass-loss rate.

R. Humphreys: My question is really more about stellar evolution than it is about Wolf-Rayet stars. Can I ask that? Ok, in fact it's about stars that *don't* become Wolf-Rayet stars. I have a question about the evolution of what might be called intermediate mass supergiants. In your scenario you put forward today you have just about everything more massive than, say, 23 solar masses becoming a Wolf-Rayet star and then going into the supernova stage. We do observe, or we think we observe that red supergiants become type II supernovae—at least some red supergiants do. We have seen one blue supergiant become a type II supernova. I was noticing in your tracks for the lower metallicity systems that they did not go back to the blue. Those tracks which would presumably be appropriate for the Magellanic Clouds do not return to the blue and therefore would not be producing the progenitor of Supernova 1987A. I am sure that by tweaking various parameters—metallicity, mass-loss rate and so on—you can send a star back to the blue. How would you address the problem of SN1987A as a supernova progenitor and red supergiants exploding as supernovae in the context of your models?

Maeder: It seems that the blue progenitor of SN1987A was enriched in helium to some extent. And it's clear that the tracks I have shown don't produce such a large increase in the helium content. If there is a high surface enrichment in helium somewhere in the evolution, it is very easy to have the blue location. We essentially have 2 explanations now for the blue progenitor of SN1987A. 1) Saio *et al.* proposed that mixing of the outer layers during the C-burning produces a helium enrichment at the surface, which results in a blue location of the SN progenitor. 2) Langer obtains the blue location and the surface N-enrichment by a finely tuned combination of a semi-convective parameter and rotational mixing. In that context, I may only say that I get very close (by $\Delta M/M \simeq 0.02$) external and intermediate convective zones in the C-burning phase in my models, which tends to make the interpretation by Saio *et al.* very plausible. However, the consequences of additional mixing processes are also very worthwhile to explore, and this is why I undertook models including rotational diffusive mixing some years ago. Whether SN1987A was a general case or just an exception—I don't know. Maybe Norbert has something to add.

N. Langer: What I would like to add is basically given in my poster contribution. I would prefer the point-of-view that the evolution of the progenitor might have been normal. All we can say is that stellar evolution models are not quite in the stage that you can predict every single event before it happens.

Maeder: If it is the normal case, how do you make a red supergiant explode as a type II supernova?

N. Walborn: There's plenty of stars between 8 solar masses and 20 solar masses.

Maeder: Let me perhaps just add one more specific comment on that point. Every kind of evolutionary models shows the extreme sensitivity of these tracks in the blue and in the red phase to the mass-loss rate or to the mixing. And on the other hand, we know that stars are rotating at different rates. Maybe one single evolutionary track with one given mass-loss rate, one kind of mixing or without mixing is able to reproduce one group of stars but not the whole spectrum of properties associated with rotation. Probably one should have models for 20 solar mass stars with zero rotation, 50 km/s, 100 and so on, in order to be able to reproduce the observed variety of rotational velocities.

A. Moffat: Can I say something about masses? I've been working on binaries, as you perhaps know, especially in the Small Magellanic Cloud. In double line systems we can really get the mass, especially since now we have polarization data to get the orbital inclinations. It's almost embarrassing to say that the masses that come out are quite small. There is a WO binary system with an O4 companion in the SMC and the mass of the W-R component is around 10 $M_\odot$. This is much less than the 20–30 $M_\odot$ your models predict. So either the observations are wrong or maybe there's some tweaking to do in the models. I don't know. This is a trend we're finding in the SMC in general, I think.

Maeder: It is true that the mass you are quoting is lower than the lower mass limit expected from single star evolution at low Z. I think one has to examine the whole expected mass spectrum of W-R stars at various Z and then one can make comparisons with observations. Maybe binary evolution and mass transfer has led to a higher mass loss in binary systems. What initial mass, for example, do you assign to the progenitor of the LBV η Carinae? Do you also have something smaller than 60?

A. Moffat: We have probably four components there, and I think the most massive one is below 90. I don't know the actual value quoted by Weigelt *et al.*

Maeder: The LBV phase is very short. They cannot be all four in the LBV phase. So there is probably only one in the LBV phase even if there is a group of four.

A. Moffat: I agree, but when you have a dynamic mass, you can't argue too much with it.

C. Leitherer: You adopted a scaling law of the form $\dot{M} \sim Z^{0.5}$ to obtain mass-loss rates as a function of Z. If I understood you correctly this has been done for the main sequence, blue supergiant, LBV (if applicable) and red supergiant phases. When one plots $\dot{M}$ as a function of time along your tracks, it is evident that the highest mass-loss rates occur in the LBV and RSG phases whereas $\dot{M}$ on the main sequence is relatively unimportant. The relation $\dot{M} \sim Z^{0.5}$ may be valid on the main sequence but what observational support do we have that this relation also holds during the LBV and RSG phases where larger amounts of mass are removed in your models? Personally I don't think that it is proven that radiation pressure *is* responsible for the large outbursts in LBV's with $\dot{M}$ of $10^{-1} - 10^{-2}$ $M_\odot$ yr^{-1}.

Maeder: You are quite right in pointing that out. This phase affects the most massive ones, and there's no reason to scale that with Z. Except the fact that we cannot introduce many, many parameters in the model—there are already enough. My hope is that within 10 years we may have mass-loss rates integrated self-consistently in the evolutionary models. The only thing we can reasonably do at the moment—except for LBV's—is to take the observed mass-loss rates and apply the proposed scaling relation. But I'm sure in the future we'll speak about complete stellar models, interior and atmosphere and wind in order to have an entirely consistent model.

Let me add that I also tried various such relationships, namely $\dot{M}$ vs. Z^α, with $\alpha = 1$, 0.5 and 0.25 and used the comparisons with number statistics to see what fits best. It is clear from these trials that the value $\alpha = 0.5$ as predicted by radiation driven wind theory by Kudritzki and coworkers gives the best agreement.

C. Leitherer: Would it be possible to turn the argument around? Do you think that mass loss is the most important ingredient which governs your evolutionary tracks as compared to overshooting and mixing processes so that we could use your predictions of ratios of red to blue supergiants or Wolf-Rayet to O stars, to constrain the mass-loss rates?

Maeder: I would say so, yes. In my opinion, overshooting is not negligible but with respect to the mass loss it is a secondary parameter.

THE OBSERVATIONAL H-R DIAGRAM AND IMF OF MASSIVE STARS

Catharine D. Garmany
Joint Institute for Laboratory Astrophysics
University of Colorado and National Institute of Standards and Technology
Boulder, CO 80309-0440
USA

1. INTRODUCTION: STARS AND THE STARBURST PHENOMENA

The starburst phenomenon in external galaxies is generally interpreted as representing a region in which the star formation rate has been temporarily elevated. The observations that form the basis of this conclusion are all global: the distances of these galaxies are too great for individual stars to be examined. Nevertheless, there is some observational evidence, reviewed by Scalo (1989) that the initial mass function (IMF) in starburst galaxies differs from the "normal" one, either by an excess of high mass stars or a deficiency of low mass stars. Generally such an effect is referred to as an IMF that is top heavy, or flatter than normal.

Before we can hope to understand top heavy IMF's in starbursts, we need to ask: How is the IMF determined locally for massive stars, and what is its accuracy? Are there data supporting variations either spatially or temporarily? How does the IMF as computed directly from star counts compare with more indirect results such as Hα flux? Some related problems include the actual determination of mass for these stars and the upper luminosity limit in the H-R diagram, since the appearance of the upper H-R diagram is a function of the IMF, star formation rate (SFR) and the results of stellar evolution. It is the purpose of this review to discuss these issues.

2. IMF FOR MASSIVE STARS: HOW IT'S DONE DIRECTLY

The galactic IMF for stars of all masses has been beautifully reviewed by Scalo (1986). However, it is appropriate to begin with a caveat from a more recent, related paper (Scalo 1989): "In order to study the spectrophotometric and chemical evolution of galaxies, it is necessary to understand the form of the stellar IMF and whether and how it depends on physical conditions. Although there have been numerous speculations concerning this question, it is no exaggeration to say that we know next to nothing about the answer."

Following Scalo, we define the IMF as the number of stars born per unit time per unit log mass interval per unit area. This function is taken to be a power law with an index which is called the slope of the IMF. Scalo denotes this index by Γ, where:

$$\Gamma = d \text{ (mass function)}/ d \log \text{mass}$$

As Scalo points out, this form is adopted more for computational convenience than from any fundamental understanding of the star formation process. Further discussion of this is given by Zinnecker (1984). Reference is also often made to the Salpeter IMF (1955) which is defined as the number of stars born per unit time per unit mass. The Salpeter mass function index is thus $\Gamma - 1$, and Scalo denotes this by γ. Another important review of the connection between stellar and galactic evolution is by Tinsley (1980): in her notation $x = -\Gamma$.

Scalo (1986) also discusses the difference between field (or composite) IMF's and cluster IMF's. In a cluster one assumes the star formation was more or less instantaneous, but in a field IMF one must correct for the relative ages of the stars. Such corrections for massive stars rely on ages derived from stellar evolution calculations.

A direct determination of the IMF will be taken to be a complete census of all stars within some given mass range and within some specified volume of space. In the galaxy, this is often arrived at from the luminosity function, using the mass-luminosity relation and a correction for evolved stars. This assumes we understand the stellar evolution history of the stars well enough to assign ZAMS masses to the evolved stars, and therefore that we know the evolutionary stage of all the stars. By contrast, indirect methods include IMFs determined from measures of integrated colors, UV fluxes, or Hα measures.

By massive stars we mean O- and B-type stars, and their descendents, with derived parameters in the following range:

Mass range: 10 $M_\odot$ and greater
Main sequence temperatures: 20,000 to 45,000 K
Bolometric magnitudes: -5 to -12
Lifetimes: 10^7 to 10^6 yr
Formation sites: OB associations and clusters in galactic spiral arms, with typical dimensions of 100 pc, which translates to many degrees in angular diameter for the nearest ones, but many arc minutes for associations in the Magellanic Clouds.

For these stars, the visual luminosity becomes less and less sensitive to the mass with increasing temperature. At spectral type O9, corresponding to a ZAMS mass of about 20 $M_\odot$, the bolometric correction is comparable in magnitude to the visual magnitude itself. A slice in absolute visual magnitude across an H-R diagram includes stars with masses from 20 to 10 $M_\odot$ and lower. This effect becomes more pronounced with increasing mass. An excellent example of this effect can be seen in Massey (1985), Figure 3, which shows the difference between an H-R diagram plotted in M_v and M_{bol}. A further problem with some IMF determinations that include massive stars is small number statistics: the larger the mass the fewer the number of stars. The correction for evolved stars is not always straightforward either. For example, in the original MK system, there is no luminosity classification for stars earlier than O9.

Although Scalo (1986) used the luminosity function to determine the IMF of massive stars, most other recent IMF determinations have started by first converting the stellar parameters for a statistically complete sample of stars to luminosity and temperature.

This requires accurate temperatures and bolometric corrections, the latter derived from stellar models. Photometry is not sufficient for the derivation of temperatures for stars earlier than B0. The color temperatures are a very poor thermometer because most of the observable wavelength lies on the Rayleigh-Jeans tail of the energy distribution, as pointed out empirically by Massey (1985) and shown theoretically by Hummer et al. (1988). Therefore the observational data must include spectra as well as photometry. The data, converted to temperature and luminosity, are plotted in an H-R diagram. Theoretical evolutionary tracks are adopted to define the mass of the stars (and bin them by mass), as well as to assign mean ages to the stars. This approach was, as far as I know, first used by Lequeux (1979) for stars brighter than apparent magnitude of 7.5 and with masses between 2.5 and 100 $M_\odot$. He found $\Gamma = -1.3$ in this study, which is almost identical to the Salpeter IMF. Garmany, Conti and Chiosi (1982), using a volume limited set of O-type stars, argued for $\Gamma = -1.6$, with the possibility that the IMF slope was flatter in the inner galactic arm than the outer one. Humphreys and McElroy (1984) included catalogs of both O and B stars and got $\Gamma = -2.2$. These and other studies are discussed by Scalo (1986) and are summarized in Figure 1. The variation in Γ from -1.3 to -2.4, based on what should be essentially the same data, hardly inspires confidence that we can determine an unambiguous value of this function. Part of the reason for this dismaying lack of unanimity may be the incomplete status of the stellar census, to be examined in the next section.

2.1. Problem: Completeness of the Data for Galactic Stars

The IMFs discussed in the previous section are referred to as field IMFs, as opposed to cluster IMFs. But what do we mean by "field" IMF for massive stars? It has been taken to include all stars, both field and cluster members, in a certain volume of the galaxy. Figure 2 shows that even those stars that are not clearly members of OB associations are clustered nearby in the plane of the galaxy. The distinction between true field stars and those which are members of clusters or OB associations is not very great. While this distinction is not important in a field IMF, a complete stellar census is very important. Let us examine how well this is known in the OB associations.

An examination of the literature on the OB associations listed by Ruprecht, Balazs and White (1981) reveals that only a small fraction of them have been studied individually. In order to better define the observational parameters of massive stars we have carried out a review of all the OB associations between $l = 50°$ to $150°$ (Garmany and Stencel, in preparation). The first question we ask is how the stellar membership in an association was settled and how the distance was derived. In many cases the majority of the members are O stars and evolved B stars, and the distance of the association is based on the average spectroscopic parallax of these presumed members, rather than on ZAMS fitting in a color magnitude diagram as is customarily done for open clusters. The spectral type-absolute magnitude calibration for O stars and B supergiants has a fairly large uncertainty, generally quoted to be as large as 0.5 mag (Conti et al. 1983; Garmany 1990). Not only does this raise questions about the accuracy of the distance modulus, but it also raises the completeness issue. Where are the B main sequence members? Some studies have concluded that they have not yet reached the main sequence or do not exist, while others dispute this (Chini and Wargau 1990). However, if the association has not been studied in detail, these stars may simply have been overlooked. And, as already pointed out, photometric studies cannot separate early B stars from O stars. Figure 3 shows examples of both an OB association that has been well

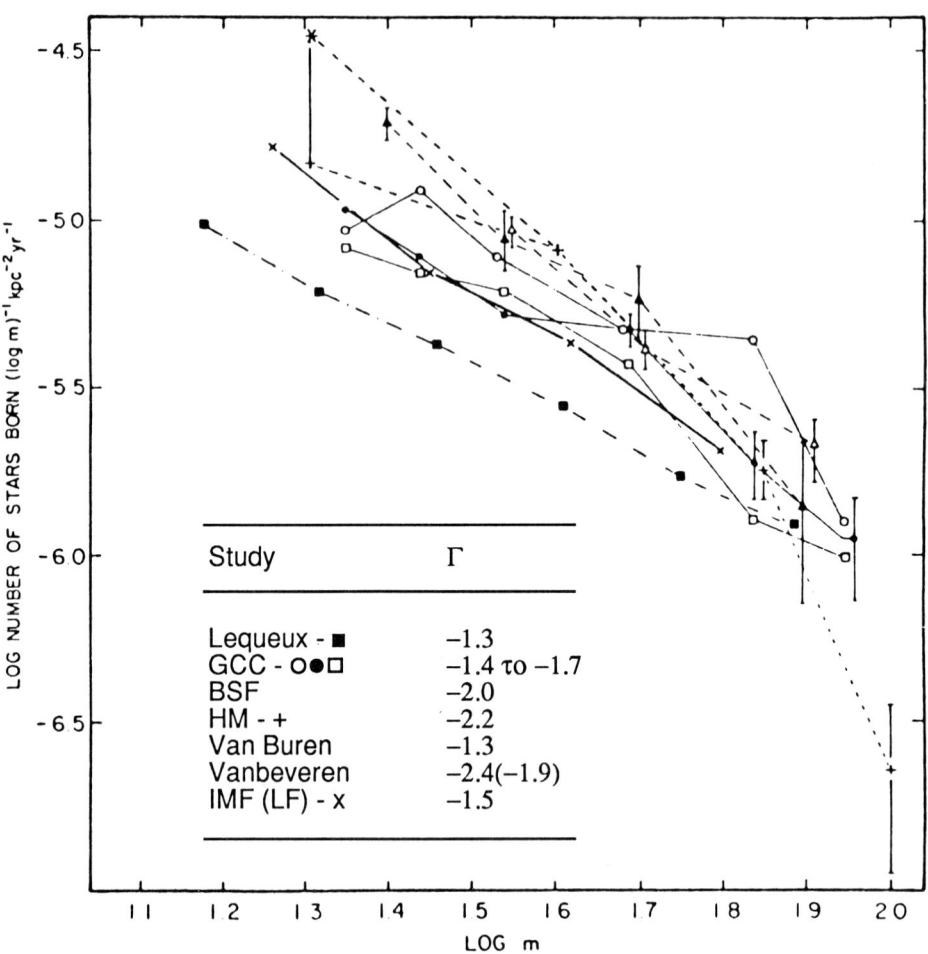

Figure 1. *Summary and comparison of different determinations of the galactic IMF for massive stars (from Scalo, 1986).*

studied and one that has not. In the latter case the distance modulus, as determined from ZAMS fitting, is uncertain by at least 1 magnitude. This uncertainty is in turn related to the apparent paucity of B main sequence stars.

Our reexamination of galactic OB associations shows that of 22 catalogued by Ruprecht et al. (1981), data for only 11 include a significant number of B main sequence members, and thus somewhat reliable distances. With all of the uncertainties in the galactic data it should come as no surprise that the computed IMFs show such a wide range, each depending on different corrections for incompleteness, temperature calibrations, etc. More observations are needed before we can confidently compute the IMF for solar neighborhood O and B stars.

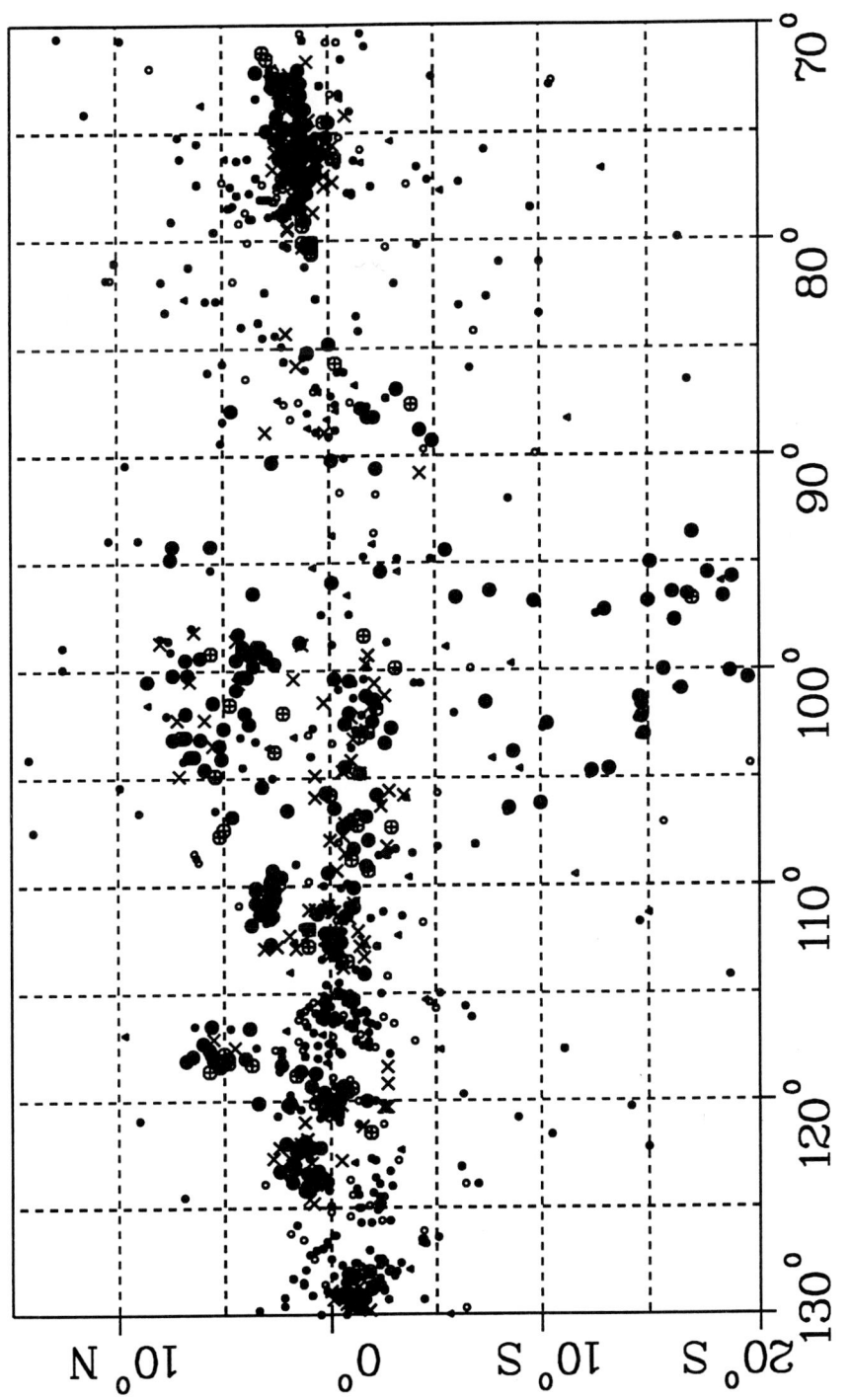

Figure 2. *Massive stars between galactic longitude $70°$ and $130°$. Stars believed to be members of OB associations are shown as large symbols, other stars by small symbols. O stars are crossed circles, B stars are filled circles, and late type supergiants are indicated by crosses.*

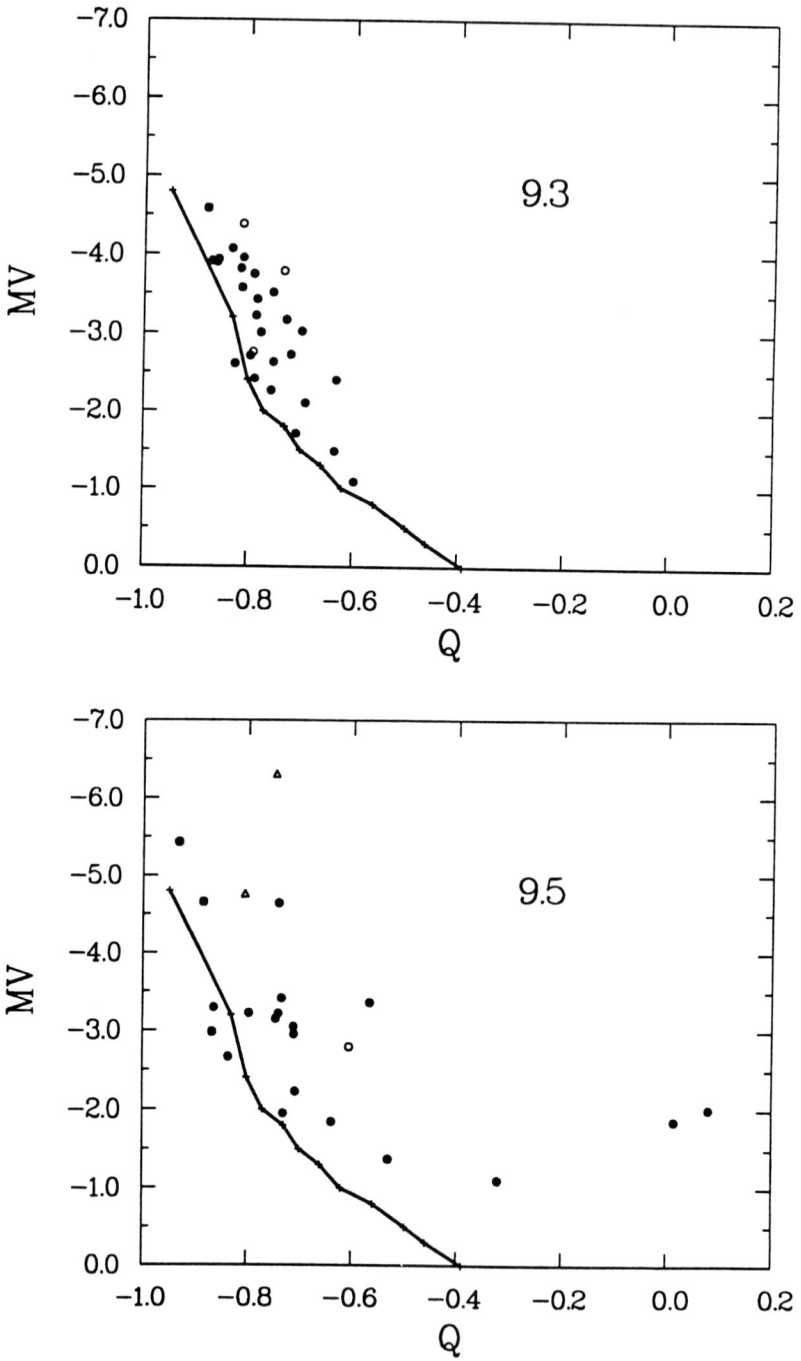

Figure 3. *Color magnitude diagram for two OB associations. The top, Cep OB3, is sufficiently well populated with main sequence B stars that a well determined cluster ZAMS fit is possible with a DM = 9.3. The lower one, Cyg OB7, is defined mainly by O stars and B supergiants, and no clear ZAMS is seen. An uncertain DM of 9.5 is adopted.*

3. MAGELLANIC CLOUDS

The only other place where the IMF can be studied directly is in the Magellanic Clouds. The Magellanic Clouds are especially well placed for the study of massive stars: the angular size of the entire galaxy is comparable to galactic OB associations, and the angular size of OB associations is comparable to the linear size of current CCD chips projected on the sky. Surveys of all bright blue stars, including moderate resolution spectroscopy of them, are almost complete and in-depth studies of associations and clusters are under way.

3.1. OB Associations

Although catalogs of OB associations in the Magellanic Clouds have been available for some time (Lucke and Hodge 1970; Hodge 1985) only recently has the stellar content been investigated in detail. In a series of papers on Magellanic Cloud OB associations, Massey and collaborators are using DAOPHOT (Stetson 1987) to get UBV photometry from mosaics of CCD frames, then selecting candidates for spectroscopy based on their colors (Massey et al. 1989: LH 117 (NGC 2122) and LH 118; Massey, Parker and Garmany 1989: NGC 346). In these three associations the ZAMS is populated from about 60 to 5 $M_\odot$. See Figure 4. There is no comparable study of galactic associations because their large angular size has precluded doing photometry of all objects to some limiting magnitude. The results for these three associations yield $\Gamma = -1.9$, consistent with a "universal" IMF and consistent with the galactic results. This result is also consistent with Mateo's (1988) results for Magellanic Cloud star clusters with masses in the range 1 to 10 $M_\odot$. However, work in progress (Massey and Garmany: NGC 602 (SMC); Parker, Garmany and Massey: LH 10; Parker, Massey and Garmany: 30 Dor) suggests that some associations have flatter IMFs. While few people would be surprised to learn that 30 Dor has a flat IMF, NGC 602 is a very unassuming region with none of the characteristics of a starburst region. These studies still represent only the tip of the iceberg, as there are over 100 OB associations catalogued in the LMC alone.

3.2. Field Stars

As was pointed out by Humphreys and Davidson (1979), above 10^4 K the maximum luminosity of early type stars increases with temperature. This is generally assumed to represent the locus where the stellar evolutionary tracks reverse direction in the H-R diagram, at an effective temperature that depends on the mass of the star. This luminosity limit, sometimes referred to as the HD limit, has been observed to be identical in the galaxy, the LMC and the SMC. Recent estimates are given by Humphreys (1987) and Fitzpatrick and Garmany (1990); see Figure 5. Thus the observation that this limit is apparently not dependent on metallicity (SMC versus the galaxy) is significant. It is also to be noted that although the HD limit was first noted in the location of the most extreme luminous blue stars, as observational material on more normal blue supergiants has become available they too obey this limit.

Although efforts have been made to derive a field IMF for the Clouds, Figure 5, which includes all of the data for LMC stars hotter than about $\log T = 3.9$, illustrates the incompleteness problem for main sequence stars, which should form the bulk of the data set. Humphreys and McElroy (1984), recognizing the incompleteness of the data

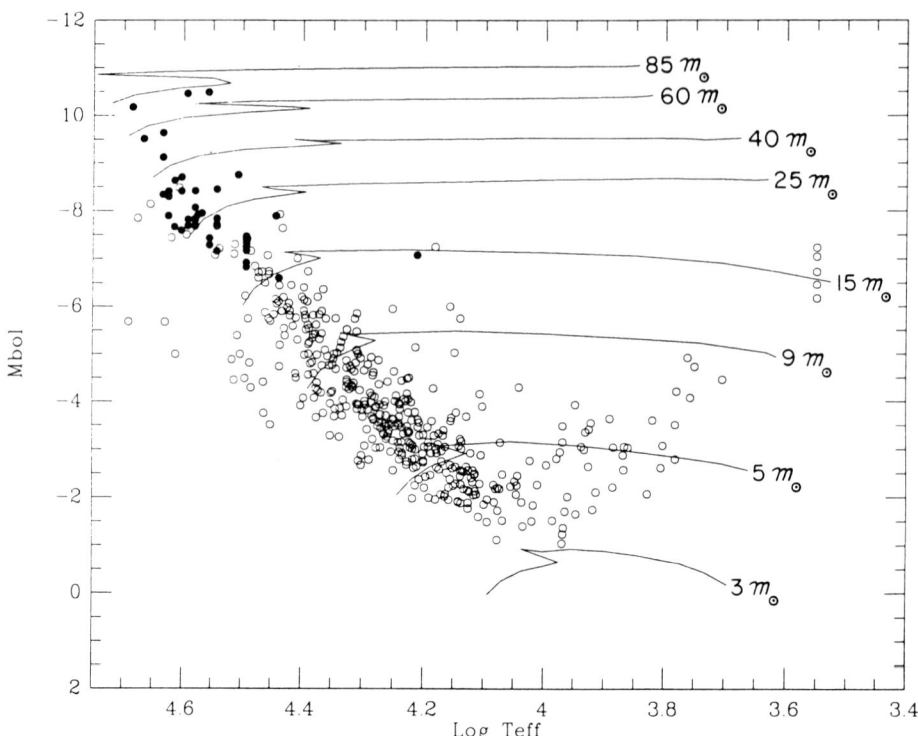

Figure 4. *The H-R diagram of NGC 346 in the SMC (from Massey et al. 1989). Filled symbols are stars with spectra and photometry, open symbols are stars with only UBV photometry.*

for stars less massive than 30 $M_\odot$, applied a simple correction factor to the numbers and concluded that the IMF in the Clouds is consistent with the galaxy. In the LMC they derived $\Gamma = -2.3$; in the SMC $\Gamma = -2.1$. Even their estimate of the completeness limit may be optimistic, as the work on OB associations discussed in the previous section indicates that many more O-type stars have yet to be uncovered. Further work by Blaha and Humphreys (1989) on the luminosity function for the galaxy and the Magellanic Clouds, which they converted to a present-day IMF, are consistent with $\Gamma = -2$. They also found that the inclusion of stars from the 30 Dor region flattened the IMF.

4. HOW ARE THE MASSES OF THE STARS DETERMINED?

There are three independent methods by which masses can be determined: eclipsing binaries, spectroscopic analysis, and comparison of stellar temperature and luminosity with the predictions of evolutionary models. As the last method is the only reasonable procedure to use for large numbers of stars, let us compare the results of the first two methods with evolutionary models.

The most fundamental way to measure a stellar mass is from an eclipsing, spectroscopic binary. Although the binary frequency of massive stars differs little from more common stars, there are relatively few good studies. Popper (1980) discussed 20 systems and Hilditch and Bell (1987) extended the sample to 31 systems. However, these compilations of mass, radii, temperatures and luminosities include no stars greater than

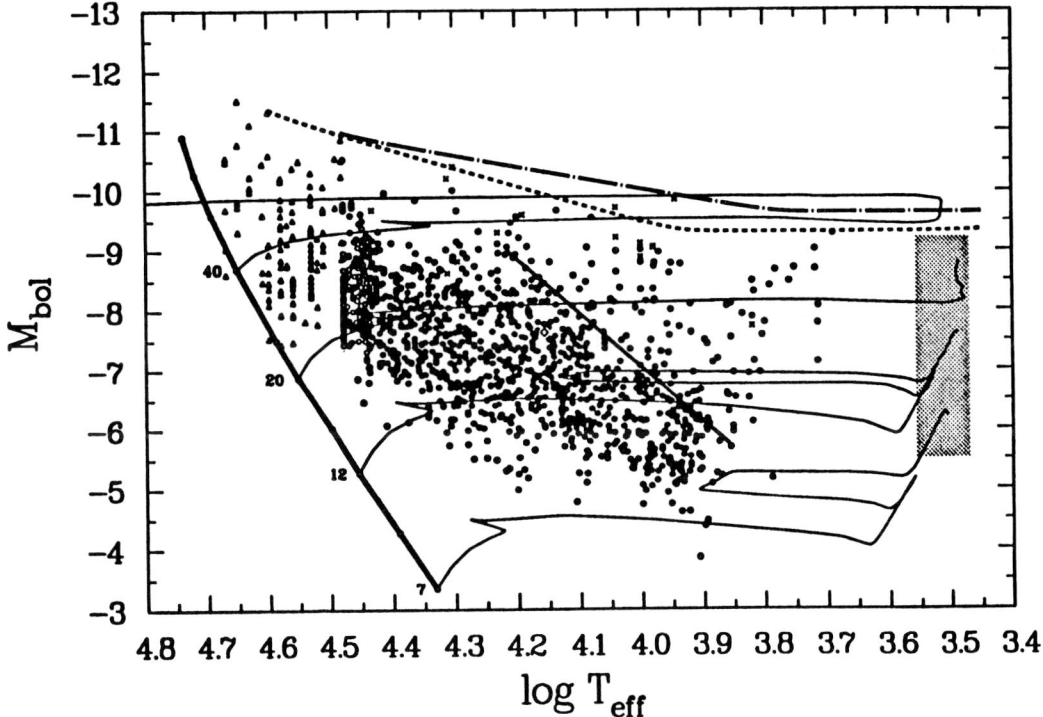

Figure 5. *Observed upper limit in the H-R diagram for LMC stars (from Fitzpatrick and Garmany 1990). Upper limit derived by Humphreys (1987) is shown as a dashed-dot line.*

40 $M_\odot$.

Masses can also be derived by either continuum or line fitting, depending on the temperature range. For supergiants cooler than 30,000K, Fitzpatrick (1987), has fit line blanketed, LTE Kurucz models at lower gravities than the published model grid, and finds that they fit the observations very well. He derives T_{eff}, $\log g$, radii, luminosity, mass: for B2I and earlier, the masses agree well with the predictions of evolutionary models but for late B supergiants, he finds much lower masses and gravities.

As discussed earlier, continuum fitting for O-type stars is not very useful because temperature, gravity and mass loss rates influence the shape. In addition, unless the individual reddening law is known, the extinction correction is uncertain. Instead, as shown by Auer and Mihalas (1972) the T_{eff} and $\log g$ can be obtained from model fitting of line wings and cores of H I, He I, and He II. Recent work includes studies by Kudritzki, Simon and Hamann (1983), Voels et al. (1989), and Bohannan, et al. (1990). Voels et al. note that their derived spectroscopic masses are only about 60% of the evolutionary masses, a discrepancy which is not yet understood.

In Figure 6 we compare the results from eclipsing binaries, continuum fitting and line fitting with a set of evolutionary tracks by Maeder and Meynet (1989). Although Fitzpatrick suggests that his masses for B supergiants may only be correct within a factor of 2, the stars closer to the ZAMS should be more reliable, and it is interesting to note that almost all of these masses are systematically smaller than Maeder's models predict. One should also take into consideration the difference between various evolutionary models: de Loore (1988) has compared a number of them, and concludes that even the ZAMS points differ by several tenths mag in M_{bol} and 0.02 in $\log T_{eff}$. Until these differences are reconciled, it might be concluded that absolute masses are not well known.

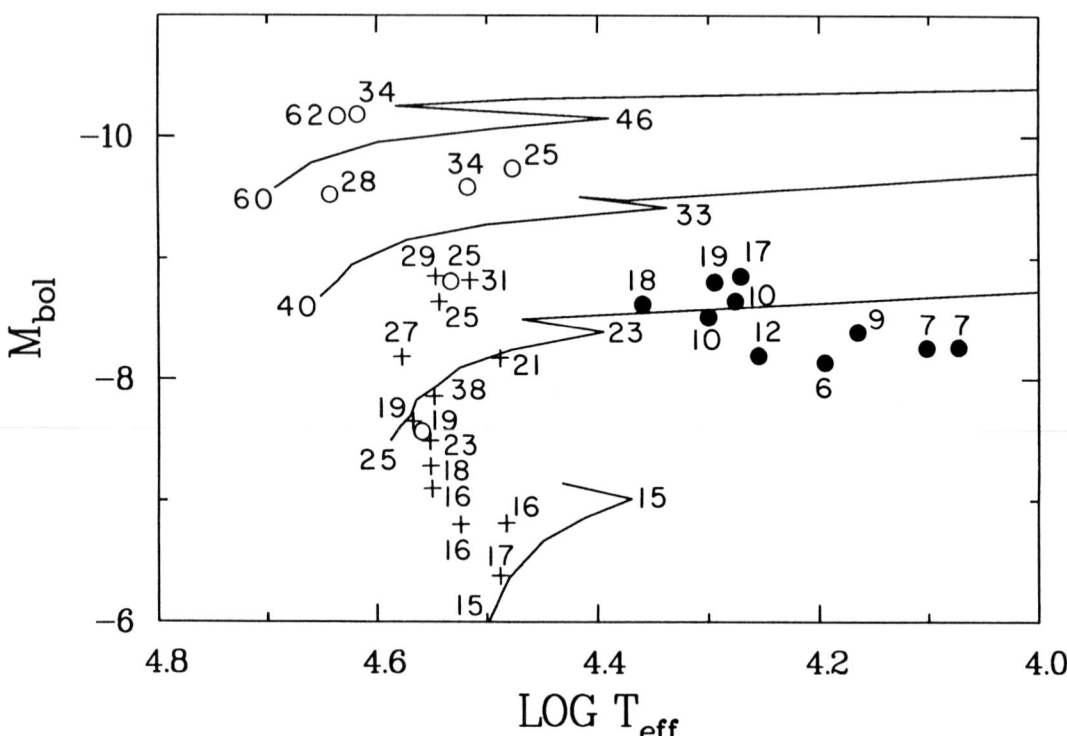

Figure 6. *Individual stellar mass determinations compared with evolutionary tracks. The open circles are stars with masses derived from line profile fitting (Bohannan et al. 1990), filled circles are masses derived from continuum fitting (Fitzpatrick 1987), and crosses are masses derived from double line spectroscopic binaries (Popper 1980, Hilditch and Bell 1987). The tracks, which include mass loss, are by Maeder and Meynet 1989 and include the ZAMS and TAMS mass.*

5. THE MOST MASSIVE INDIVIDUAL STARS

What is the most massive star known? Not surprisingly, candidates are rare and distant, and often difficult to resolve among their fainter neighbors. The case of R136 is a good example: as the resolution of 30 Doradus improved the mass of this object dropped from 2000 to 250 $M_\odot$ (Walborn 1984). As also pointed out by Humphreys and Aaronson (1987), the candidate brightest blue supergiants in external galaxies often are actually compact H II regions, blue clusters or composite spectra. New image deconvolution techniques, both at the telescope and in the reduction procedure, are producing exciting results. However, the final mass of the brightest star in such resolved clusters depends greatly on the adopted temperature and bolometric correction. A good example of this is the LMC star Sk−69°253, which was described by Walborn (1977) as a "compact pentagonal, quintuple visual system" and classified as B0.7-1 I. Heydari-Malayeri, Magain and Remy (1989) have presented a reconstructed image showing at least 14 components. Component A is Sk−69°253, and they estimate its magnitude to be 12.0 based on the deconvolution and the aperture photometry of Ardeberg et al. (1972) for the composite system of $V = 11.23$, $B - V = -0.02$ and $U - B = -0.81$. They conclude that from the original unresolved image photometry and spectral type the derived mass is greater than 120 $M_\odot$, but from their image the mass is only about 70 $M_\odot$. In contrast, Fitzpatrick and Garmany (1990), using the Ardeberg et al. photometry, adopted $\log T = 4.229$ and $M_{bol} = -8.962$. From Maeder and Meynet's (1989) evolutionary tracks, this implies a mass less than 40 $M_\odot$, not 120 $M_\odot$ as estimated by Heydari-Malayeri et al.

Another good example of how different techniques lead to different final masses is the star NGC 346, no. 1, which was classified O4 III (f) by Walborn (1978). Massey, Parker and Garmany (1989), noting that this is a multiple image, derived UBV colors from CCD frames and separately classified this star as O5.5 If. Adopting a temperature of 38,900 K, a bolometric correction of −3.65 and absolute magnitude of −6.85, the mass, from Maeder and Meynet's (1989) evolutionary tracks, is between 80 and 85 $M_\odot$. Kudritzki et al. (1989), using the same CCD photometry and high dispersion spectra, derived a temperature for this star from line analysis of 43,500 K, and from their derived $\log g$, a mass of 113 $M_\odot$.

6. RELATION OF NEBULAR Hα FLUX TO STELLAR CONTENT

For distant galaxies, the integrated Hα luminosity is one of the main measures of recent star formation. How does this measure compare with cases in which the stellar content is known? This is a difficult comparison to make in the galaxy because distances and stellar content of associations are so uncertain.

A good comparison can be made in the Magellanic Clouds. Kennicutt and Hodge (1986) have measured the integrated Hα luminosity for all of the HII regions, which not surprisingly are coincident with the OB associations. Massey et al. (1989) have compared the implied Lyman continuum photon flux with that predicted from summing the contribution from all of the hottest stars in LH 117 and 118. Table 1 has been adapted from Massey et al. to include the number of ionizing photons per star, based on stellar models giving the expected Lyman photons per second (Panagia 1973) and radii computed from the temperatures and bolometric magnitudes. The total of all ionizing photons from Table 1 is about a factor of 1.7 greater than what Kennicutt and Hodge detect, which is larger than the expected errors. This may indicate that the H II region is density, rather than ionization bounded. It should be noted that the majority of the stellar radiation is contributed by the three hottest stars in Table 1. If the most luminous star, which is a bit overluminous for a spectral type of O4:III, were actually a binary of, say, two O6 III stars, then the total ionizing flux from the system would drop to 4.7×10^{49}, and the flux from the stellar census would only be 1.3 times that detected by Kennicutt and Hodge. Table 1 also illustrates how insensitive the Hα flux is to the actual slope of the IMF.

TABLE 1. *Ionizing photons per star in LH 117*

Sp. Type	$\log T_{eff}$	M_{bol}	Photons s^{-1} · 10^{-49}
O4:III(f)	4.65	−10.8	8.9
O 3-4	4.68	−9.7	3.5
O 3-4	4.68	−9.7	3.5
B2 I	4.29	−8.9	0.0
O7V	4.60	−8.7	1.0
O6V	4.62	−8.5	1.0
O9.7	4.48	−8.4	0.2
O6.5	4.62	−8.3	0.7
O9	4.55	−7.8	0.2
O9V	4.55	−7.7	0.2
O9V	4.55	−7.6	0.2
O9	4.55	−7.5	0.2
O8	4.58	−7.2	0.2
O9	4.55	−7.2	0.2
O9.5	4.54	−7.0	0.2
B0	4.47	−6.6	0.0
B0	4.47	−6.3	0.0
B0	4.47	−6.1	0.0

7. CONCLUSIONS

It does not appear that IMF studies of massive stars offer any crisp, definitive predictions that can be applied to the starburst phenomenon yet. Within the admittedly large uncertainties, the slope of the global IMF in the galaxy, the LMC and the SMC appear similar, as does the upper luminosity limit. There is increasing evidence that the IMF may be flatter in some associations in the Clouds, and a reexamination of galactic OB associations is in order. The answer to the very interesting question about the existence of lower mass cutoff is still lost in incomplete data samples and unquestioned assumptions about star formation rates. Even our assignment of masses to individual stars may have systematic errors. Nevertheless, the questions that need to be answered are clear, and their applications to extragalactic issues are urgent.

This work was partially supported by NSF grant AST88-06594 to the University of Colorado.

REFERENCES

Ardeberg, A., Brunet, J. P., Maurice, E., and Prevot, L. 1972, *Astr. Ap. Suppl.*, **6**, 249.
Auer, L. H., and Mihalas, D. 1972, *Ap. J. Suppl.*, **24**, 153.
Blaha, C., and Humphreys, R. M. 1989, *A. J.*, **98**, 1598.
Bohannan, B., Voels, S. A., Hummer, D. G., and Abbott, D. C. 1990, *Ap. J.*, in press.
Chini, R., and Wargau, W. F. 1990, *Astr. Ap.*, **227**, 213.
Conti, P. S., Garmany, C. D., de Loore, C., and Vanbeveren, D. 1983, *Ap. J.*, **274**, 302.
de Loore, C., 1988, *Astr. Ap.*, **203**, 71.
Fitzpatrick, E. L. 1987, *Ap. J.*, **312**, 596.

Fitzpatrick, E. L., and Garmany, C. D. 1990, *Ap. J.*, in press.
Garmany, C. D. 1990, in *Properties of Hot Luminous Stars*, ed. C. D. Garmany, Astron. Soc. Pacific Conf. Series vol. 7, p. 16.
Garmany, C. D., Conti, P. S., and Chiosi, C. 1982, *Ap. J.*, **263**, 777.
Heydari-Malayeri, M., Magain, P. and Remy, M. 1989, *Astr. Ap.*, **222**, 41.
Hilditch, R. W., and Bell, S. A. 1987, *M.N.R.A.S.*, **229**, 529.
Hodge, P. 1985, *Pub. A.S.P.*, **97**, 530.
Hummer, D. G., Abbott, D. C., Voels, S. A., and Bohannan, B. 1988, *Ap. J.*, **328**, 704.
Humphreys, R. M. 1987, in *Instabilities in Luminous Early-Type Stars*, ed. C. de Jager and H. J. G. L. M. Lamers (Dordrecht: Reidel), p. 3.
Humphreys, R. M., and Aaronson, M. 1987, *Ap. J.*, **318**, L69.
Humphreys, R. M., and Davidson, K. 1979, *Ap. J.*, **232**, 409.
Humphreys, R. M., and McElroy, D. B. 1984, *Ap. J.*, **284**, 565.
Kennicutt, R., and Hodge, P. R. 1986, *Ap. J.*, **306**, 130.
Kudritzki, R. P. Cabanne, M. L., Husfeld, D., Niemela, V. S., Groth, H. G., Puls, J., and Herrero, A. 1989, *Astr. Ap.*, **226**, 235.
Kudritzki, R. P., Simon, K. P., and Hamann, W. R. 1983, *Astr. Ap.*, **118**, 245.
Lequeux, J. 1979, *Astr. Ap.*, **80**, 35.
Lucke, P. B., and Hodge, P. W. 1970, *A. J.*, **75**, 171.
Maeder, A., and Meynet, G. 1989, *Astr. Ap.*, **210**, 155.
Massey, P. 1985, *Pub. A.S.P.*, **97**, 5.
Massey, P., Garmany, C. D., Silkey, M., and DeGioia-Eastwood, K. 1989, *A. J.*, **97**, 107.
Massey, P., Parker, J. W., and Garmany, C. D. 1989, *A. J.*, **98**, 1305.
Mateo, M. 1988, *Ap. J.*, **331**, 261.
Panagia, N. 1973, *A. J.*, **78**, 929.
Popper, D. 1980, *Ann. Rev. Astr. Ap.*, **18**, 115.
Ruprecht, J., Balazs, B., and White R. E. 1981, *Catalogue of Star Clusters and Associations*, ed. B. Balazs (Akademiai Kiado, Budapest).
Salpeter, E. E. 1955, *Ap. J.*, **121**, 161.
Scalo, J. 1986, *Fund. Cosmic Phy.*, **11**, 1.
_____. 1989, *Windows on Galaxies*, ed. A. Renzini, G. Fabbiano, and J. S. Gallagher (Dordrecht: Kluwer), p. 125.
Stetson, P. B. 1987, *Pub. A.S.P.*, **99**, 191.
Tinsley, B. M. 1980, *Fund. Cosmic Phy.*, **5**, 287.
Voels, S. A., Bohannan, B., Abbott, D. C., and Hummer, D. G. 1989, *Ap. J.*, **340**, 1073.
Walborn, N. R. 1977, *Ap. J.*, **215**, 53.
_____. 1978, *Ap. J.*, **224**, L133.
_____. 1984, in *Structure and Evolution of the Magellanic Clouds*, IAU Symp. 108, ed. S. van den Bergh and K. S. de Boer (Dordrecht: Reidel), p. 243.
Zinnecker, H., 1984, *M.N.R.A.S.*, **210**, 43.

DISCUSSION

A. Maeder: Among all the predictions of the stellar evolution theory, the M-L relation for main-sequence stars is the safest one and I am not ready to accept a change to it. There are large differences between observational and evolutionary masses but this is true for the entire range of stellar masses. Even between 1 and 2 solar masses you have big difficulties in such comparisons. One certain reason for that is tidal interaction in the case of spectroscopic binaries. The evolution of those stars would be different from the single-star evolution in the sense that they would be mixed and overluminous for their mass. This is just what you have: your binaries are overluminous for their mass in comparison with the calculation. Is there any indication that the difference with respect to the theoretical relation is just connected to the distance between the two components? Certainly, single-star evolution and binary evolution are not the same, even if there is no mass transfer between the components.

Garmany: That may well be true. We have only very few data points, and I'm simply pointing this out as a caution. These binaries, as I tried to stress, have been chosen to be as non-interacting as possible. I think certainly if you have two visual binaries, and could observe them for 100,000 years you wouldn't argue that they should be treated any differently than single stars. Obviously, spectroscopic binaries have some kind of interaction, but I still find it curious because these values also agree with the masses for the single stars that Hummer and co-workers have been determining.

R. Kudritzki: Just a comment on these masses. I think a very good way to test evolutionary masses would be a comparison with v_∞-masses. At the moment I can't say much about the discrepancy. I think there is a clear discrepancy for extreme Of stars or supergiants. I don't know if such a discrepancy also exists for the main sequence but that would be an ideal test case.

A. Maeder: Of course—but then one should also see to what extent the masses derived from the v_∞ are affected by rotation. How reliable are they for rotating stars since the wind models apply to non-rotating stars?

E. Fitzpatrick: I'd like to comment on the interpretation of the masses I got from these supergiants. I believe that the low masses that resulted from my fitting of the UV continua of the late B supergiants with atmosphere models is a reflection of the inappropriateness of the atmosphere models for these types of stars—rather than an indication that these evolutionary masses are in error. I have a healthy skepticism of some of the details of stellar evolution calculations, but I think that most of the factor of 2–3 mass discrepancy results from plane-parallelism or LTE in the atmospheres or from inadequacies in the atmospheric opacities.

C. Norman: I have a question referring to the last part of your talk. When you were comparing indirect methods, which starburst people use, and the more direct methods that you are using, the conclusion you came to at least in this talk was a little bit depressing for us starburst people. When is the situation going to change and how is it going to change?

Garmany: With the advent of CCD's we have been able to study this in the Magellanic Clouds, and with the expected advent of really large CCDs we hope to return to the galaxy and study entire OB associations down to a few solar masses, so in principal, the answer is "soon". Of course, more than just spectrophotometric data would be helpful, and I had great hopes for HIPPARCOS. We probably need a better understanding of massive star evolution for checking the direct–indirect methods.

I'd also mention that even direct determinations of the IMF are rather a function of author. It might be interesting to send a data set to all derivers of this parameter, and let everybody return their answer in a sealed envelope.

C. Norman: What about the *lower* cut-off mass of the IMF? When will that situation improve? That is crucial also.

Garmany: You mean the lower cut-off in the B-star regime? I guess it depends on how much evidence you're going to want, and I think it will turn out that there'll be some regions that have lower cut-offs, and others that don't. The Scorpius/Centaurus OB association has no sign of a lower mass cut-off. I didn't bring a transparency of that but it has been studied. The stars just keep on all the way down. On the other hand, in Cepheus OB3 you only see everything down to B2. Yet I'm wondering now that the work that was done in the 60's was very careful when everyone said there was no evidence for membership below about B2. Maybe it is correct. Maybe there really isn't anything, but again I think we've got to look deeper. Unfortunately, you probably need some astrometric work, and if HIPPARCOS will provide data, that will be an important step.

C. Blaha: I'd like to make a cautionary note. It's easy to get a flatter IMF by missing some of the late O and early B stars. Roberta Humphreys and I have analyzed the impact of the 30 Doradus stars on the IMF and luminosity functions of the entire LMC, and we found several results. Adding the 30 Doradus sample to the LMC left the U, B, and V luminosity functions essentially unchanged. The 30 Doradus sample flattened the slope of the IMF from -2.0 to -1.5 but this 30 Doradus data set was markedly incomplete for late O and early B stars. Using recent estimates of the number of 30 Dor stars in this regime results in a slope for the IMF closer to -2.0.

Garmany: You're absolutely right. I think that's one of the reasons why Joel Parker's work when it's finally complete will be important. It will go down to 10 solar masses and we will be able to answer this question directly.

S. Heap: I wanted to ask about the uncertainties in Γ, the slope. You've shown that you need a wide baseline and temperature to determine the slope. Once you go to the higher temperatures you're dealing with very early O stars, which have no He I. I don't know how you get your temperatures, therefore I don't know how you get your bolometric correction, therefore I don't know how you get your luminosities, therefore I don't know how you get your masses. I was wondering, you get -1.9 for Γ but what is the error bar?

Garmany: The formal error is like 0.2 for those associations. In those stars we see He I *and* He II so that we can classify them in the Walborn system and assign an effective temperature, bolometric correction, etc.

S. Heap: But even the best of the non-LTE spectroscopic analysis has at least a 10% uncertainty in the temperature.

Garmany: That's true and we haven't tried to fold that into the IMF. However, I think that the bins in which we place the masses are large enough that they easily accommodate a 10% effect in the temperatures. Of course, if there is a systematic effect, *e.g.*, if all these temperatures are systematically too high or too low, the whole IMF will be changed.

S. Heap: I suppose it is not so much the temperature by itself, but the indirect effect on the bolometric correction?

Garmany: Yes, exactly. This influences where the stars end up in the HR diagram. Although the modelers would claim that they're better than 10% on their temperature scale now.

N. Scoville: I'm a little bit confused about your answer to Colin's question. The slope of the upper end of the initial mass function is important for starburst models, but I think probably more critical, in fact, is whether there are any low-mass stars. I'm not quite sure whether you're saying there are some clusters for which you see only OB stars, and you could measure the low-mass stars if they existed. That is, there aren't any or whether it's impossible to measure them. I would have thought you could take a nearby cluster with proper motions and measure all the way down to 1 solar mass.

Garmany: In principle you could, but it hasn't been done.

N. Scoville: Why not?

Garmany: These things are enormous. It hasn't been a popular thing to do for the past 10–15 years and it's been hard to get observing time.

N. Scoville: Physically it's more important to measure the low-mass stars than the high mass stars because I think we have a fairly good handle on the overall luminosity which is dominated by the highest mass stars that you get. The high-mass end is reasonably well determined, but the overall mass which is being used up in that population is the most critical parameter for understanding the ages of starbursts.

Garmany: You really need something like HIPPARCOS. Proper motions for O stars are very difficult, and the ones that have been done from the ground are very uncertain. I did some for my thesis 20 years ago, and I think they're garbage.

N. Scoville: I don't understand. I'm talking about proper motions of a few solar mass stars in the nearest OB association.

Garmany: Oh, stars with masses of a few solar masses? We don't see them yet. We haven't gone down that far. The nearest OB association is Orion at 500 parsecs, but there they haven't reached the main sequence. If they're there, they're still contracting, they're still T Tauri stars.

N. Scoville: That's fine. You would at least get some feeling for the distribution of final masses. In the case of the Orion region I would have thought you could use the cloud as the screen to cut off any more distant background stars.

Garmany: But you have things along the line of sight, and you have too many other stars confusing the issue.

A. Campbell: The question of the lower limit of the IMF is very important. Rebecca Elson and I have data for 30 Doradus that should reach to $\sim 1.5\ M_\odot$. We hope to have a preliminary answer in the next few months. I have a second comment on the question of using nebular emission lines to measure IMF parameters: Katy said someone had tried to use the Hα luminosity, although I have never heard of this being done. But you can do quite a good job by looking at the relative strengths of other emission lines, *e.g.*, [OII]λ3727 and [OIII]$\lambda\lambda$5007, 4363. It's not easy because you have all the complications of the nebular geometry, but several groups are attempting to do it, and hopefully we will have an answer for H II galaxies and giant H II regions in the near future.

V. Robledo-Rella: How much do you think metallicity (Z) can affect the IMF? i) There have been suggestions that in regions of low Z a tendency to form stars of large mass could be present. ii) Direct estimations of the IMF *depend* also on the recent SFR, and this in turn may depend on Z by the self-regulating effect that stars may have, since stellar evolution depends on Z.

Garmany: I have approached this from the observational point-of-view. It appears that there is no evidence for IMF differences, or differences in the most massive star population, between the local Galaxy, LMC and SMC. The SMC, with the lowest metallicity of the three, may actually have fewer massive stars, but this is probably because of the overall smaller mass of the SMC.

IR– AND MM– OBSERVATIONS OF REGIONS OF MASSIVE STAR FORMATION

Ian Gatley, K. M. Merrill, and A. M. Fowler
Kitt Peak National Observatory
P.O. Box 26732
Tucson, AZ 85726
USA

Abstract. The advent of large format infrared array detectors gives us the tool necessary to determine the distribution of young stars within their natal molecular cloud. Such studies (the most detailed of which is the investigation of L1630 by E. Lada *et al.*) amply confirm the notion that stars form in clusters within dense molecular clumps.

In order to assess the prospects for the determination of the Initial Luminosity Function in regions of massive star formation we present large area ($>$ 100 square arcminute) infrared J, H, and K band images of M17 and M42 (the Orion Nebula) taken at KPNO using a 256×256 Platinum Silicide array detector. In M17—the more luminous of our targets—we find a cluster of OB stars with more than 100 members, every one of which has excess infrared emission suggestive of a surrounding "cocoon" of material left over from the star formation process. The luminosity function of this OB cluster is consistent with the distribution described by the Salpeter Initial Mass Function (IMF). Near the faint limit of the data there is a hint of a departure from the Salpeter IMF, in the sense of a deficiency of stars less massive than the Sun. In earlier work (DePoy *et al.* 1990) we found conclusive evidence for a similar deficit in NGC2023—NGC2023 is some four times closer than M17; that is, at the same distance as M42.

Our observations of the Trapezium cluster in M42 reveal two surprises. First, the luminosity function is steeper than that expected from the Salpeter IMF. Second, the percentage of cluster members with excess infrared emission is rather small. A plausible explanation for these results is that the cluster is not coeval; in such a model the massive stars with infrared excesses are very young, while the remaining cluster members are older. The extraordinarily high stellar density of the Trapezium Cluster has been noted in the literature for more than fifty years; perhaps repeated bursts of star formation at the same site are responsible.

1. INTRODUCTION

Massive stars form in clusters within giant molecular clouds; the formation process is inefficient, and the dust in the inevitable remnant cloud material shrouds the young cluster. Star formation occurs on a large scale; nearby examples, within a few kiloparsecs of the sun, typically have sizes of hundreds of square arcminutes. Here we review the results of a large-scale survey of such a cloud, L1630, both in CS (which traces the density distribution of the cloud) and in the near infrared (which reveals the embedded population of young stars). The large size and heavy local extinction associated with massive star formation render such regions natural candidates for study with infrared arrays. The potential of this kind of investigation, presently in its infancy, is very great. The results presented here offer a glimpse of these future possibilities, and show the prospects for eventual precise determination of the Initial Luminosity Function in regions of massive star formation.

2. GLOBAL STAR FORMATION IN THE L1630 CLOUD

In her survey of the L1630 molecular cloud, Elizabeth Lada measured 13,000 spatial positions on a 1 arcminute grid in the J=2-1 transition of CS, using the AT&T Bell Labs 7m telescope. The total cloud area studied was 3.6 degrees, but CS was detected in only 10% of the positions, confined in 42 dense cores. An infrared K band (2.2 micron) survey of a large portion of L1630—3,000 fields of one square arcminute—covering a total area of 0.8 square degrees, was also executed by Lada using the 1.3m telescope at KPNO. The infrared measurements included locations both with and without CS emission. The star formation in L1630 occurs exclusively in the dense gas. In fact the majority of the young stars are confined to the three most massive cores, which together account for less than 8% of the total molecular gas mass in the area surveyed. Figure 1a shows the CS(2-1) map with the boundaries of the K survey superposed. Figure 1b shows the distribution of infrared sources brighter than K = 12 mag.

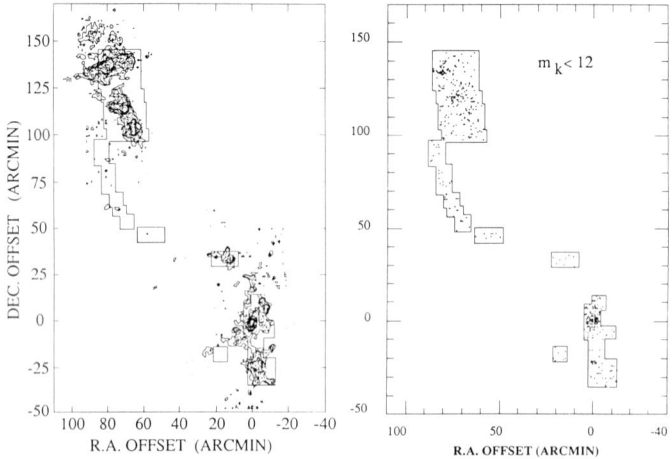

Figure 1. *From Lada (1990), used by permission.*

3. STAR FORMATION IN M17

The HII region M17 is partially embedded in its parent molecular cloud, and so exhibits a very striking "blister" morphology (Icke, Gatley and Israel, 1980). The far infrared luminosity of this nebula is 6×10^6 $L_\odot$ (Harper et al. 1976; Gatley et al. 1979). Optical searches failed to identify the OB cluster members because of large local extinction, but near infrared searches have been more successful (Chini, Elsasser and Neckel, 1980). Subsequent photometry of the brighter members of the cluster by Chini and Krugel (1985) have shown these stars to have infrared excesses among the largest of any optically identified stars studied to date; this infrared excess is thermal emission from a dust "cocoon" left over from the star formation process, and is therefore a signpost of extreme stellar youth.

An early example of a K band array image of M17 was given by Gatley (1988). The M17 data to be discussed here appeared on the cover of Science magazine (2 Dec 1988; vol 242, pp1217-1348) as a color composite image, which was encoded by wavelength as J=blue, H=green, K=red; that is, as M17 might appear if the human eye responded to infrared (rather than visible) light.

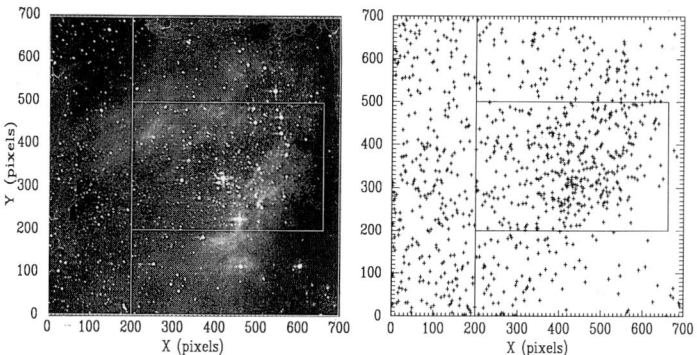

Figure 2.

Figure 2a shows a monochrome image of M17 at wavelength K in a region approximately 10 × 10 arcminutes in size; the positions of all stars brighter than K=13 mag (the 6 sigma per pixel limiting magnitude of the K image) are indicated by a cross in Figure 2b. The limit K=13 mag corresponds to an A type main sequence star at the distance of M17 (2.2 kpc). For the purposes of analysis we identify three distinct areas within the field of Figure 2. First, a strip down the left side of the frame where the extinction is known to be small, and where the nebula is absent; this "Left field" will provide a check on our ability to subtract background stars. Second, a region equal in area to the Left field, centered near the nominal position of the stellar cluster in M17 (Icke, Gatley and Israel, 1980); this "Center field", which contains much of the OB cluster, suffers only moderate extinction at infrared wavelengths. Third, a region of higher extinction coincident with the denser portions of the molecular cloud, which is especially evident to the lower right of Figure 2; because this region is opaque even in the infrared, nothing further can be said of its stellar population from these data. The locations of both Left and Center fields are indicated in Figure 2b; each has an area of 24.7 square arcminutes.

Figure 3a shows a histogram of cumulative number of stars versus K magnitude (that is, number of stars brighter than a given magnitude) for the Left field. The curve displayed through these data is the prediction of the Bahcall-Soneira model of the Galaxy. It is evident that the model fits the data very well, as expected. Figure 3b shows a similar plot for the Center field; here the number of stars is well in excess of the model—again as expected—because these supernumary stars are the members of the OB cluster. Subtraction of the data for the Left field from that of the Center field (Figure 3c) reveals that there are 99 cluster members within the Center field.

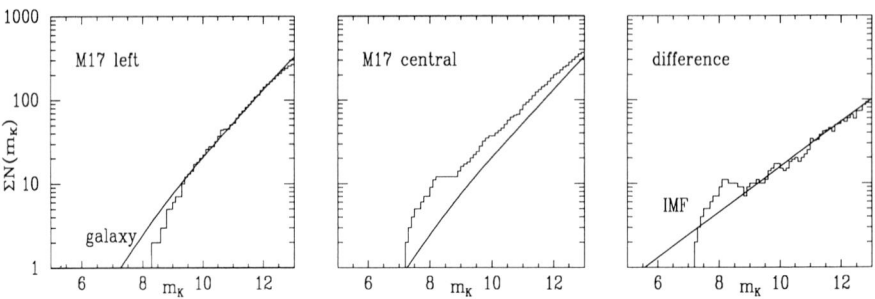

Figure 3.

The curve (labelled "IMF") displayed with the data in Figure 3c shows the distribution of stars expected from the form of the Salpeter Initial Mass Function (IMF); a simple exercise in dimensional analysis avoids the necessity to specify any strict correspondence between observed brightness and mass, as follows:

For OB stars, wavelength K lies within the domain of the Rayleigh-Jeans approximation. Therefore the observed K brightness depends only on the area and temperature of the stellar photosphere. For O-F main sequence stars, the radius scales as approximately the mass to the 0.8 (0.76) power and the temperature scales as approximately the mass to the 0.5 (0.54) power. Therefore the K brightness scales approximately as the mass squared (2.06). This relationship between K brightness and mass allows us to convert the Salpeter IMF from its usual form, number of stars versus mass, to the form plotted in Figure 3c. The agreement between this curve and the data in Figure 3c shows at once that the distribution of stellar brightness in M17 generally conforms to that expected from the Salpeter IMF.

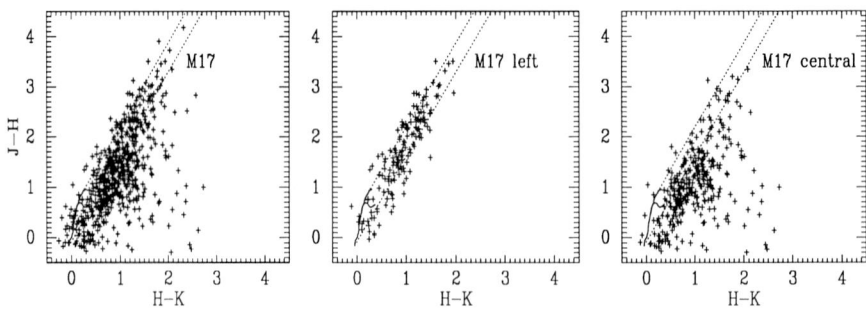

Figure 4.

The results of this simple K band star-counting experiment can profitably be complemented by analysis of the infrared colors of the stars. Figure 4a shows a color-color diagram (J-H versus H-K) for the whole of the M17 image (Figure 2), for all stars detected at J, H and K. The locus of the main sequence is shown as a light line in the lower left, and the envelope of reddened main sequence stars is shown by the lines extending away from the main sequence locus to the upper right. Stars down and to the right of this envelope have an "infrared excess" (see, for example, Gatley et al. 1981)—a phenomenon indicative of stellar youth. Such stars have often been described as "protostars". Figure 4a reveals at once that the image of M17 contains both a large number of reddened stars and a large number of stars with an infrared excess.

In the analysis of the stellar colors the sample within the Left field (defined above) again provides a useful control, because (by virtue of its precise fit to the Bahcall-Soneira model) it contains very few cluster members. Figure 4b shows the color-color diagram for the Left field; the data are distributed along the reddening vector. If these measurements are projected along the reddening vector onto the line J-H=zero (that is, are dereddened from their J-H color) they all exhibit a residual "dereddened H-K color" of less than 0.4 mag; this numerical result provides a suitable criterion by which to select objects with an unequivocal infrared excess. The color-color diagram for the Center field is shown in Figure 4c. Comparison of Figures 4b and 4c shows that there are clearly many stars with excesses in the Center field. Application of the numerical criterion that the "dereddened H-K color" exceed 0.4 mag reveals that there are 116 objects in the Central field with an infrared excess. This is comparable to the number of cluster members found in the star counting experiment, and therefore demonstrates that every member of the OB cluster in M17 has an infrared excess.

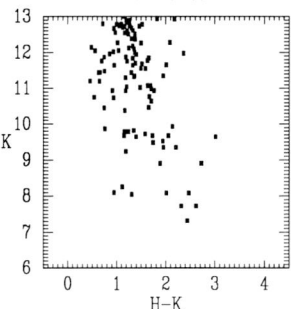

Figure 5.

The infrared excess exhibited by bright OB stars in M17 has been noted previously, and has been interpreted as "cocoons" of material left over from the star formation process (Chini and Krugel; 1985). The color-magnitude diagram (Figure 5) for the much larger sample of stars studied here shows that the color excess increases with brightness for this cluster. Interpretation of the physical origins of this result must await further measurements at longer infrared wavelengths. With only three wavelengths measured we cannot uniquely correct both for infrared excess and for interstellar extinction in our K band star counting experiment. In fact these two phenomena enter our analysis with opposite sign (the excess making a given star brighter, and the extinction making it fainter). It is, however, much more important to notice—as a corollary to the dimensional analysis argument given earlier—that the IMF analysis is affected only by the *differences* in excess and extinction from one star to another.

Because the cluster members can be uniquely identified by their infrared excess, it is possible to plot the spatial location of the cluster. In Figure 6a we indicate the locations of those stars which lie within the reddening envelope ("dereddened H-K

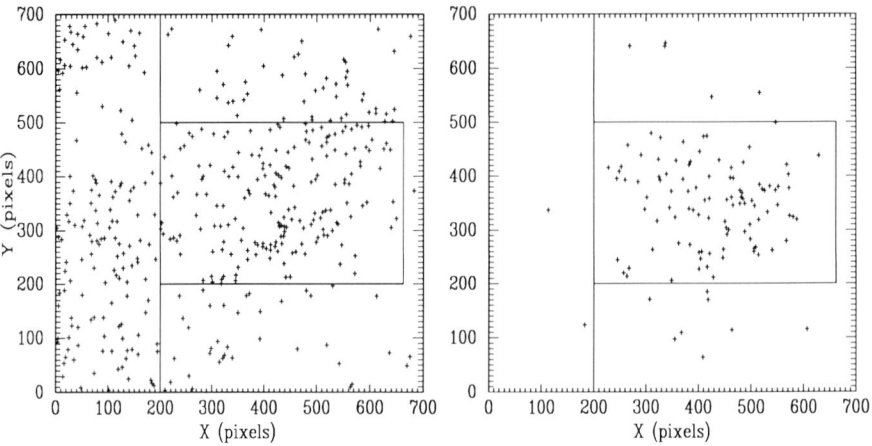

Figure 6.

color" less than 0.4 mag); these are the stars within the image which are not members of the M17 cluster. In Figure 6b we indicate the locations of those stars with infrared excesses ("dereddened H-K color" exceeds 0.4 mag); these are the cluster members.

The limiting magnitude of the K band image, K=13 mag, corresponds to a main sequence star of early A spectral type. Even with the most optimistic correction for the large infrared excess measured here, the spectral type deduced remains earlier than mid A. To pursue the question of formation of lower mass stars in this cluster, fainter limiting magnitudes are necessary. As it happens, the particular detector used in this experiment has a very low quantum efficiency, only 0.5 percent, at wavelength K. The quantum efficiency at wavelength H is an order of magnitude better. Unfortunately the extinction at H is higher, and no longer can be ignored in the Central field (as we did for the K star counting experiment). After a correction corresponding to about 4 magnitudes of visual extinction, however, it is possible to construct from the H image a version of the cluster membership histogram in reasonable agreement with the corresponding K diagram (Figure 3c, above); this H histogram shows a tendency to depart from the Salpeter IMF curve, in the sense of a deficit of stars at about spectral type G. Because of the limitations of the analysis, the onset of confusion due to crowding in the field, and the absence of color (J, H, and K) measurements for these fainter stars, this suggestion of a deficit of low mass stars must be regarded as tentative. It is, however, quite straightforward to investigate such trends in clusters closer to the sun; we therefore turn our attention to star forming regions in Orion (at a distance of approximately 500 parsecs).

4. STAR FORMATION IN NGC 2023

Before the availability of infrared array detectors, Sellgren (1983) demonstrated that the stars in the reflection nebula NGC 2023 were distributed in accordance with the Salpeter IMF down to a limiting magnitude of K=12 mag. Extrapolation of this relationship to K=15 mag suggests that approximately 50 stars of spectral type G through K might be present. A search to this limit revealed no additional cluster members (DePoy et al.; 1990). There is therefore a very significant dirth of detected stars in NGC 2023 of one solar mass and below. There may well be no such stars in this cluster.

One possible alternative—that the lower mass stars are still heavily enshrouded in a "cocoon" so dense that it is not penetrated by infrared radiation—has yet to be investigated. As DePoy et al. suggest, if such embedded stars are present, a straightforward search at 10 microns will probably detect them.

Approximately 50 percent of the stars detected in NGC 2023 have an infrared excess.

5. STAR FORMATION IN THE ORION NEBULA

In Figure 7 we compare the positions of stars in our wide field infrared image of the OMC1/OMC2 region (left) with the ^{13}CO map of Kutner et al. (1976, right, used by permission). A tendency for the stars to follow the distribution of dense molecular gas—similar to that demonstrated explicitly for L1630 above (Figure 1) may be noted.

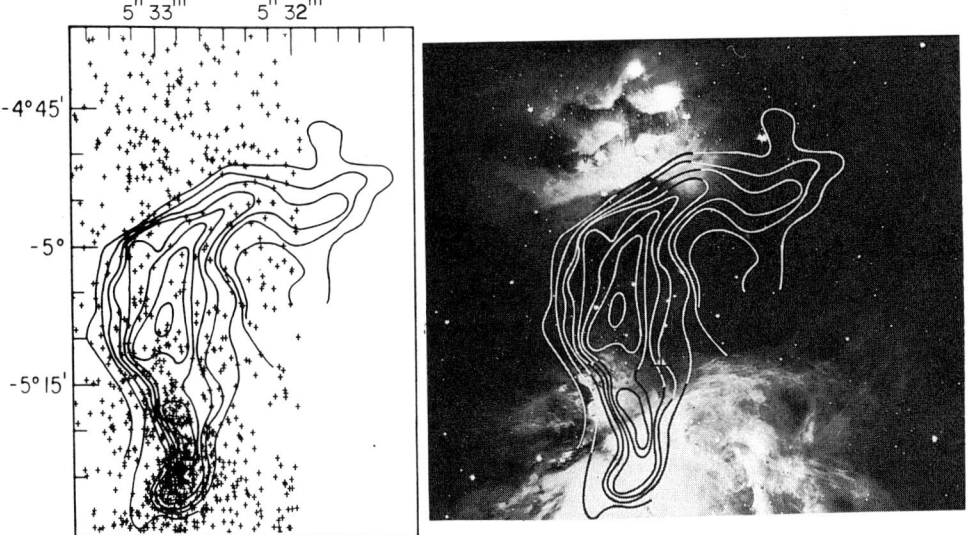

Figure 7.

Figure 8a shows a histogram of cumulative star counts (number of stars brighter than a given magnitude) at wavelength K for the Trapezium cluster in the Orion nebula; this sample consists of all stars within a region of 51.5 square arcminutes brighter than K=12 mag. The stellar density in this cluster is so high than contamination of the sample by Galactic field stars is minimal; the result of the Bahcall-Soneira model calculation is dispayed with the data in Figure 8a. Two results are immediately evident from inspection of Figure 8a. First—in sharp contrast to the results in M17 and NGC 2023—the histogram for the Trapezium cluster has a slope steeper than the Salpeter IMF for the more massive stars. Second, there is a deficit of low mass stars similar to that seen in NGC 2023.

The steep slope of the luminosity function in the Trapezium cluster can be most easily understood via the assumption that the cluster has a range of stellar ages. Evidence

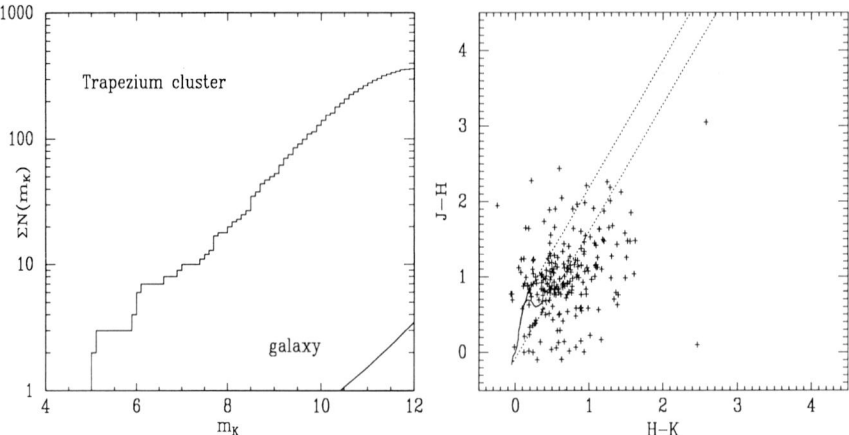

Figure 8.

for this hypothesis is provided by the color-color diagram for the cluster, displayed in Figure 8b; unlike the case of M17, where all the cluster members exhibit an infrared excess, only a minor fraction of the Trapezium cluster members have an excess.

The majority of the members of the cluster are not, in fact, particularly young; the slope of the luminosity function in a cluster steepens with time because the most massive members evolve away most rapidly. It is noteworthy that the slope of the Trapezium luminosity function parallels that of the Galactic population for K = 8 - 10 mag. The more luminous Trapezium members exhibit a different effect: for $K < 8$ mag the slope of the luminosity function is shallower, and so more like that of the Salpeter IMF; there are members of the cluster of spectral type as early as O5, which are therefore very young; there are cluster members with infrared excesses. Clearly, then, there are both old and young members of the Trapezium cluster.

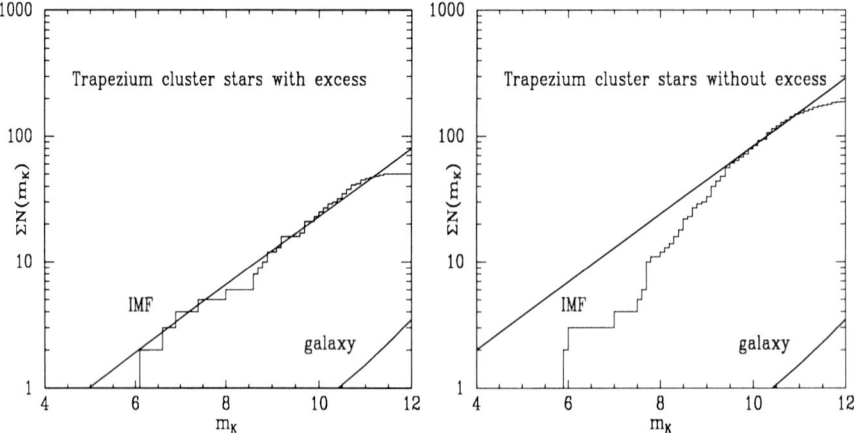

Figure 9.

The infrared colors provide a mechanism for separation of the younger and older populations within the Trapezium cluster. The histograms of cumulative star counts (number brighter than a given magnitude) for those members with infrared excesses and without infrared excesses, respectively, are plotted in Figures 9a and 9b. The (younger) stars with excesses fit the form of the Salpeter IMF; the (older) stars without excesses fit the form of the Galactic field population through approximately 1 $M_\odot$, and then exhibit the deficit also observed in NGC 2023. Therefore the Trapezium cluster exhibits a range of stellar ages. The notoriously high stellar density within this cluster (Baade and Minkowski 1937) may result from repeated episodes of star formation at the same site.

6. CONCLUSIONS

1) Star formation in L1630 occurs exclusively in the dense gas.
2) The OB stars in M17 follow the Salpeter IMF.
3) Every OB star in M17 has an infrared excess.
4) The brighter the star, the bigger the excess.
5) NGC2023 has a deficit of low mass stars.
6) The Trapezium cluster is not coeval.

We are grateful to Pat Britt, Nick Buchholz, Lindsey Davies, Darren DePoy, Tetsuo Hasegawa, Jim Herring, Jim Gates, Cathy and Sam Gatley, Dick Joyce, Charlie Lada, Elizabeth Lada, John Little, Ron Probst, Frank Stuart, and Sidney Wolff for their help and support.

REFERENCES

Baade, W. and Minkowski, R., 1937, *Ap. J.*, **86**, 119.
Chini, R., Elsasser, H., and Neckel, Th., 1980, *Astr. Ap.*, **91**, 186.
Chini, R., and Krugel, E., 1985, *Astr. Ap.*, **164**, 175.
DePoy, D., Lada, E., Gatley, I., and Probst, R., 1990, *Ap. J. (Letters)*, In Press.
Gatley, I., Becklin, E., Sellgren, K., and Werner, M., 1979, *Ap. J.*, **233**, 575.
Gatley, I., 1988, *in NATO ASI "Galactic and Extragalactic Star Formation"*, Kluwer, eds Pudritz and Fich.
Gatley, I., DePoy, D., and Fowler, A., 1988, *Science*, **242**, 1264.
Gatley, I., Becklin, E., Hyland, A., and Jones, T., 1981, *M.N.R.A.S.*, **197**, 17p.
Harper, D. A., Low, F., Rieke, G., and Thronson, H., 1976, *Ap. J.*, **205**, 136.
Icke, V., Gatley, I. and Israel, F. 1980, *Ap. J.*, **236**, 808.
Kutner, M. L., Evans, N. J., II, and Tucker, K. D., 1976, *Ap. J.*, **209**, 452.
Lada, E., 1990, PhD Dissertation, University of Austin at Texas.
Sellgren, K. 1983, *A. J.*, **88**, 985.

DISCUSSION

P. Conti: I was very taken with the way you got the stars with the infrared excess out of there, and therefore derive an IMF. But, I'll tell you what's bothering me a little about this. How come stars all have an infrared excess and you must be looking at mostly B stars, A stars, and F and G stars. Do you have any good physical explanation why they all have an infrared excess? Remember they have very different contraction times?

Gatley: Yes, I agree. I think that's an important question—the work on low mass star formation. Actually let me just say I was very amused that you thought Orion was puny and we should talk about more massive stars because when I talk to purely infrared audiences, they're always upset that Orion's got two massive stars in it. And they want to talk about Taurus or something. Lots of them have been working on that problem. Steve Strom especially has been looking at T Tauri stars. They find excesses in T Tauri stars younger than about three million years, and they think that it's a disc. And that the dissipation of that disc is the physics behind your question. I think I'm quite surprised to find excesses like that in the OB stars in M17. But this is not a new result, this was found originally by Chini who went and looked in the optical and found some B stars in M17. And then looked at them in the thermal infra-red. And there's a nice paper in which he describes the cocoon that he feels is still surrounding those stars. Evidently, in this very early phase of star formation, that there is leftover material from the star formation process, and the precise physics of that dissipates I think is one of the challenges that faces us all and to understand it. And I don't know, I've been at some pains today just to use it as a diagnostic of youth. But I think it has to do with dissipation at the disc.

R. Kudritzki: I'm extremely impressed by these pictures, but the next step would, of course, be to take spectra in the infrared. What can you say about this?

Gatley: Yes actually I completely agree with you. As I was telling you, these cameras were only built within the last couple of years. I'm happy to tell you now that the National Observatories both at Kitt Peak and at CTIO have spectrographs based on the same technology. One such spectrograph which works at a maximum resolution of about a thousand. We have just made a major improvement in the sensitivity of the chips which is reported in the April 10th *Astrophysical Journal*, and you can see an example there of the kind of spectra we can take. But it's safe to say that with that instrument working on the 4 meter telescope you can easily take spectra of any of these targets, and in fact, with regard to what Peter was saying this morning, I would say that the sample of a hundred objects in M17 is the classic place to begin. I don't know how many of those are going to turn out to be Wolf-Rayet stars or what they are. Their literature says the earlier stars in there are O4. And I think it's a glorious opportunity to go measure spectra of all of them, and I expect and hope that that will get done real soon.

T. Heckman: I'm a little bit confused about how you went from the number counts to an IMF. There must be a lot of corrections that you have to put into that.

Gatley: Fortunately not. Although there are a couple of assumptions. All I said there was a functional form, so I haven't made any attempt to make a direct conversion from observed brightness to particular mass. Only to a functional form. As you know, the Salpeter IMF is written as a power law. I want to take advantage, in this case, of the thing that Katy described as a disadvantage, namely that we're well in the Rayleigh-

Jeans approximation. The brightness of these stars depends only on the area of the photosphere and the temperature. So it goes like R^2T. For the main sequence stars, at least, each of those quantities can be written as a power law. The radius goes like the mass to the power 0.8, the temperature goes like the mass to the power 0.4. So the area temperature product goes like the mass squared. So you can just make a simple change of variables in the Salpeter formalism, and that's all I drew on there.

I deliberately excluded the stars that are heavily buried in the cloud. I estimate the main extinction to those to be about 4/10ths of a magnitude at 2 microns. But they all suffer it, and so because I'm only plotting a functional form through there, it's just a shift in where I plot the line on the graph. Same thing for the excess, if they all suffered the same excess it would just offset the coordinates. It's actually a very robust kind of analysis, but of course you might want to argue with the assumption that these are main sequence stars. I don't know that that's true at all, but I don't know where else to begin. But I think the attraction of it is that it's a nice simple dimensional analysis kind of argument, and I think it's a good place to begin. And for me the real surprise was that I took this data and did nothing to them other than what I described to you today—slapped on the Salpeter IMF and it went right through the data in both cases. I think it's always hard to judge whether that's an important datum or not, and when it fits your expectations. I didn't know where else to go with this analysis and it seems to me that it probably is important that you get exactly what you do get here.

N. Scoville: I would have thought actually that the assumption that there were main sequence objects would be pretty poor once you get far down the main sequence. Because the pre-main sequence times are of the order of 10^7 years, so a few solar masses—I think.

Gatley: Yes and I think that criticism applies more strongly to this apparent rollover at low masses, I don't think it has much to say, I don't think it's much of a problem with the OB star part of it. But certainly when you get to low masses some of the points that were being made this afternoon about contraction on to the main sequence may be important.

N. Scoville: The specific question I actually had was that in Orion, I guess there were two populations, which seem to imply different lengths of ...

Gatley: Well I split them into two, I'm not sure that that's a very good idea. I mean earlier today someone said that rather than, I think Roberta Humphreys said rather than two particular ages it might be better to imagine them as a spread of ages. That was more for illustration.

N. Scoville: Ok, but the high mass population is presumably 10^6 years or younger and what was the time for that assuming that they originally had high mass stars?

Gatley: Well again, I think that one would be talking up several million years and there abouts. It's difficult, I'll be happy to discuss it with you in detail and we can take it apart, but I don't have a specific model for that, just simply to show that some of them are old enough that they no longer have this infrared excess.

N. Scoville: No, but translating that cutoff in K magnitude into a mass, what mass was implied? That's what I didn't understand.

Gatley: They were all off of K=12 is at about one solar mass. I'm not sure which cut off you're talking about.

N. Scoville: Wait, I'm talking about high mass end of the extended population in the cloud. There was a termination at some value.

Gatley: I think it's slightly arbitrary how you make that division, but as I say I'll be happy to discuss it with you. The young population contains θ^1 C Ori. Of course, quite what you do with the aged part of the population I'm not very sure.

30 DORADUS, STARBURST ROSETTA

Nolan R. Walborn[1]
Space Telescope Science Institute[2]
3700 San Martin Drive
Baltimore, Maryland 21218
USA

Abstract. The 30 Doradus region of the LMC is a Rosetta Stone for the interpretation of similar, more distant regions, since it is accessible to spatially resolved investigation of its components with modern instrumentation. Following an introductory review of the various scales and populations relevant to the region, several current programs which promise improved understanding will be reviewed. They include extensive new optical spectroscopy and photometry of the central cluster, which will provide substantial new information about the stellar content and IMF; infrared surveys which will clarify questions about ongoing star formation and the lower main sequence; and high-energy phenomena in the surrounding interstellar medium, as revealed by high-resolution nebular spectroscopy and X-ray observations.

1. INTRODUCTION

An understanding of the structure and content of 30 Doradus is an essential prerequisite to the interpretation of more distant starburst regions. Paradoxically, we already know far more about 30 Doradus than can be reviewed in the present time and space, and yet several of its most fundamental aspects are only now being revealed by investigations in progress. The principal reason is that, although 30 Doradus is much more accessible to detailed study than any other starburst, it still represents a difficult observational problem at the limit of a given instrumentation, so that any advance in resolution, signal-to-noise, and/or wavelength coverage immediately yields significant new information. Therefore, rather than attempting to cover all recent developments, I shall, following a brief survey of the relevant spatial scales and stellar populations, concentrate on three current areas of research which in my judgment will provide substantial new insights into the nature of this spectacular region.

---

[1] Visiting Astronomer, Anglo-Australian Telescope.
[2] Operated by the Association of Universities for Research in Astronomy, Inc., under contract with the National Aeronautics and Space Administration.

2. SCALES AND POPULATIONS

In comparing 30 Doradus with more distant objects, it is important to distinguish, in descending order of size, the scales of the 30 Dor *region, nebula, ionizing cluster*, and *central object* (R136). The region is usually taken to encompass Henize (1956) N158/N159/N160 to the south, with a scale of about 1 kpc, although on many photographs the twice greater region of N135 appears as a coherent entity containing 30 Dor (N157) and the other nebulosities (Hodge and Wright 1967, Elliott *et al.* 1977). More than half of the LMC W-R stars listed by Breysacher (1981) are within N135, as are large numbers of B and M supergiants, minus one which was the SN 1987A progenitor. Of course, some of this noncoeval population will be projected onto, or even contained within, the 30 Dor Nebula, which has a diameter of about 200 pc (Walborn 1984). The 30 Dor ionizing cluster, defined by the distribution of the very early O stars (Melnick 1985, Walborn 1986), extends about 3' or 40 pc, which also corresponds to the extent of the brightest Tarantula filaments. The luminous central object R136 (Feast, Thackeray, and Wesselink 1960) is now well established as the dense core of the ionizing cluster, containing 27 resolved components (Walborn 1973; Walker and O'Donoghue 1984; Moffat, Seggewiss, and Shara 1985; Weigelt and Baier 1985; Neri and Grewing 1988); note the striking correspondences between the maximum-entropy structure of Walker and O'Donoghue and the speckle-masking components of Neri and Grewing. Figure 1 is a shallow cut from a remarkable CCD image with $0\rlap{.}''6$ resolution obtained at the ESO 2.2m and kindly made available by Dr. G. Meylan, which shows by inspection the outer structure of R136a previously seen only in the maximum-entropy and speckle analyses: supermassive stars have five points! One awaits with anticipation even better images from the ESO NTT and the HST. Such dense cores may be a characteristic of the formation of massive starbursts, as evidenced by the related galactic objects NGC 3603 (Walborn 1973, Moffat *et al.* 1985) and W49 (Welch *et al.* 1987). These various scales are summarized in Table 1, both in linear measure, and in angular for the LMC and greater distances. Perhaps the most noteworthy entry implies that the intricacies of the 30 Dor ionizing cluster, to be further discussed below, will subtend less than $1''$ in the Virgo Cluster, not to mention even more distant systems.

TABLE 1. *Characteristic dimensions*

	Linear	Angular		
		52.5 kpc	1 Mpc	10 Mpc
LMC	5 kpc	5°	15'	1.5'
30 Doradus Region	1 kpc	1°	3'	20''
30 Doradus Nebula	200 pc	15'	50''	5''
30 Doradus Cluster	40 pc	3'	10''	1''
R136	2.5 pc	10''	0.5''	0.05''
R136a	0.25 pc	1''	0.05''	0.005''

Table 2 summarizes the outstanding properties of four distinct evolutionary stages of massive young regions. The 30 Dor ionizing cluster definitely corresponds to the Carina phase, but all four populations are present in the field of the 30 Dor Nebula, and the question is, which if any of the other three are generically related to the ionizing cluster? As discussed in Section 4 below, there is now very strong evidence for sequential formation of a physically associated Orion phase; on the other hand, a physically

Figure 1. *Ten-sec B CCD exposure of the 30 Dor cluster from the ESO 2.2m by Dr. G. Meylan. North is at top and east to the left; the extent in declination is $2'$ and the resolution is $0''.6$. Note the structure of R136a, the bright composite object at the center.*

associated h and χ Per phase can be excluded, as was already discussed by Walborn (1984). The relatively numerous late-O/early-B supergiants (Sco OB1 phase) present a more difficult problem, however; as shown by Walborn (1986) and discussed by Moffat et al. (1987), they are less concentrated toward R136 than the earlier types, but further work is required for a definitive answer. The two WC objects within the area of the ionizing cluster are an additional puzzle; they may be associated with the Sco OB1 population, or alternatively they might be very massive binary mass-transfer systems. With regard to the interpretation of the integrated spectra of distant starbursts, it is worth noting that the emission-line flux of a single early WC can equal the sum of those of 20 late WNs (Smith 1990).

TABLE 2. *Morphology of massive young regions*

Region	Visually Brightest Stars	MS Turnoff Spectrum	~Mass	~Age	H II, Dust	Red Sg
Orion Nebula	ZAMS O, (IR)	(PMS)	—	$< 10^6$ yrs	Yes	No
Carina Nebula	O3, WN	O3	100M$_\odot$	3×10^6	Yes	No
Scorpius OB1	OB Sg	O6-7	60	$4\text{-}5 \times 10^6$	No	No
$h + \chi$ Persei	A Sg	O9-B0	20	$8\text{-}9 \times 10^6$	No	Yes

3. OPTICAL SPECTROSCOPY AND PHOTOMETRY: THE CURRENT GENERATION

Spectroscopic knowledge of the 30 Dor ionizing cluster as of that time (to about 14th mag) was summarized by Melnick (1985) and Walborn (1986), who showed that it constitutes the most spectacular concentration of hot, massive stars for which detailed information is available. The cluster includes five examples of the remarkable, rare O3 If/WN-A supergiant category, with apparent magnitudes similar to those of the brightest unresolved speckle components within R136a, which are probably similar objects and strong candidates for the most massive stars known. In collaboration with C. Blades, I have undertaken to extend this information to about magnitude 15.5, by means of the AAT multiple-object, fiber-optics system, in order to investigate whether later spectral types or somewhat more reddened very early types predominate at that level. The results of a first-pass classification of these data are summarized in Table 3, which shows that, along with the significant number of later types found, the number of known very early types has been doubled by this study. Most or all of the evolved B- and A-type objects are probably nonmembers of the ionizing cluster. In addition to the 111 OBA types, 11 W-Rs exclusive of R136 (Moffat et al. 1987) and 13 late-type spectra (a mixture of LMC supergiants and foreground dwarfs/giants) are known in the field. However, Figure 1 shows many crowded objects of similar magnitudes which have not yet been observed spectroscopically, and R136 contains at least 27 components brighter than 16th mag which are unknown spectroscopically, except for two additional W-Rs.

In addition to the statistical information, as is usually the case in studies of this kind, the new observations have revealed a number of individual objects of special astrophysical interest. They include some probable very early evolutionary stages (Walborn and Blades 1987; Section 4 below), very early-O giants, the first O(n)(f)p spectrum in the LMC, a new heavily-reddened Of supergiant (Walborn 1990), B supergiants with CNO anomalies, and a variable B-type spectrum with an apparent He II $\lambda 4686$ emission line (Section 5 below).

Currently, several CCD programs are underway which will soon provide accurate photometry for most stars in the field for the first time. Perhaps the most extensive is by J. Parker, C. Garmany, and P. Massey, but other studies are also in progress (separate private communications from A. Campbell, G. Meylan, and A. Moffat). In addition to permitting evaluation of the reddening and absolute magnitudes of the spectroscopically classified stars, their work will contribute vital information about the IMF and the lower main sequence (if it is observable), although for the former purpose combination with the spectral types is essential (Massey, Parker, and Garmany 1989). A discussion of 30 Doradus in relation to the LMC blue globular clusters has been

TABLE 3. *The known OB stellar content of 30 Doradus*

Spectral Types	Melnick, Walborn	Walborn & Blades	TOTAL
O3-6 V	14	17	31
O3-6 III	3	3	6
O3 If*	5	0	5
O6-9 V	4	11	15
O6-9 III	5	10	15
O6-8 Iaf	1	1	2
O9 I	2	1	3
B0-1 V	0	1	1
B0-1 III	0	7	7
B0-1 I	11	3	14
B1-3 III	0	2	2
B1-3 I	1	2	3
B5-A I	4	3	7
TOTAL	50	61	111

presented by Kennicutt and Chu (1988), and to the integrated LMC IMF by Blaha and Humphreys (1989).

The hydrogen-ionizing luminosity of the spectroscopically classified O and W-R stars in the 30 Dor cluster (exclusive of R136) has been computed following the procedures of Walborn (1984), the result being 2.8×10^{51} sec^{-1}. Fortuitously, this result is essentially identical to that obtained in the earlier analysis based primarily on approximate photometry alone, although that estimate involved 136 stars while the present one contains 88; the reason is that some of the former turned out to be of type B, while others are of earlier O types than had been expected. The earlier estimate of 2×10^{51} sec^{-1} for R136a cannot yet be improved upon. For comparison, the H II luminosity within a 140 pc diameter is 8.7×10^{51} sec^{-1} (Mills, Turtle, and Watkinson 1978). Several sources for the "missing" ionization can be readily suggested: early-type members of the central cluster not yet classified spectroscopically (see above); higher than assumed ionizing luminosities for the W-R stars; early-type stars in the periphery of the Nebula, not included in the central cluster count (*e.g.*, R130, R144, R146, R147, Breysacher 73; Testor and Schild 1990, Schild *et al.* 1990, and Lortet *et al.* 1990 have just announced the discovery of several early-O spectra including a new O3 If/WN-A near the latter, emphasizing anew that the census of even the most spectacular stellar component in this region remains incomplete); and shocks or soft X-rays (Section 5 below). Discussions of the global energetics of 30 Doradus have recently been published based upon the UV (Smith, Cornett, and Hill 1987), Hα (Kennicutt and Hodge 1986), IRAS (Schwering 1989), CO (Cohen *et al.* 1988), and radio continuum (Klein *et al.* 1989).

4. INFRARED: NEW BEGINNINGS

The dominant optical components of the 30 Doradus Nebula and ionizing cluster are evolved: the most massive stars have reached the W-R stage, implying ages of two to three million years (Maeder 1983), and the nebulosity appears to be expanding away from the cluster (*e.g.,* Walborn 1980). Nevertheless, an H_2O maser source has been discovered in the northeast quadrant of the Nebula (Whiteoak *et al.* 1983, Whiteoak and Gardner 1986). Walborn and Blades (1987) reported two early O-type objects in dense nebular knots, one within the maser positional error circle and the other nearby within the brightest Tarantula filament, and suggested that they may represent a significantly earlier evolutionary stage. There are several other prominent nebular knots in this sector which may contain fainter embedded stellar objects. Subsequently, a third essentially identical stellar/nebular knot has been found within the second-brightest Tarantula filament immediately west of R136; it is identified in Figure 2.

In a remarkable IR imaging survey of the 30 Dor Nebula, Hyland *et al.* (1990; see also Hyland and Jones 1990, Jones *et al.* 1986) have found definitive evidence for current massive star formation in the form of four luminous protostars, three of which are located in the northeast quadrant near the maser and optical knots, while the fourth is just a few arcsec from the new stellar/nebular knot west of R136, as also shown in Figure 2. The morphology strongly suggests that the energetic activity of the central ionizing cluster has initiated a new stellar generation in these surrounding areas, and this region now becomes one of the strongest cases for sequential star formation known. The protostars have extremely red J-K colors and featureless, red 2μ spectra, and they are significantly above the survey magnitude limit, suggesting a possible deficiency of lower masses. These results presage a substantial advance in our understanding of 30 Doradus and massive star formation; one awaits with anticipation further images from the new IR arrays.

5. HIGH ENERGY: LAST THINGS

The morphology of massive young regions implies that rich OB clusters disperse their remanent gas and dust on the evolutionary timescale of the most massive stars (Table 2). The presumed agents of this dispersal are the high-energy stellar radiation, winds, and supernova events. Direct evidence of this activity is provided by high-velocity nebular kinematics and soft X-ray emission.

The moderate-velocity, large-scale expansion(s) of 30 Doradus (of order 10 km/sec) have been discussed by, *e.g.,* Cox and Deharveng (1983), Meaburn (1984), and Clayton (1987). However, there is also evidence for numerous smaller expanding shells with much higher velocities (of order 100 km/sec) throughout the Nebula (Meaburn 1984, 1990; Chu and Kennicutt 1988, 1990; Chu, private communication); one striking example has been discussed in detail by Meaburn (1988). The latter object is centered on the small h and χ Per-type cluster, containing B, A, and M supergiants, 3' northwest of R136. In the AAT stellar spectroscopic study (Section 3 above), a variable B-type spectrum with probable He II $\lambda 4686$ emission has been found in this cluster; it may be considered a candidate for the (binary) stellar remnant of a SN which produced the high-velocity shell.

The Einstein Observatory detected 30 Doradus as an extended, soft X-ray source with remarkable structural correspondences to the optical nebular morphology, as shown in Figure 3 (see also Harnden and Seward 1988); detailed analyses of these data have just recently appeared, yielding a luminosity of order 10^{37} ergs/sec and a total thermal energy of 10^{52} ergs (Chu and MacLow 1990a,b; Wang and Helfand 1990a,b; Helfand 1990). The X-ray emission and the high-velocity gas motions are undoubtedly related,

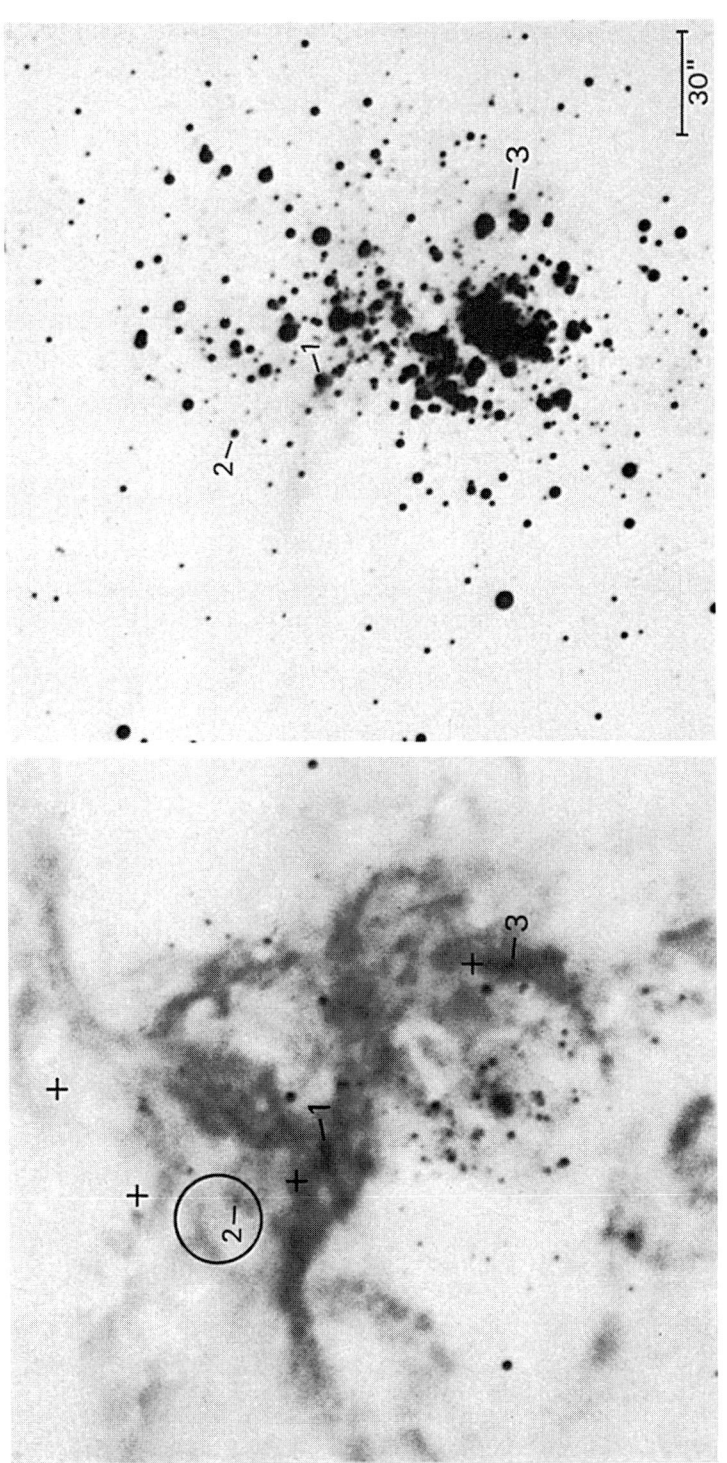

Figure 2. [OIII] (left) and blue continuum (right) images of 30 Dor, described by Walborn and Blades (1987). The numbers identify nebular knots with embedded early O stars, the crosses represent approximate locations of the IR protostars, and the circle denotes the H_2O maser position.

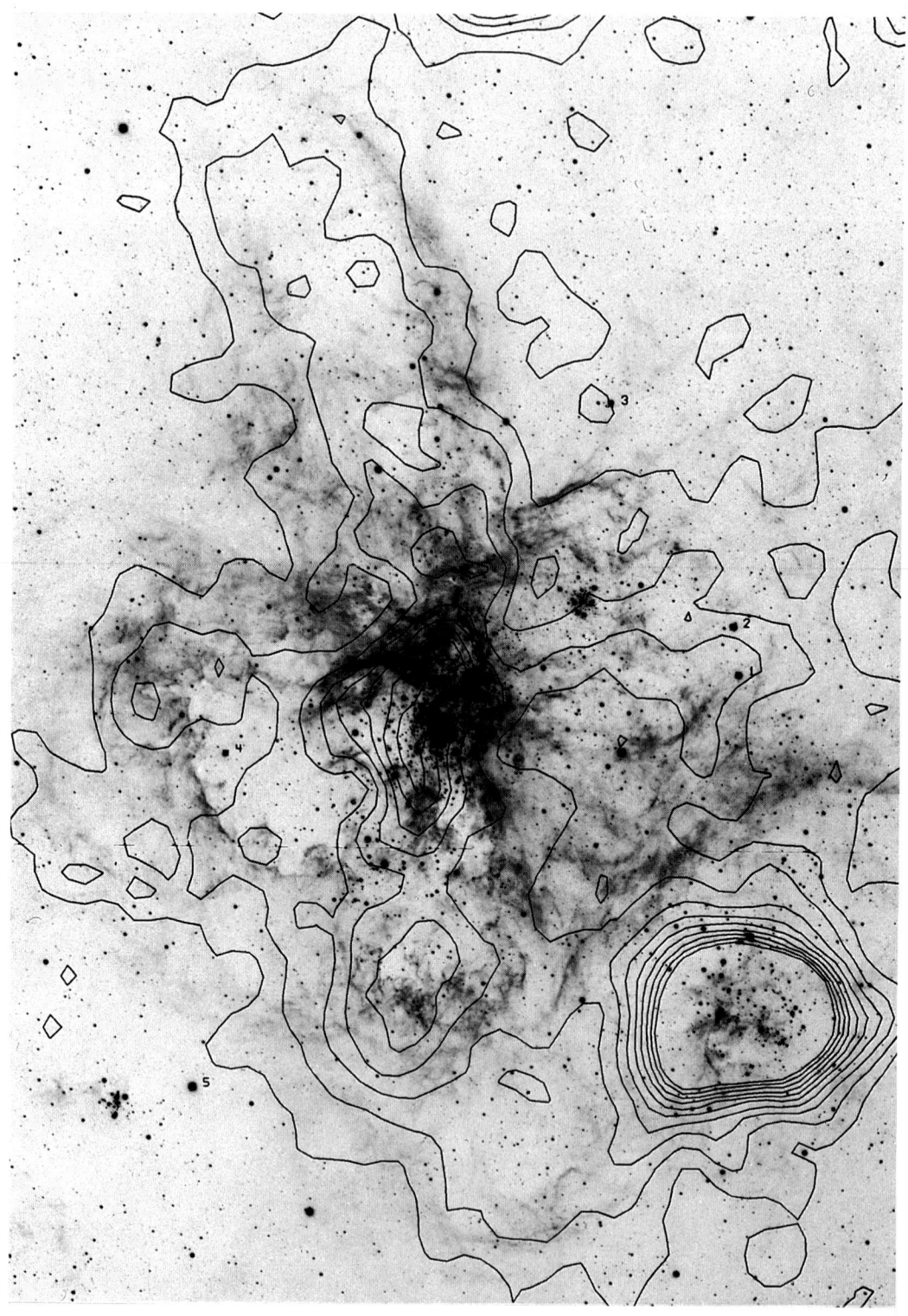

Figure 3. *Einstein Observatory IPC X-ray contours provided by Dr. F. Seward, superposed on the [S II] optical image of 30 Dor described by Walborn (1984).*

since they have similar energy contents, as is the case in the Carina Nebula (Seward and Chlebowski 1982, Walborn and Hesser 1982, Walborn 1982). Following the procedures in the latter reference, the kinetic energy from the stellar winds of the spectroscopically classified O and W-R stars in 30 Dor has been estimated as 10^{52} ergs; galactic wind parameters were used since extensive determinations as a function of spectral type for LMC stars are not yet available, but on the other hand the R136 components and other unclassified members of the ionizing cluster are not included. As is the case in Carina, the stellar winds are probably capable in principle of generating the high-energy phenomena observed in the Nebula, but it is likely that embedded SNR contribute as well in 30 Doradus.

REFERENCES

Blaha, C. and Humphreys, R. M. 1989, *A. J.*, **98**, 1598.
Breysacher, J. 1981, *Astr. Ap. Suppl.*, **43**, 203.
Chu, Y.-H. and Kennicutt, R. C., Jr. 1988, *A. J.*, **96**, 1874.
_____. 1990, in IAU Symp. 148, *The Magellanic Clouds*.
Chu, Y.-H. and MacLow, M.-M. 1990a, *Ap. J.*, in press.
_____. 1990b, in IAU Symp. 148, *The Magellanic Clouds*.
Clayton, C. A. 1987, *Astr. Ap.*, **173**, 137.
Cohen, R. S., Dame, T. M., Garay, G., Montani, J., Rubio, M., and Thaddeus, P. 1988, *Ap. J. (Letters)*, **331**, L95.
Cox, P. and Deharveng, L. 1983, *Astr. Ap.*, **117**, 265.
Elliott, K. H., Goudis, C., Meaburn, J., and Tebbutt, N. J. 1977, *Astr. Ap.*, **55**, 187.
Feast, M. W., Thackeray, A. D., and Wesselink, A. J. 1960, *M.N.R.A.S.*, **121**, 337.
Harnden, F. R., Jr. and Seward, F. D. 1988, in Fourth George Mason Astrophysics Workshop, *Supernova 1987A in the Large Magellanic Cloud*, ed. M. Kafatos and A. G. Michalitsianos (Cambridge), p. 408.
Helfand, D. J. 1990, in IAU Symp. 148, *The Magellanic Clouds*.
Henize, K. G. 1956, *Ap. J. Suppl.*, **2**, 315.
Hodge, P. W. and Wright, F. W. 1967, *The Large Magellanic Cloud* (Smithsonian Publ. 4699).
Hyland, A. R. and Jones, T. J. 1990, in IAU Symp. 148, *The Magellanic Clouds*.
Hyland, A. R., Straw, S., Jones, T. J., and Gatley, I. 1990, in preparation.
Jones, T. J., Hyland, A. R., Straw, S., Harvey, P. M., Wilking, B. A., Joy, M., Gatley, I., and Thomas, J. A. 1986, *M.N.R.A.S.*, **219**, 603.
Kennicutt, R. C., Jr. and Chu, Y.-H. 1988, *A. J.*, **95**, 720.
Kennicutt, R. C., Jr. and Hodge, P. W. 1986, *Ap. J.*, **306**, 130.
Klein, U., Wielebinski, R., Haynes, R. F., and Malin, D. F. 1989, *Astr. Ap.*, **211**, 280.
Lortet, M. C., Schild, H., and Testor, G. 1990, in preparation.
Maeder, A. 1983, *Astr. Ap.*, **120**, 113.
Massey, P., Parker, J. W., and Garmany, C. D. 1989, *A. J.*, **98**, 1305.
Meaburn, J. 1984, *M.N.R.A.S.*, **211**, 521.
_____. 1988, *M.N.R.A.S.*, **235**, 375.
_____. 1990, in IAU Symp. 148, *The Magellanic Clouds*.
Melnick, J. 1985, *Astr. Ap.*, **153**, 235.
Mills, B. Y., Turtle, A. J., and Watkinson, A. 1978, *M.N.R.A.S.*, **185**, 263.

Moffat, A. F. J., Niemela, V. S., Phillips, M. M., Chu, Y.-H., and Seggewiss, W. 1987, *Ap. J.*, **312**, 612.

Moffat, A. F. J., Seggewiss, W., and Shara, M. M. 1985, *Ap. J.*, **295**, 109.

Neri, R. and Grewing, M. 1988, *Astr. Ap.*, **196**, 338.

Schild, H., Lortet, M. C., and Testor, G. 1990, in IAU Symp. 143, *Wolf-Rayet Stars*.

Schwering, P. B. W. 1989, *Astr. Ap. Suppl.*, **79**, 105.

Seward, F. D. and Chlebowski, T. 1982, *Ap. J.*, **256**, 530.

Smith, A. M., Cornett, R. H., and Hill, R. S. 1987, *Ap. J.*, **320**, 609; err. ibid. **355**, 746 (1990).

Smith, L. F. 1990, in IAU Symp. 143, *Wolf-Rayet Stars*.

Testor, G. and Schild, H. 1990, *Astr. Ap.*, in press.

Walborn, N. R. 1973, *Ap. J. (Letters)*, **182**, L21.

_____. 1980, *Ap. J. (Letters)*, **235**, L101.

_____. 1982, *Ap. J. Suppl.*, **48**, 145.

_____. 1984, in IAU Symp. 108, *Structure and Evolution of the Magellanic Clouds*, ed. S. van den Bergh and K. S. de Boer (Reidel), p. 243.

_____. 1986, in IAU Symp. 116, *Luminous Stars and Associations in Galaxies*, ed. C. W. H. de Loore, A. J. Willis, and P. Laskarides (Reidel), p. 185.

_____. 1990, in ASP Conf. Ser. Vol. 7, *Properties of Hot Luminous Stars*, ed. C. D. Garmany (BookCrafters), p. 23.

Walborn, N. R. and Blades, J. C. 1987, *Ap. J. (Letters)*, **323**, L65.

Walborn, N. R. and Hesser, J. E. 1982, *Ap. J.*, **252**, 156.

Walker, A. R. and O'Donoghue, D. E. 1984, *Astr. Express*, **1**, 45.

Wang, Q. and Helfand, D. J. 1990a, preprint.

_____. 1990b, in IAU Symp. 148, *The Magellanic Clouds*.

Weigelt, G. and Baier, G. 1985, *Astr. Ap.*, **150**, L18.

Welch, Wm. J., Dreher, J. W., Jackson, J. M., Terebey, S., and Vogel, S. N. 1987, *Science*, **238**, 1550.

Whiteoak, J. B. and Gardner, F. F. 1986, *M.N.R.A.S.*, **222**, 513.

Whiteoak, J. B., Wellington, K. J., Jauncey, D. L., Gardner, F. F., Forster, J. R., Caswell, J. L., and Batchelor, R. A. 1983, *M.N.R.A.S.*, **205**, 275.

DISCUSSION

D. Weedman: Note the X-ray image of the LMC. By far the brightest source is LMC X-1—a massive binary. Understanding the X-ray properties of starburst galaxies—which is critical to understanding the cosmological X-ray background—comes down to the question of understanding the evolution of binary stars!

T. Heckman: It would be extremely interesting to compare the overall spectral energy distribution of 30 Dor (from X-ray through radio) to those of starburst galaxies. Have you (or has someone else) done this?

Walborn: It could be done, but would require considerable circumspection for a meaningful comparison with the distant starbursts, in view of the relatively greater spatial resolution in 30 Doradus. For example, as pointed out by Weedman and Griffiths, the X-ray emission of starbursts is likely to be dominated by massive X-ray binaries, with young SNR as additional contributors. There are such objects in the greater 30 Doradus region, but the emission of the Nebula itself is dominated by a soft, diffuse component.

C. Norman: When and how can we expect to learn the IMF, the lower and upper mass cut-offs, in 30 Doradus. How did the environment trigger the burst?

Walborn: Preliminary IMF results are presented by J. Parker in a poster at this meeting. More definitive results will be available later this year from a combination of their photometry and our spectral types. The origin of the burst is of course a matter of speculation; one recent suggestion is a collision of two supergiant shells currently observed as extended filamentary arcs enveloping the region.

A. Moffat: While there appear to be some very luminous O stars (compared to the Galaxy) in 30 Dor, the descendant WNL population in 30 Dor does not seem to be any brighter *on average* than the WNL stars in the rest of the LMC (Moffat 1990, *Ap. J.*), or even the Galaxy. Thus, one wonders if the *average* (main-sequence) O-star population *really* is overluminous (*i.e.*, are selection effects operating here among the O stars in that one tends to study the apparently brighter stars first: are these also the intrinsically brightest?).

Walborn: The most luminous stars in 30 Dor are not main-sequence, but O3 If/WN-A; the brightest are 1 mag brighter than the O3 If and WN-A's in the Carina Nebula. I think there is a physical selection effect in 30 Dor: since it is much more massive than the galactic regions, the high-mass tail of the IMF is observed further up.

PROPERTIES OF GIANT H II REGIONS

Robert C. Kennicutt, Jr.
Steward Observatory
University of Arizona
Tucson, Arizona 85721
USA

Abstract. Giant H II regions offer the opportunity to study a starburst at close range, and to critically evaluate the diagnostic methods which are used to study more distant starburst galaxies. This paper reviews what is known about the physical characteristics of nearby giant H II regions, with special emphasis on those areas which are most relevant to the starburst phenomenon.

1. INTRODUCTION

Giant H II regions in external galaxies (often called GEHRs) are important stepping stones between the study of stellar populations in our own Galaxy and the luminous starburst galaxies. GEHRs such as 30 Doradus are starbursts in every sense of the term, with hundreds to thousands of massive stars embedded within gas clouds of up to several million solar masses. They provide information on the structure, dynamics, and evolution of local starbursts over a wide range of physical conditions and galactic environments. In particular, observations of nearby H II regions are especially useful for the following:

1. *Diagnostics of Starbursts*: Nearby GEHRs serve as the testing grounds for the various diagnostic methods which are used to measure the star-formation rates and initial mass functions in galaxies. In most starburst galaxies these quantities must be inferred indirectly, from modelling of the integrated spectra of entire star-forming regions. The same techniques can be applied to GEHRs, but their stellar contents can also be studied directly, and used to test the nebular-derived parameters.
2. *Comparison with Normal Galaxies*: Although GEHRs are rare phenomena in the Galaxy, they are a dominant mode of star formation in late-type galaxies (Section 3). One of the most fundamental unanswered questions about starbursts is whether they represent an entirely distinct mode of star formation, unique to the starburst galaxies, or whether they merely represent an amplified level of (otherwise normal) star formation. Comparing the properties of starbursts with nearby GEHRs offers the best means of answering this important question.
3. *Dynamics and Regulation of Star Formation*: GEHRs are the optimal laboratories

for studying the interactions and feedback between young massive stars and the interstellar medium. They possess an abundance of massive stars and supernovae, a dense ambient interstellar medium, and an ionized, line-emitting gas, which enables one to derive detailed information on the densities, temperatures, pressures, kinematics, and energetics of the star-forming regions. This potential for investigating the physical triggering and regulation of the starbursts has not yet been fully exploited, but over the long term this may be the most significant contribution that observations of nearby GEHRs can make toward understanding the starburst phenomenon.

In this paper I review what is known about the physical properties of giant H II regions, with special emphasis on the applications described above. I also refer the reader to recent reviews by Dinerstein (1990) and Shields (1990), which emphasize the abundance properties of extragalactic H II regions.

2. PHYSICAL CHARACTERISTICS OF GIANT H II REGIONS

2.1. Range of Physical Properties

As with any class of object whose definition is based on the adjective "giant," there is a wide disparity in what is considered a giant H II region. Indeed, the literature includes frequent references to "giant," "supergiant," "hypergiant," "starburst," and even "jumbo" H II regions, with virtually no correlation between the nomenclature and the actual size or luminosity of the objects in question. For the purposes of this discussion I will consider a giant H II region to be one which is large enough so that its physical properties are dominated by the composite properties of its embedded stellar population, rather than by its most massive few stars.

The most comprehensive catalogs of H II regions in external galaxies are based on Hα emission-line imaging surveys, mainly by Hodge and co-workers (*e.g.*, Hodge 1969, 1974; Hodge and Kennicutt 1983), and by a large group at the Observatoire de Marseille (*e.g.*, Pellet *et al.* 1978, Bonnarel *et al.* 1986, Courtès *et al.* 1987, Deharveng *et al.* 1988). The widespread availability of CCD cameras has revolutionized the field, and now it is relatively straightforward to obtain deep images of galaxies in Hα or other strong lines down to limiting emission measures of order $10\ cm^{-6}\ pc$ in exposures of less than an hour.

The most massive single stars produce ionizing luminosities of several times $10^{49}\ s^{-1}$, so in quantitative terms GEHRs as defined above are characterized by luminosities of order $10^{50}\ s^{-1}$ and above, or luminosities in the Hα emission line of order $10^{39}\ erg\ s^{-1}$ and higher. Table 1 lists the actual order-of-magnitude ranges in diameter, luminosity (corrected for extinction), equivalent stellar contents, and ionized gas masses for GEHRs in nearby spiral and irregular galaxies (Kennicutt 1984, 1988). Also listed for comparison are the Orion Nebula (Kennicutt 1984) and the prototype nuclear starburst galaxy NGC 7714 (Weedman *et al.* 1981).

Compared to most ordinary star-forming regions in the solar neighborhood, such as the Orion Nebula, the GEHRs are enormous, typically 100–10,000 times the luminosity and mass. The Milky Way may contain a handful of bona fide GEHRs, with the best candidates being the highly obscured W49 and NGC 3603 regions, or perhaps the Carina complex (Kennicutt 1984). None of these regions, however, approaches the scale of 30 Doradus, or the brightest regions in M101, M82, or NGC 2403. At the other extreme, there are no known nearby ($D \leq 20\ Mpc$) GEHRs which approach the most luminous

TABLE 1. *Range of physical properties*

	GEHRs	Orion	NGC 7714
Diameter (pc)	100–1000	10	600
Log L(Hα) ($erg\ s^{-1}$)	38–41	37	42
Log Q(H) ($photons\ s^{-1}$)	50–53	49	54
Ionizing O Stars	10–10,000	6	>10,000
Log M(H II) (M$_\odot$)	3–7	2–3	7

starburst nuclei (*cf.*, Kennicutt, Keel, and Blaha 1989). Extremely luminous GEHRs have been reported in several distant "clumpy irregular" galaxies (*e.g.*, Benvenuti *et al.* 1982), and in such extreme cases GEHRs as bright as L(Hα) = 10^{42} $erg\ s^{-1}$ may form.

2.2. Physical Description

In addition to the large differences in global properties between GEHRs and small Galactic H II regions, there are important structural differences. For example, the gas densities in GEHRs are usually 1–2 orders of magnitude lower ($N_e \sim$ 10–100 cm^{-3}) than in Galactic regions like the Orion Nebula (*e.g.*, Israel 1976, O'Dell and Castañeda 1984, Kennicutt 1984, Copetti 1990). Most GEHRs also possess less visual extinction than typical Galactic star-forming regions, as evidenced from both radio-continuum observations (*e.g.*, Israel and Kennicutt 1980, Caplan and Deharveng 1986, van der Hulst *et al.* 1989) and infrared data (*e.g.*, Strom *et al.* 1974, Skillman and Israel 1988). There are notable exceptions, including the ultraluminous IRAS galaxies (*e.g.*, M82), and the nuclear regions of many nearby galaxies (*e.g.*, M83, NGC 253, IC 342), but most extranuclear disk regions exhibit visual extinctions of only 0–3 mag. GEHRs are also distinct in terms of their internal kinematics, with supersonic velocity dispersions ($\sigma_v \sim$ 15–40 $km\ s^{-1}$), and often with localized regions with much higher velocities (see discussion in Sec. 5). Many of the large H II regions also exhibit diffuse X-ray and/or nonthermal radio emission. These observations provide evidence of a much more complex and dynamic environment than an ordinary small H II region.

The composite nature of the stellar population in GEHRs has important effects on their structure and evolution. For example, several observations suggest that the age range among GEHRs is considerably greater than the main-sequence lifetime of the most massive individual stars. This evidence includes the presence of supergiants (*e.g.*, McGregor and Hyland 1981, Massey *et al.* 1989b) and supernova remnants (*e.g.*, Long *et al.* 1981, Chu and Kennicutt 1990) in nearby GEHRs, and the observation of a bright, supergiant-dominated continuum in the integrated spectra of GEHRs (*e.g.*, Searle 1971, Shields and Searle 1978, McCall *et al.* 1985). A practical consequence is that the ionization structure and integrated continua of the GEHRs are dominated not by the most massive single star, as is often the case in small H II regions, but rather by the age and IMF of the embedded stellar association or cluster (*e.g.*, Shields and Tinsley 1976, Dottori and Bica 1981, Copetti *et al.* 1985). This may make it possible to use the integrated spectra of GEHRs to probe the behavior of the upper IMF, as discussed later.

The high concentration of massive stars in the cores of GEHRs has other important consequences. In the central region of a luminous H II region such as 30 Doradus, the expected interval between successive supernovae ($\sim$ 100–1000 yr) is probably much shorter than the lifetime of an individual supernova remnant, and consequently supernovae will tend to occur inside existing remnants or stellar-wind-blown bubbles. This presents ideal conditions for the development of collective dynamical structures such as supershells.

In summary, although GEHRs are often characterized either as scaled-up versions of the Orion Nebula or as large clusters of such regions, neither of these descriptions is accurate. The most realistic characterization of a GEHR is as a microcosm of a galaxy's entire interstellar medium, containing all of the its principal phases, in a dynamic, nonequilibrium configuration.

These distinctions can have important practical consequences. A good example is the long-debated question of whether GEHRs are ionization-bounded or density-bounded objects. The question is relevant to several applications, including the measurement of star-formation rates in galaxies, modelling the ionization and structure of the gaseous halos and disks of galaxies, and the ionization of the intergalactic medium at early cosmological epochs. Although arguments abound in the literature for both alternatives, I suspect that a hybrid description is most accurate. In a highly filamentary H II region many lines of sight from the central ionizing stars will be optically thick to UV radiation, and the gas in those regions will dominate the integrated nebular spectrum. Consequently, this spectrum will exhibit all of the earmarks of a classical, ionization-bounded H II region. On the other hand, supernovae and stellar winds almost certainly excavate holes which will be optically thin to ionizing radiation, and hence an H II region which for spectroscopic applications can be treated as a radiation-bounded volume may provide a copious source of ionizing radiation and particles to the surrounding galaxy.

3. SYSTEMATICS OF H II REGION POPULATIONS IN GALAXIES

Hα imaging surveys of nearby galaxies show that GEHRs occur almost exclusively in late-type (Sc–Irr) galaxies. This is part of a general trend in the luminosity and mass functions of H II regions with Hubble type (Kennicutt 1988; Kennicutt, Edgar, and Hodge 1989). Early-type galaxies generally exhibit lower levels of massive star formation per unit area overall (*e.g.*, Kennicutt 1983), and photometry of the H II regions shows that this is due to both a smaller number of H II regions per unit area, regardless of nebular luminosity, and a significantly steeper nebular luminosity function.

This Hubble-type dependence is most pronounced for the brightest H II regions. For example, Table 2 lists the frequency of occurrence of GEHRs with Hα luminosities above 3×10^{39} *erg s*$^{-1}$ (roughly the observed luminosity of 30 Doradus), for different galaxy types, based on data from Hunter and Gallagher (1985), Kennicutt and Chu (1988), Kennicutt *et al.* (1989), and Caldwell *et al.* (1990). The frequency per unit galaxy luminosity (expressed in units $M_B = -19.5$) decreases by nearly a hundredfold from types Irr I to Sb, and to zero in Sa galaxies. Indeed, the brightest H II region in the Sa sample is an order of magnitude fainter than 30 Doradus.

The physical cause for this trend is not fully understood. To the extent that the Hα luminosities of bright H II regions scale with the luminosity and mass of the embedded star clusters and associations, the trend must reflect a systematic gradient in the masses of star clusters and their parent atomic/molecular clouds along the Hubble sequence (Kennicutt 1988, Kennicutt *et al.* 1989). This interpretation is supported by the obser-

TABLE 2. *Frequencies of giant H II regions in galaxies*

Type	N(Galaxies)	L_B	N(GEHRs)	N(GEHR)/L_B
Sa	12	26.6	0	0.00
Sb	10	43.1	1	0.02
Sc	15	22.9	25	1.09
Irr	22	3.1	7	2.28

vation of a similar increase in the blue luminosities of the brightest stellar associations with Hubble type (Wray and de Vaucouleurs 1980). However, other effects, such as a gradient in extinction and/or the upper IMF with Hubble type, cannot be ruled out. Radio observations of the H II regions (e.g., Israel and Kennicutt 1980, Caplan and Deharveng 1986, Kaufman et al. 1987, van der Hulst et al. 1988) show that extinction variations among different galaxies are too small to explain all of the gradient in Table 2, but they may contribute partially to the trend. Virtually nothing is known about the importance of IMF variations along the Hubble sequence, though there is circumstantial evidence from spectra of GEHRs for some systematic variation in the IMF with metallicity, as discussed below.

4. STELLAR CONTENT AND INITIAL MASS FUNCTION

One of the recurring questions at this workshop is whether the IMF in starbursts is radically different from that in normal galaxies. Measurements of the IMF in nearby GEHRs are an important element in resolving this controversy. All of the various claims of abnormal IMFs in starburst galaxies have been based on modelling of the integrated optical and/or infrared spectra of the regions, and the nearest GEHRs provide us with a means of testing the reliability of these models from first principles. Moreover, if the IMF in the starburst galaxies is abnormal, it is important to test whether this is a property of all regions of massive star formation, or whether it is unique to the extreme starburst environment.

During the past decade there have been extensive observations of the brightest stars in nearby GEHRs, as reviewed by Rosa and D'Odorico (1986). The 30 Doradus region has attracted the most attention (see Walborn's review in this volume), and studies of resolved stars, especially Wolf-Rayet stars, have been extended to other galaxies in the Local Group. Unfortunately, surveys of the visually brightest stars in galaxies tend to undersample the most massive, blue stars, which preferentially reside in H II regions, and IMFs derived from such data are subject to large systematic errors (Massey 1985, Blaha 1989). In order to circumvent this problem, Massey, Garmany, and their collaborators have obtained UBV photometry and followup classification spectra for complete samples of stars in a large sample of H II regions in the Magellanic Clouds, and these data are used to construct complete upper H-R diagrams and mass functions. To date three H II regions in the LMC and SMC have been measured (Massey et al. 1989a, b), and all are consistent with a power-law mass function with slope $d\psi(m)/dm = -2.8 \pm 0.2$ over the mass range 10–85 $M_\odot$. This is close to the IMF found for OB stars in the solar neighborhood (Garmany et al. 1982, Scalo 1986). The interpretation of these data is somewhat dependent on the assumed ages of the ionizing associations; if the clusters contain a significant population of stars older than 3 *Myr*, then the true

IMF will be shallower than the present-day mass function measured by Massey *et al.* Preliminary observations of 30 Doradus by Melnick (1985) and Parker (this meeting) also suggest that the IMF in this region may be considerably shallower.

For more distant GEHRs the individual ionizing stars cannot be resolved, and any information on the IMF must be extracted from the composite spectrum. The information which is derived in this way is much less detailed, but regions covering a much wider range of abundances and galactic environments can be studied. The stellar mix in the H II region influences several properties of the integrated spectrum, including the shape of the stellar continuum longward of the Lyman break, especially in the UV (Rosa *et al.* 1984, Rosa and Arsenault 1990), the equivalent widths (EWs) of the Balmer emission lines, which measure the ratio of stellar ionizing photons shortward of the Lyman limit to the flux in the visible spectrum, and the degree of ionization in the nebula, which provides information on the shape of the ionizing spectrum in the 13.6–54 eV (228–912 Å) region.

It has long been recognized that GEHRs in disk galaxies exhibit not only well-known excitation gradients, but also parallel gradients in the ionization and in the Balmer-line EWs (*e.g.*, Searle 1971, McCall *et al.* 1985). The excitation gradient is primarily caused by an underlying heavy-element gradient, through its effect on the cooling of the H II regions, but the ionization and EW gradients must be caused by some other mechanism, because they involve processes which are at energies more than an order of magnitude larger than the excitation or Boltzmann energies in the ionized gas. The most likely mechanism is a decrease in the mean color temperature of the ionizing star clusters at higher metallicity. The required temperature gradient is quite large; for example, in M101 the mean ionization temperature decreases from roughly 50,000 K in the outermost H II regions ($Z \simeq 0.2 Z_\odot$) to 37,000 K in the inner regions, with $Z \simeq 1$–$2 Z_\odot$ (*e.g.*, Shields and Searle 1978).

The observational support for this result has been strengthened considerably in the past few years. The original interpretation of the nebular ionization gradient (Shields 1974, Shields and Tinsley 1976) was based on the behavior of O^{++}/O^+ alone, and as pointed out by Evans and Dopita (1987), these gradients could be produced by changes in nebular structure, specifically the nebular "ionization parameter," with little or no need for a change in the stellar ionizing spectrum. However, a more rigorous treatment of the ionization, based on observations of O^{++}/O^+ and S^{++}/S^+, confirms the presence of a decrease in mean ionization temperature with increasing metallicity (Shields and Searle 1978, Vilchez and Pagel 1988, Garnett 1989). Campbell (1988) has modelled the oxygen line ratios and electron temperatures for a sample of low-abundance H II galaxies, and reaches similar conclusions.

The physical origin of this change in the stellar radiation field is less clear, however. Shields and Tinsley (1976) suggested that it is caused by a systematic decrease in the stellar mass limit with increasing metallicity, from approximately 100 $M_\odot$ in the outer disk of M101 to $\sim$25–40 $M_\odot$ in the inner, metal-rich disk. Other causes cannot be ruled out, however. These include an increase in ultraviolet extinction with metallicity (Sarazin 1976), an increase in stellar line blanketing (Balick and Sneden 1976), or a reduction in the effective temperatures or main-sequence lifetimes of massive stars with increasing abundance (Rosa and Arsenault 1990). The latter may be the most serious competitor to the variable-IMF picture. Rosa and Arsenault (1990) have used Maeder's most recent evolutionary tracks to model the evolution of massive star clusters with different compositions and ages, and they conclude that most of the changes in nebular equivalent widths and UV–optical continuum colors can be reproduced with a metallicity-independent IMF. It is not yet clear whether this model can reproduce all of the spectral gradients seen in GEHRs, but it demonstrates the need for including the effects of metallicity on evolution in any attempted analysis of the IMF in these regions. Whatever their interpretation, the observed spectral gradients imply that star formation rates derived from nebular spectra are sensitive to metallicity at a significant level.

Before leaving the subject of the stellar content of GEHRs, I briefly mention one other area of current interest, namely the discovery of nebular He II $\lambda 4686$ emission in several nearby GEHRs, including 3 regions in the Magellanic Clouds (Pakull and Angebault 1986, Stasinska et al. 1986, Garnett et al. 1990), one in IC 1613 (Garnett et al. 1990), and the main H II complex in the dwarf compact galaxy I Zw 18 (Skillman and Kennicutt, in preparation). He II $\lambda 4686$ emission has occasionally been reported in the spectra of several high-excitation GEHRs and blue compact dwarfs, but most of the previous observations were made at moderate to low dispersion with 1D detectors, and in such cases it is difficult to distinguish nebular features from stellar He II emission (Bergeron 1977). With the use of long-slit CCD spectroscopy or interference-filter images, it is now possible to unambiguously test for nebular emission, and to determine precisely the spatial extent and location of the high-ionization regions.

The nature of the ionizing source for the He II regions is not completely clear. Normal stars ($T_{eff} \leq 50,000 K$) produce negligible flux below the He II edge, so the observation of doubly ionized helium implies the existence of very hot stars or some nonstellar ionization process. If the gas is photoionized by hot stars, the observed level of helium ionization requires stars with temperatures of at least 70,000 K (Stasinska et al. 1986, Garnett et al. 1990). Wolf-Rayet stars are prime candidates; two of the H II regions above contain luminous W-R stars, and spectroscopic surveys of H II regions surrounding isolated W-R stars have revealed at least 6 examples of He II emission (Pakull 1990, Niemela 1990). However, stellar photoionization may not be the sole source of the emission. At least one of the regions (N159F in the LMC) appears to be photoionized by radiation from an X-ray source (Pakull and Angebault 1986, Pakull and Motch 1989), and shock ionization is yet another possibility in some regions. Regardless of the explanation, however, these observations demonstrate the existence of very-high-ionization regions in at least some normal GEHRs. It is interesting to note that all of the regions found so far possess metal abundances well below solar, and it would be interesting to test whether low metallicity is a necessary condition.

5. KINEMATICS, DYNAMICS, EVOLUTION

As discussed in the introduction, GEHRs offer ideal environments for studying the interaction and feedback between massive stars and the interstellar medium. Smith and Weedman (1971) first demonstrated that the GEHRs in the Magellanic Clouds are characterized by highly supersonic velocity dispersions, much higher than in most small H II regions. Melnick (1977) extended this work to GEHRs in other galaxies, and discovered correlations between the velocity dispersions and the diameters and luminosities of the H II regions. The possibility of using these relations as extragalactic distance indicators has stimulated a large body of observations (e.g., Terlevich and Melnick 1981, Hippelein 1986, Roy and Arsenault 1986). The rms dispersions of the largest H II regions are typically 15–40 $km\ s^{-1}$, and this supersonic turbulence probably accounts for the highly filamentary structure of most GEHRs.

The energy source for these motions, and the physical origin of the L-σ and D-σ

correlations, are not completely understood. Terlevich and Melnick (1981) originally argued that the dispersions reflect virial motion of H II regions and their associated star clusters, while others (*e.g.*, Gallagher and Hunter 1983) have argued that large-scale gas flows driven by stellar winds from the OB associations offer a more plausible explanation.

Clues to the nature and origin of the motions come from spatially resolved maps of the velocity structure of individual GEHRs. These show direct evidence of stellar-wind and supernova-driven flows (*e.g.*, Gallagher and Hunter 1983, Meaburn 1984, Chu and Kennicutt 1990). The most revealing observations are high-dispersion echelle and Fabry-Perot spectroscopy of the H II regions in the Magellanic Clouds. These show that a GEHR is typically characterized by a series of discrete shells, with velocities of order 10–30 km s^{-1} (Meaburn 1984). The largest shells often coincide with holes around OB associations in the GEHRs (*e.g.*, Meaburn *et al.* 1989).

The most fascinating features are high-velocity shells ($v_{exp} = 100$–300 km s^{-1}), which kinematically resemble fast stellar-wind-blown bubbles or supernova remnants (Chu and Kennicutt 1988). The core of 30 Doradus contains a large number of such shells, which occur both as isolated bubbles and as large, composite structures. There is considerable evidence linking these fast shells to young stellar-wind-driven structures, supernova remnants, or combinations of these. The largest fast shells in 30 Doradus, for example, coincide with *Einstein* X-ray sources and nonthermal radio sources (Chu and Kennicutt 1990, Ye 1989). Their kinetic energies are sufficiently high that they are best explained as arising from the stellar winds and supernovae from several stars. These may well represent the nearest examples of the "supershells" predicted by Mac Low and McCray (1988) and others.

There is less direct evidence for significant numbers of remnants in other GEHRs. Multi-frequency radio-continuum observations of large numbers of GEHRs in M51 and M33 show that the average spectral index is significantly steeper than expected for pure nebular free-free emission, probably the result of unresolved supernova remnants inside the H II regions (van der Hulst *et al.* 1988; van der Hulst, private communication). Ye (1989) identified radio SNR candidates within several H II regions in the LMC and SMC, and an unusually luminous nonthermal source (several hundred times brighter than Cassiopeia A) has been identified in NGC 5471, the most luminous GEHR in M101 (Skillman 1986). The radio source coincides with an unusually luminous, fast-expanding shell in Hα (Chu and Kennicutt 1986).

The total kinetic energy in the shells is comparable to that in the low-velocity "turbulent" component, and hence it is possible that massive stars provide the bulk of the kinetic energy in the surrounding H II region. The near coincidence between this energy and the gravitational binding energy of the complex, pointed out by Terlevich and Melnick (1981), is intriguing, as it suggests that 30 Doradus is nearly in a self-regulating star-formation regime, similar to that postulated for galactic disks by Dopita (1985) and others.

This work was supported in part by NSF Grant AST-8996123. I also wish to acknowledge the many contributions from my collaborators in this work, especially Paul Hodge, You-Hua Chu, Howard French, and Evan Skillman.

REFERENCES

Balick, B., and Sneden, C. 1976, *Ap. J.*, **208**, 336.
Benvenuti, P., Casini, C., and Heidmann, J. 1982, *M.N.R.A.S.*, **198**, 825.
Bergeron, J. 1977, *Ap. J.*, **211**, 62.
Blaha, C. A. 1989, Ph.D. thesis, University of Minnesota.
Bonnarel, F., Boulesteix, J., and Marcelin, M. 1986, *Astr. Ap. Suppl.*, **66**, 149.
Caldwell, C. N., Kennicutt, R. C., Phillips, A., and Schommer, R. A. 1990, *Ap. J.*, submitted.
Campbell, A. 1988, *Ap. J.*, **335**, 644.
Caplan, J., and Deharveng, L. 1986, *Astr. Ap.*, **155**, 297.
Chu, Y.-H., and Kennicutt, R. C. 1986, *Ap. J.*, **311**, 85.
_____. 1988, *A. J.*, **95**, 1111.
_____. 1990, in *The Magellanic Clouds and Their Dynamical Interaction with the Milky Way*, (Dordrecht: Kluwer), in press.
Chu, Y.-H., and Mac Low, M.-M. 1990, *Ap. J.*, in press.
Copetti, M. V. F. 1990, *Astr. Ap.*, **229**, 533.
Copetti, M. V. F., Pastoriza, M. G., and Dottori, H. A. 1985, *Astr. Ap.*, **152**, 427.
Courtes, G., Petit, H., Sivan, J. P., Dodonov, S., and Petit, M. 1987, *Astr. Ap.*, **174**, 28.
Deharveng, L., Caplan, J., Lequeux, J., Azzopardi, M., Breysacher, J., Tarenghi, M., and Westerlund, B. 1988, *Astr. Ap. Suppl.*, **73**, 407.
Dinerstein, H. L. 1990, in *The Interstellar Medium in External Galaxies*, ed. H. A. Thronson and J. M. Shull (Dordrecht: Kluwer), in press.
Dopita, M. A. 1985, *Ap. J. (Letters)*, **295**, L5.
Dottori, H. A., and Bica, E. L. D. 1981, *Astr. Ap.*, **102**, 245.
Gallagher, J. S., and Hunter, D. A. 1983, *Ap. J.*, **274**, 141.
Garmany, C. D., Conti, P. S., and Chiosi, C. 1982, *Ap. J.*, **263**, 777.
Garnett, D. R. 1989, *Ap. J.*, **345**, 282.
Garnett, D. R., Kennicutt, R. C., Chu, Y.-H., and Skillman, E. D. 1990, in preparation.
Hippelein, H. H. 1986, *Astr. Ap.*, **160**, 374.
Hodge, P. W. 1969, *Ap. J. Suppl.*, **18**, 73.
_____. 1974, *Ap. J. Suppl.*, **27**, 113.
Hodge, P. W., and Kennicutt, R. C. 1983, *A. J.*, **88**, 296.
Hunter, D. A., and Gallagher, J. S. 1985, *Ap. J. Suppl.*, **58**, 533.
Israel, F. P. 1976, Ph.D. thesis, Leiden University.
Israel, F. P., and Kennicutt, R. C. 1980, *Astrophys. Lett.*, **21**, 1.
Kaufman, M., Bash, F. N, Kennicutt, R. C., and Hodge, P. W. 1987, *Ap. J.*, **319**, 61.
Kennicutt, R. C. 1983, *Ap. J.*, **272**, 54.
_____. 1984, *Ap. J.*, **287**, 116.
_____. 1988, *Ap. J.*, **334**, 144.
Kennicutt, R. C., and Chu, Y.-H. 1988, *A. J.*, **95**, 720.
Kennicutt, R. C., Keel, W. C., and Blaha, C. A. 1989, *A. J.*, **97**, 1022.
Kennicutt, R. C., Edgar, B. K., and Hodge, P. W. 1989, *Ap. J.*, **337**, 761.
Long, K. S., Helfand, D. J., and Grabelsky, D. A. 1981, *Ap. J.*, **248**, 925.
Massey, P. 1985, *Pub. A.S.P.*, **97**, 5.

Massey, P., Garmany, C. D., Silkey, M., and DeGioia-Eastwood, K. 1989a, *A. J.*, **97**, 107.
Massey, P., Parker, J. W., and Garmany, C. D. 1990, *A. J.*, **98**, 1305.
Mac Low, M.-M., and McCray, R. A. 1988, *Ap. J.*, **324**, 776.
McCall, M. L., Rybski, P. M., and Shields, G. A. 1985, *Ap. J. Suppl.*, **57**, 1.
McGregor, P. J., and Hyland, A. R. 1981, *Ap. J.*, **250**, 116.
Meaburn, J. 1984, *M.N.R.A.S.*, **211**, 521.
_____. 1988, *M.N.R.A.S.*, **235**, 375.
Meaburn, J., Solomos, N., Laspias, V., and Goudis, C. 1989, *Astr. Ap.*, **225**, 497.
Melnick, J. 1977, *Ap. J.*, **213**, 15.
_____. 1985, in *Star Forming Dwarf Galaxies*, ed. D. Kunth, T. X. Thuan, and T. T. T. Van (Gif sur Yvette: Editions Frontieres), p. 171.
Niemela, V. S. 1990, in *Wolf-Rayet Stars and Interrelations with Other Massive Stars in Galaxies*, ed. K. A. van der Hucht and B. Hidayat, (Dordrecht: Kluwer), in press.
O'Dell, C. R., and Castañeda, H. O. 1984, *Ap. J.*, **283**, 158.
Pakull, M. W. 1990, in *Wolf-Rayet Stars and Interrelations with Other Massive Stars in Galaxies*, ed. K. A. van der Hucht and B. Hidayat, (Dordrecht: Kluwer), in press.
Pakull, M. W., and Angebault, L. P. 1986, *Nature*, **322**, 511.
Pakull, M. W., and Motch, C. 1989, *Nature*, **337**, 337.
Pellet, A., Astier, N., Vialc, A., Courtes, G., Maucherat, A., Monnet, G., and Simien, F. 1978, *Astr. Ap. Suppl.*, **31**, 439.
Rosa, M., and Arsenault, R. 1990, private communication.
Rosa, M., and D'Odorico, S. 1986, in *Luminous Stars and Associations in Galaxies*, ed. C. W. H. De Loore, A. Willis, and P. Laskarides (Dordrecht: Reidel), p. 355.
Rosa, M., Joubert, M., and Benvenuti, P. 1984, *Astr. Ap. Suppl.*, **57**, 361.
Roy, J.-R., and Arsenault, R. 1986, *Ap. J.*, **302**, 579.
Sarazin, C. L. 1976, *Ap. J.*, **208**, 323.
Scalo, J. S. 1986, *Fund. Cosmic Phys.*, **16**, 1.
Searle, L. 1971, *Ap. J.*, **168**, 327.
Shields, G. A. 1974, *Ap. J.*, **193**, 335.
_____. 1990, *Ann. Rev. Astr. Ap.*, **28**, 525.
Shields, G. A., and Searle, L. 1978, *Ap. J.*, **222**, 821.
Shields, G. A., and Tinsley, B. M. 1976, *Ap. J.*, **203**, 66.
Skillman, E. D., and Israel, F. P. 1988, *Astr. Ap.*, **203**, 226.
Smith, M. G., and Weedman, D. W. 1971, *Ap. J.*, **169**, 271.
Stasińska, G., Testor, G., and Heydari-Malayeri, M. 1986, *Astr. Ap.*, **170**, L4.
Strom, S. E., Strom, K. E., Grasdalen, G. L., and Capps, R. W. 1974, *Ap. J. (Letters)*, **193**, L7.
Terlevich, R., and Melnick, J. 1981, *M.N.R.A.S.*, **195**, 839.
van der Hulst, J. M., Kennicutt, R. C., Crane, P. C., and Rots, A. H. 1988, *Astr. Ap.*, **195**, 38.
Vilchez, J. M., and Pagel, B. E. J. 1988, *M.N.R.A.S.*, **231**, 257.
Wang, Q., and Helfand, D. 1990, *Ap. J.*, in press.
Weedman, D. W., Feldman, F. R., Balzano, V. A., Ramsey, L. W., Sramek, R. A., and Wu, C.-C. 1981, *Ap. J.*, **248**, 105.
Wray, J. D., and de Vaucouleurs, G. 1980, *A. J.*, **85**, 1.
Ye, T. 1989, Ph.D. thesis, University of Sydney.

DISCUSSION

R. Kudritzki: I have two comments: (1) much of your analysis depends on the "hardness ratio" of the ionizing continuum as inferred from the strengths of the corresponding recombination lines. The calculations that I have presented do now indicate that this ratio might depend also on metallicity. For instance, the number of photons ionizing O II does depend on metallicity, at least for $T_{eff} \lesssim 42{,}000$ K. (2) You mentioned that now 5 cases are known, where He II $\lambda 4686$ is seen as a recombination line and that this would require $T_{eff} \gtrsim 73{,}000$ K for the ionizing stars. I don't think that you need temperatures that high, because our new "Unified Model Atmospheres" predict a factor of 200 to 1000 more He II photons shortward of 228 Å than the older hydrostatic model atmospheres. It would be interesting to carry out careful, narrow-band $\lambda 4686$-imaging of those H II regions that are well known to contain O3 stars.

N. Panagia: I believe that dust absorption cannot have any major effect on the hardness of the radiation field in H II regions. In fact, one should require that the dust absorption cross section increase steeply with frequency shortward of the Lyman edge and there is no material of any size which can do that. On the other hand, a higher metal abundance can reduce the hardness of the UV spectrum in three independent ways: (1) line blanketing, which is most important shortward of about 600 Å; (2) higher interior opacity, which makes the stellar radius bigger for equal luminosity and mass; (3) upper cutoff of the IMF, which is predicted to decrease with metal abundance. This latter effect seems to be the dominant one in our Galaxy (Panagia 1979, in *Radio Recombination Lines*, Reidel, pp. 99–105).

S. Heap: Do you observe any systematic differences between the H II regions in flocculent and grand-design spirals?

Kennicutt: In galaxies with strong grand-design patterns, the H II regions in the spiral arms tend to be brighter than those located between the arms. However, the global levels of Hα emission and star formation do not appear to be significantly different in the two types of spirals.

J. Young: You showed that the luminosity in giant H II regions per unit blue luminosity increases from Sa $\to$ Sc, but the L_B's include light from the bulge. In order to determine the variations in star-forming properties of disks, it is important to normalize by a disk luminosity or disk property. We have done this in normalizing global SFR's by the available H$_2$ mass, and find no variation in the mean SFR per unit M(H$_2$) for Sa $\to$ Scd.

Kennicutt: You are correct in that the bulge luminosity is included in the normalizations. However, normalizing to the disk luminosity alone reduces the gradient with type only by a factor of two or so, much less than the hundredfold gradient in GEHR frequency.

SUPERNOVAE AND SUPERNOVA REMNANTS IN STARBURSTS

Roger A. Chevalier
University of Virginia
Department of Astronomy
P.O. Box 3818
Charlottesville, VA 22903
USA

Abstract. Type Ib and Type II supernovae are of special interest for starburst galaxies because they are thought to have massive star progenitors. Type II supernovae generally have red supergiant progenitors (SN 1987A being an exception), while Type Ib's have more compact progenitors so that they have characteristic circumstellar surroundings and characteristic radio emission. Starburst galaxies are thought to have high supernova rates, but these events are difficult to observe optically because of high dust obscuration and high background. These problems are reduced at radio wavelengths and, on the basis of radio evolution, it is likely that there was a massive Type II supernova in the center of M82 in the mid-1950's and a Type Ib event in late 1980. The high supernova rate generates a high interstellar pressure in nuclear starbursts, so that the evolution of the circumstellar medium around massive stars is modified. In particular, the red supergiant wind is stopped by the interstellar pressure at a diameter of a few 0.1's of a pc, where a dense shell is formed. The interaction of the supernovae with such dense shells may give rise to the compact radio sources that are frequently observed in starburst galaxy nuclei. Because the filling factor of cool gas is small, supernovae are most likely to deposit their energy in hot, low density gas that is easily able to escape from the galactic nucleus in a wind. There is good evidence for such a wind in M82. The high pressure in the nuclear region can drive a shock front into the surrounding galactic disk. The ring of neutral gas observed in M82 may be formed in this way; the expansion of the ring may be accompanied by a propagating starburst.

1. INTRODUCTION

A starburst galaxy is one with a region of strong star formation, including massive stars; the deaths of the more massive stars are expected to result in supernovae. An example is the nuclear region of M82, where Rieke *et al.* (1980) have estimated a supernova rate of 0.3 yr^{-1}. Despite the high expected supernova rates, very few optical supernovae have been discovered in starburst regions. This can be attributed to two factors: high dust obscuration and the difficulty in detecting a variable source against a high surface brightness background. These problems do not exist in the radio regime, so that this regime appears to be a particularly favorable observational handle on supernova properties. Section 2 reviews the properties of the basic supernova types, with an emphasis on their radio emission. The discussion includes young supernova remnants. The interaction of supernovae with their surroundings for the special conditions present in starburst galaxies are discussed in section 3. The expected and observed properties of supernovae in starburst galaxies are discussed in section 4. Section 5 deals with the effects of supernovae on the large-scale flows in starburst galaxies. In this paper the value of the Hubble constant is taken to be 75 km s^{-1}Mpc^{-1}.

2. SUPERNOVAE AND YOUNG SUPERNOVA REMNANTS

There are now known to be three major classes of supernovae: Type Ia (SN Ia), Type Ib (SN Ib), and Type II (SN II). The distinction between the SN I and the SN II is that the latter show hydrogen lines in their spectra, while the former do not. The distinction between SN Ia and SN Ib near maximum light is more subtle and was not clearly recognized until the mid-1980's (Wheeler and Levreault 1985). The primary spectroscopic difference near maximum light is the lack of the $\lambda 6150$ feature (presumably due to Si II $\lambda 6355$) in the spectra of SN Ib. At late times (≥ 6 months), the spectra are very different: the SN Ia show strong emission lines of FeII and FeIII, while the SN Ib show emission lines of OI, OII, CaII, NaI, MgI, and Fe group elements. The late spectra of SN Ib resemble the late spectra of SN II, but without the hydrogen lines. SN 1987K appeared to be a normal SN II near maximum light, but at late times, the hydrogen lines disappeared and the spectrum was like that of a SN Ib (Filippenko 1988).

It appears that SN Ia are associated with an old stellar population, so they are not of particular interest with regard to starburst galaxies. Before the recognition of SN Ib, there did appear to be a correlation of SN I with star formation rates in galaxies; however, it now seems likely that the correlation is due to SN Ib and that the SN Ia rate is relatively independent of the spiral galaxy type (van den Bergh and Tammann 1991). The explosion mechanism for SN Ia appears to be the Carbon deflagration of a massive white dwarf in a binary (Woosley and Weaver 1986), but the evolution leading up to the explosion is poorly understood. It is possible that some relatively young binary systems can produce SN Ia. SN Ia are optically the brightest type, on average, with $M_B = -18.9$. They have not produced detectable radio sources (Weiler et al. 1986).

SN Ib and SN II both show a clear correlation with HII regions and with star formation rates in galaxies. Van den Bergh and Tammann (1991) find that the rates for both increase towards late type spirals at a constant ratio of 5 for the SN II to the SN Ib rate. The ratio is higher than other estimates because of the recognition of a population of faint SN II. The conclusion is that SN Ib are massive stars that have lost their hydrogen envelopes prior to the explosion; this could occur either as result of strong mass loss by a single star, or mass loss and transfer in a close binary system. Studies of stellar evolution with mass loss, combined with empirical information, indicate that stars more massive than about 25 $M_\odot$ lose their hydrogen envelopes during their lifetimes (Chiosi and Maeder 1986). Stars with masses above 60 $M_\odot$ remain blue supergiants throughout their life, while those in the range 25 – 60 $M_\odot$ become red supergiants before ending their lives as blue supergiants. While these stars are attractive candidates for the SN Ib, model explosions of massive cores give broader light curves than are observed (Ensman and Woosley 1988). Nomoto et al. (1990) find that the light curves are best fit in a model with a 3.7 $M_\odot$ core ejecting 2 $M_\odot$ in an explosion. The original stellar mass in this case is 13 $M_\odot$, so that the mass loss probably occurs in a binary system. Such an explosion can plausibly produce the 0.2 $M_\odot$ of ^{56}Ni that is needed for the light curve of a SN Ib. For lower mass stars (about 10 $M_\odot$), very little ^{56}Ni may be formed in the explosion; this may also occur for a high mass star (K. Nomoto, private communication). Since the light emitted near maximum comes from radioactivity for the explosion of a compact star, such an event would be faint. Initial discussions of the explosion of massive stars without their envelopes were based on this assumption (Chevalier 1976) and it is possible that such explosions do occur but have not yet been observed. The Cas A supernova is a candidate.

SN Ib show a spread of about 2 magnitudes in their absolute magnitudes near maximum light, but the statistics are too poor to tell whether this is due to variable extinction or to intrinsic differences between the events. SN II clearly show a range of properties, with M_B varying from -14.3 for SN 1987A to -19.8 for SN 1979C. The light curve shapes also show considerable variation, although most SN II light curves show late time tails that imply the ejection of about $0.1 - 0.6$ $M_\odot$ of ^{56}Ni (Young and Branch 1989). The reason for the low luminosity of SN 1987A appears to be the small radius of the progenitor star; the light curves of other SN II are best modeled by the explosions of red supergiants. While the explosion of SN 1987A as a blue supergiant may be related to the low metallicity in the LMC, a detailed understanding of this type of SN II progenitor is lacking.

TABLE 1. RADIO SUPERNOVAE

SN	Type	Ratio to Cas A at 6 cm	α	$\dot{M}/v_w$ $(M_\odot/\text{yr})/(\text{km/s})$
1979C	II	~ 250	-0.72 ± 0.05	1×10^{-5}
1980K	II	~ 15	-0.50 ± 0.06	3×10^{-6}
1983N	Ib	~ 125	-1.03 ± 0.06	5×10^{-7}
1986J	II	~ 700	-0.67 ± 0.06	2×10^{-5}
1987A	II	~ 0.02	-1.0 ± 0.1	1×10^{-8}

Radio emission from supernovae provides useful constraints on their properties. The circumstellar interaction model for the emission (Chevalier 1982), which was developed for the SN II events SN 1979C and SN 1980K, has been strengthened by the observation of other events. For SN 1979C and SN 1980K themselves, observations at X-ray, ultraviolet, and infrared wavelengths also show evidence for strong circumstellar interaction (for reviews, see Fransson 1986, Chevalier 1990). Table 1 gives a summary of the basic data on the radio supernovae for which it is possible to estimate the density of the circumstellar matter from the turn-on of the radio flux. The data are from Weiler et al. (1986), Weiler, Panagia, and Sramek (1990), Chevalier (1990), and references therein. The spectral index in the optically thin phase, α, is defined by flux $\propto \nu^\alpha$. In general, the derived mass loss for the SN II progenitors is consistent with strong mass loss from a red supergiant star. The low radio luminosity and early turn-on of the radio emission from SN 1987A can be attributed to the low density wind surrounding a blue supergiant star. SN 1986J is currently the brightest radio supernova. It was not observed optically near maximum light, but is now seen as a faint source with emission lines, especially Hα (Rupen et al. 1987). The radio emission in this case did not show the exponential rise seen in other radio supernovae, but showed a more gradual, perhaps power law, increase (Weiler, Panagia, and Sramek 1990). The sharp rise can be attributed to free-free absorption external to the emitting shell; the more gradual rise implies that the absorbing material is mixed into the emitting region. One possibility is that the forward shock front is irregular so that different parts of the shell become visible at different times. VLBI observations of the supernova show that the large scale structure is not spherically symmetric (Bartel, Rupen, and Shapiro 1989).

While SN II show a large dispersion in radio luminosity (SN 1986J at maximum was roughly a factor 10,000 more luminous than SN 1987A), the SN Ib show a relatively small dispersion. SN 1983N (Sramek, Panagia, and Weiler 1984), SN 1984L (Panagia, Sramek, and Weiler 1986), and SN 1990B (Sramek, Weiler, and Panagia 1990) showed a range of about a factor of 3 in maximum luminosity. SN 1985F was not detected as a radio source (Weiler et al. 1986), but was only observed at late time. The reason for the small dispersion is not clear; the implication is that the stellar evolution leading up to the explosion gives a similar circumstellar medium in each case.

Our current understanding suggests that dense circumstellar matter is required for a bright radio supernova. SN 1987A has been faint in the radio, but ultraviolet and optical observations show that there is dense gas within a light-year of the supernova (Fransson et al. 1989; Wampler et al. 1990). This gas is thought to have been lost in a previous red supergiant phase and some of it has been swept into a dense shell by the blue supergiant wind. When the supernova shock wave does interact with the shell in 10 to 15 years, a bright radio source is expected (Chevalier and Liang 1989). While SN 1987A appears to have exploded with most of its hydrogen envelope (Arnett et al. 1989), other stars may explode as blue supergiants as a result of strong mass loss. These Wolf-Rayet stars may have a phase of dense mass loss and shells are sometimes observed around these stars. The supernova remnant Cas A, the most luminous radio source in the Galaxy, does show evidence for shell interaction (Chevalier and Liang 1989) and it is believed that the progenitor star was a WN7–WN9 Wolf-Rayet star (Fesen, Becker, and Blair 1987). The radius of the shell surrounding the explosion was about 1.6 pc.

3. THE SURROUNDINGS OF SUPERNOVAE

The properties of a young supernova remnant, particularly its radio emission, depend on its immediate surroundings. The case of a starburst region is different from that of the interstellar medium like that in most of the Galaxy because of the large pressure generated by the high supernova rate. To be specific, I consider conditions in the central region of M82, a well observed starburst galaxy. Chevalier and Clegg (1985) made a theoretical estimate of the central conditions, based on a supernova rate of 0.3 yr^{-1}. For a freely expanding wind with an energy input of $\dot{E} = 10^{43}\alpha$ ergs s^{-1} and a mass input of $\dot{M} = \beta M_\odot \text{yr}^{-1}$, the central pressure and density are $p/k = 5.6 \times 10^7 \alpha^{1/2} \beta^{1/2} \text{cm}^{-3}\text{K}$ and $\rho = 1.2 \times 10^{-25} \alpha^{-1/2} \beta^{3/2}\text{g cm}^{-3}$. These estimates are the peak values; at the edge of the star forming region, both physical quantities drop by a factor of about 3. The pressure estimate can be compared to a Galactic value of about 3000 cm^{-3}K.

There is considerable uncertainty in the parameters that go into the theoretical estimate, so it is useful to have estimates from observations of M82. The [SII] optical doublet ratio gives a pressure close to that given above (O'Connell and Mangano 1978; Chevalier and Clegg 1985), but it could be an overestimate because of overpressure in a young HII region. X-ray observations (Watson, Stanger, and Griffiths 1984; Schaaf et al. 1989) give a similarly high pressure, but there are uncertainties in the filling factor of the X-ray emitting gas. Estimates of p/k from the ratio of [OIII] fine structure lines are in the range $10^6 - 10^7 \text{cm}^{-3}\text{K}$ (Lugten et al. 1986). In numerical estimates below, I take a value $p/k = 10^7 \text{cm}^{-3}\text{K}$.

One implication of the high pressure is that the filling factor, f, of cool gas is small. For example, Lo et al. (1987) note that the mass of molecular gas in the nuclear region is about $6 \times 10^7 M_\odot$ in the central 200×700 pc of M82, as derived from CO observations. At a temperature of 100 K, the volume of the gas is $1 \times 10^4 \text{pc}^3$, giving $f \approx 1 \times 10^{-4}$. By comparing the density needed to collisionally thermalize the rotational transition of CO with the observed average density, Nakai et al. (1987) deduce $f \leq 10^{-3}$. Lugten et al. (1986) derive similarly small filling factors for neutral and cool ionized gas. An implication of the small filling factors for the dense gas is that the supernovae are likely to explode in the hot, low density gas that fills most of the volume. Even in our Galaxy, massive stars appear able to escape the dense clouds in which they formed.

In section 2, we found that circumstellar interaction is responsible for radio emission from supernovae. A high pressure environment can modify the nature of the circumstellar medium. While on the main sequence, massive stars have fast winds that can create bubbles in the surrounding medium. Weaver et al. (1977) have described the evolution of such bubbles for the Galactic interstellar medium, showing that the bubble radii can grow to 10's of pc for typical wind parameters ($\dot{M} = 10^{-6} M_\odot \text{yr}^{-1}$ and wind velocity $v_w = 2000$ km s^{-1}). For the present case, the expanding bubble comes into pressure equilibrium with the surrounding medium within 10's of years, much less than the ages of the stars. Under these circumstances, a standing shock is present in the wind at a radius $r_{sh} = (0.881 \dot{M} v_w / 4\pi p)^{1/2}$ (Parker 1963). For the massive star wind parameters and $p/k = 10^7 \text{cm}^{-3}\text{K}$, $r_{sh} = 0.3$ pc. The wind gas is heated to a temperature of 5×10^7 K in the shock wave and its cooling time is greater than the stellar lifetime. The shocked wind properties are in fact similar to those of the major component of the interstellar medium.

When the massive star evolves to the red supergiant phase, mass loss continues, but in a slow dense wind. The wind again rapidly comes into pressure equilibrium with the surrounding medium. The shock radius is now

$$r_{sh} = 0.16 \left(\frac{\dot{M}}{5 \times 10^{-5} M_\odot \text{yr}^{-1}}\right)^{1/2} \left(\frac{v_w}{15 \text{ km s}^{-1}}\right)^{1/2} \left(\frac{p/k}{10^7 \text{cm}^{-3}\text{K}}\right)^{-1/2} \text{pc}$$

where parameters for a strong red supergiant wind have been used. The shocked gas is now able to cool on a timescale that is short compared to the stellar evolution time, so that a dense shell builds up behind the shock front. The temperature in the shell is determined by the ambient radiation field; for a temperature of 100 K, the density is 10^5cm^{-3}.

When the supernova explosion occurs, the expanding gas initially interacts with the circumstellar medium. It then interacts with the low density, high pressure medium, where it is decelerated by the high pressure. For an energy of 10^{51} ergs and the above pressure, pressure equilibrium is reached at a radius of about 16 pc (see also Schaaf et al. 1989). The supernova remnant does not enter a radiative phase before this occurs, so that the supernova energy is available to drive a galactic wind with high efficiency.

4. SUPERNOVAE IN STARBURST GALAXIES

While extinction and high background are problems for the discovery of optical supernovae in starburst galaxies, supernovae have been found in the hot spots of two galactic nuclei. Wood and Andrews (1974) describe the Type II event SN 1968L in the nuclear region of M83 (NGC 5253). The supernova could be followed for 2 months before it faded into the nuclear background. Lacques et al. (1980) found a brightening of a nuclear hot spot in NGC 2903 that was consistent with expectations for a SNII. Lebofsky and Rieke (1979) find patchy extinction toward this galactic nucleus. While Rieke et al. (1980) find an extinction of $A_V = 25$ mag toward the general galactic nucleus of M82, O'Connell and Mangano (1978) found an extinction of only $A_V = 5$ mag toward region 'A' in the nucleus; there are some holes in the thick dust layer. A supernova in region 'A' might be observed at 14 or 15 mag. While the extinction problem is small along particular lines of sight, the background problem remains. The supernova is likely to be fainter than the hot spot on which it is projected, so that accurate photometry is needed to detect the event. An alternate means of detection is the spectroscopic detection of broad lines; lines broader than several 100 km s^{-1} are likely to have a supernova origin (unless an active galactic nucleus is present). A broad Hα line is characteristic of SNII.

One way to alleviate the extinction problem is to observe in the infrared. Van Buren and Norman (1989) have explored this possibility and have suggested K-band imaging. The background problem remains and may be aggravated by the fact that supernovae emit most of their light at optical wavelengths, while starburst nuclei may have a substantial near-infrared flux from red supergiants. Van Buren and Norman find that high resolution imaging or very accurate photometry is needed. Based on observations of SN 1987A, they note the promise of infrared lines, in particular CoII $\lambda 10.5\mu m$, ArII $\lambda 7.0\mu m$, and NiII $\lambda 6.6\mu m$. The CoII line is of special interest because it is emitted by the radioactive isotope ^{56}Co and the line strength gives information on the amount of explosive processing.

The problems of observing supernovae at optical wavelengths in starbursts disappear at radio wavelengths, where the high surface brightness of nonthermal sources allows them to stand out against the background. More than 40 compact radio sources have been found in the nuclear region of M82 (Kronberg, Biermann, and Schwab 1985; Bartel et al. 1987; Kronberg 1988, and references therein) and they are thought to be associated with supernova events. The case is especially strong for two sources with a well-defined time evolution. One of these, 41.5 +59 7, is dubbed the "rapid turnoff" source by Kronberg and Sramek (1985) because of its rapid evolution. The source was observed with a 6 cm flux of 7.1 mJy in February 1981 and had dropped to less than 1.5 mJy fourteen months later (Kronberg and Sramek 1985). This evolution is consistent with the expected radio evolution of a SN Ib, if the source's age was less than about 100 days at the time of the first observation. The peak radio luminosity was then likely to be within a factor of a few of the typical peak luminosity of a SN Ib. The spectral index of 41.5 +59 7 was observed to be $\alpha = -1.1$; this steep spectral index is close to that observed for SN Ib (see section 2). The fact that the spectral index had a value characteristic of the optically thin regime implies that the supernova was at least 30 days old at the time of the radio observations. The radio evidence thus strongly suggests that a SN Ib occurred in the nuclear region of M82 sometime in late 1980 or early 1981. The fact that the SN II rate is about 5 times the SN Ib rate in spiral galaxies makes this result of particular interest.

The brightest compact radio source in M82, 41.9 +58, has been the subject of intense study, including VLBI mapping (Kronberg, Biermann, and Schwab 1985; Bartel et al. 1987). The spectral evolution has been followed since the early 1970's, when the source showed a strong low frequency turnover. The source has declined in flux since that time and the spectral turnover has moved to lower frequencies, as in the radio supernovae. Chevalier (1983) applied the circumstellar interaction model to this source and found that the evolution was roughly consistent with an age of 10 years at the time of the first observations in 1966 and a presupernova mass loss rate of $4 \times 10^{-4} M_\odot \text{yr}^{-1}$ for a wind velocity of 10 km s^{-1}. When combined with the VLBI size (Bartel et al. 1987), these mass loss properties imply a total circumstellar envelope mass of at least 6 $M_\odot$, i.e., a massive progenitor star is required. From the expansion of the VLBI image, Bartel (1988) estimated that the supernova occurred in 1955^{+10}_{-20}, consistent with the age estimate from the spectral evolution. The evolution of this source is closest to that of the radio supernova SN 1986J (Weiler, Panagia, and Sramek 1990). Neither object shows a very sharp low frequency turnover and the absorption can be attributed to free-free absorption by gas that is within the same volume as the emitting gas. Both cases are likely to be the Type II explosion of a massive star.

There is less information on the dozens of other compact sources in M82. Their diameters generally appear to be in the range of 0.1's of a pc to several pc (Kronberg, Biermann, and Schwab 1985; Bartel et al. 1987). The high radio luminosity at a substantial radius implies that there is dense gas at this radius. The cool interstellar gas is dense, but it was argued in section 3 that its filling factor is small, so a circumstellar origin is more probable given the large number of sources. Chevalier and Liang (1989) argued that interaction with a dense circumstellar shell is likely and suggested that the shell is formed by the interaction of a blue supergiant wind with the dense wind from a previous red supergiant phase; this requires that the supernovae typically explode as blue stars. The considerations of section 2 suggest that this is not necessary. A dense shell is formed naturally where the red supergiant wind is decelerated by the pressure of the interstellar medium; this occurs at about the correct radius to explain the observed structures. If the supernovae explode as red supergiants, they should be radio bright within a few years of the explosion, and then brighten again after decades when they interact with the dense shell. Most of the time would be spent in latter phase. In the case of a blue supergiant explosion, the supernova remains radio faint until the shell interaction.

It is clear that there is much to be gained from detailed radio monitoring of the compact sources in M82. While M82 provides a particularly good opportunity, the nuclei of other starburst galaxies, such as NGC3448 (Noreau and Kronberg 1987) and NGC253 (Turner and Ho 1985; Antonucci and Ulvestad 1988), are also rich in compact sources. Monitoring of a number of starburst galaxies should give a reasonable probability for the discovery of new radio supernovae. Huang et al. (1990) have recently discovered a probable radio supernova in the starburst galaxy NGC3690.

5. LARGE-SCALE FLOWS

The release of large amount of energy in a relatively small volume implies that high pressures are generated that are able to drive an outward flow. For an OB association in a galactic disk, the result is a 'superbubble' (Tenorio-Tagle and Bodenheimer 1988 and references therein). For the more extreme conditions in a starburst galaxy nucleus, it is plausible that a strong wind is driven out from the nuclear region (Chevalier and Clegg 1985). There is now a body of observational evidence that a nuclear wind is being driven out of the starburst nucleus of M82. First, observations with both the Einstein Observatory (Watson, Stanger, and Griffiths 1984; Fabbiano 1988) and EXOSAT (Schaaf et al. 1989) have shown evidence for an extended region of X-ray emission around M82. The X-ray spectrum is not well determined, but is consistent with thermal emission from gas with a temperature of a 9^{+9}_{-4} keV (Schaaf et al. 1989). Fabbiano (1988) finds that the emission extends to 8.5 kpc to one side of the nucleus of M82. The X-ray emission cannot be produced in a fast, freely expanding wind because of its low X-ray emissivity; it is plausible that the emission is produced where the wind runs into clouds (Chevalier and Clegg 1985). Radiative shocks in clouds can produce optical emission lines and there is evidence for such emission in the region surrounding M82 (McCarthy, Heckman, and van Breugel 1987). The decreasing pressure deduced from the emission line ratios is consistent with that expected in the wind model. While the wind is able to freely expand along the minor axis of M82, it is constrained by the dense gas in the galactic plane. The region of supernova activity is coincident with a region of disturbed CO emission that shows 'plumes' extending to 500 pc from the galactic plane (Lo et al. 1987; Nakai et al. 1987). On a larger scale, Bland and Tully (1988) find evidence for expanding 'walls' of a bipolar flow on both sides of the galaxy up to 3 kpc; the opening angle of the flow is about 35°. Thus, the evidence points to a well-developed bipolar flow driven from the nuclear region of M82.

Rieke, Lebofsky, and Walker (1988) have suggested that the evolution of such an energetic region can be divided into 6 phases: 1. massive stars form; 2. supernova explosions begin; 3. hot gas breaks out of the nucleus (e.g. NGC253); 4. cloud collisions and supernova shocks trigger star formation around the nucleus (e.g. M82); 5. decline in star formation; 6. starburst over, exceptionally low star formation, remnant X-ray sources (e.g. M31). The evolution through the first 3 phases has been modeled by numerical simulations (Tomisaka and Ikeuchi 1988; Mac Low, McCray, and Norman 1989), which show that the creation of a bipolar flow is plausible. The suggestion of a propagating starburst is interesting and is supported by the fact that the compact radio sources in M82 occupy a region with the same linear extent as that of the disturbed CO (Lo et al. 1987; Nakai et al. 1987).

To examine this idea more quantitatively, I examine a simple model for the propagation of the starburst region into the disk of M82. I assume that the gas density, ρ_o, is initially constant with radius and has a fixed height H. The star formation rate is initially proportional to the swept up mass, so that both $\dot{M}$ and $\dot{E}$ for the galactic wind are proportional to R^2, where R is the radius of the starburst region. The model of Chevalier and Clegg (1985) assumes spherical symmetry, so it is not exactly applicable, but the general scaling laws should still be valid. Thus the pressure, p, in the starburst region is $\propto \dot{M}^{1/2}\dot{E}^{1/2}/R^2$ and is expected to stay constant as the starburst propagates into the disk.

The interior pressure works against the ram pressure of the swept up mass and the gravitational force to build up a ring of matter at radius R. The equation of motion for the ring is

$$M_r \frac{dV}{dt} = 2\pi RH \left(p - \rho_o V^2\right) - M_r g(R) + M_r \frac{V_\phi^2}{R},$$

where M_r is the ring mass, $V = dR/dt$ is the radial velocity of the ring, V_ϕ is the rotational velocity of the ring, and g is the gravitational acceleration. The value of g can be estimated from the rotational velocities in the central region of M82; there is some gas spread through the central region. The rotation curve shows a strong gradient over the central 5 arcsec (80 pc) in radius, and is moderately flat outside of this point (O'Connell and Mangano 1978; Beck et al. 1978; Nakai et al. 1987); a rough estimat the rotational velocity, v_c, in the constant velocity region is 100 km s^{-1}. I assume a constant value of v_c throughout the central region, although this overestimates the gravitational acceleration in the very center. With $M_r = \pi R^2 2H\rho_0$, the equation of motion becomes

$$R\frac{dV}{dt} = 2\left(\frac{p}{\rho_0} - V^2\right) - v_c^2 + V_\phi^2.$$

The value of V_ϕ can be estimated from conservation of angular momentum of gas in the central region on the assumption of constant initial rotational velocity. The integrated angular momentum out to R is then $2\pi H R^3 \rho_0 v_c/3$ and the angular momentum in the ring is $\pi H R^3 \rho_0 V_\phi$; thus we have $V_\phi = 2v_c/3$. The known parameters in the equation of motion are all constant with radius, which implies that the velocity V is also a constant:

$$V = \left(\frac{p}{\rho_0} - \frac{5}{18}v_c^2\right)^{1/2}.$$

There is an upper limit to the density, $\rho_c = 18p/5v_c^2$, that allows the ring to propagate outward against the gravitational force. For $p/k = 10^7 \mathrm{cm}^{-3}$ K and $v_c = 100$ km s^{-1}, $\rho_c = 5 \times 10^{-23}$ g cm^{-3}. With $R = 350$ pc and $H = 100$ pc, the critical mass of the ring is $6 \times 10^7 M_\odot$; this radius is chosen because it characterizes the outer extent of the region of compact radio sources and disturbed CO emission. The gaseous ring in the central region of M82 is estimated to have a molecular mass of $6 \times 10^7 M_\odot$ (Lo et al. 1987) and an HI mass of $1 \times 10^7 M_\odot$ (Weliachew et al. 1984). However, the gas is not in a narrow ring at radius R, as in the model, but is spread over a range of radius; Nakai et al. (1987) find that the molecular gas covers the radius range $80 - 400$ pc from the center. The breadth of the ring could be partially due to gravitational forces. Another factor is that in regions of varying gas density, the expanding ring is expected to sweep up lower density gas, while dense clouds are left behind. Weliachew et al. (1984) note that some molecular clouds may be inside of the HI ring, and Lo et al. (1987) find that the HI ring is slightly outside of the molecular ring, although the two gaseous components are generally coincident. It appears that the gaseous ring is consistent with a propagating starburst, although a dynamical origin for the ring, as opposed to a hydrodynamic origin, cannot be ruled out.

The above model assumes that the star formation rate is proportional to the total swept up gaseous mass. The gaseous mass will eventually be depleted by star formation and by mass loss in a galactic wind; when this occurs, the pressure in the starburst is expected to drop with expansion. The depletion of gas may thus be a factor in the duration of the starburst phase. A decrease in density out from the center of the galaxy could be another factor.

The above discussion applies to M82 and to other relatively low luminosity starburst galaxies, like NGC253. More luminous starbursts involve mergers (Sanders et al. 1986) and the area of the starburst is probably determined by the area of the interaction. Lo et al. (1987) note that the infrared luminosity per unit area of starburst region is approximately constant over the whole range of starburst luminosities; this may be due to the saturation of the starburst at high star formation efficiency. In the wind model, the pressure in the starburst region is then independent of the starburst luminosity and the strength of the wind is proportional to the area of the starburst region. Heckman, Armus, and Miley (1987) have found evidence for galactic winds in infrared galaxies that are more powerful than the one in M82 by a factor 10–100. Winds of this magnitude may be present during galaxy formation and could play a role in the distribution of heavy elements surrounding galaxies. Our study of supernova-driven flows in starbursts is just beginning.

I am grateful to Robert O'Connell for useful discussions. This work was supported in part by NASA grant NAGW-764.

REFERENCES

Antonucci, R. R. J. and Ulvestad, J. S., 1988, *Ap. J. (Letters),* **330**, L97.
Arnett, W. D., Bahcall, J. N., Kirshner, R. P., and Woosley, S. E., 1989, *Ann. Rev. Astr. Ap.,* **27**, 629.
Bartel, N., 1988, in *Supernova Shells and Their Birth Events*, ed. W. Kundt (Berlin: Springer), p. 206.
Bartel, N., Rupen, M. R., and Shapiro, I. I., 1989, *Ap. J. (Letters),* **337**, L85.
Bartel, N., et al. 1987, *Ap. J.,* **323**, 505.
Beck, S. C., Lacy, J. H., Baas, F., and Townes, C. H., 1978, *Ap. J.,* **226**, 545.
Bland, J. and Tully, R. B., 1988, *Nature,* **334**, 43.
Chevalier, R. A., 1976, *Ap. J.,* **208**, 826.
—————. 1982, *Ap. J.,* **259**, 302.
—————. 1983, in *IAU Symp. 101, Supernova Remnants and Their X-Ray Emission*, ed. J. Danziger and P. Gorenstein (Dordrecht: Reidel), p. 71.
—————. 1990, in *Supernovae*, ed. A. G. Petschek (Berlin: Springer), p. 91.
Chevalier, R. A. and Clegg, A. W., 1985, *Nature,* **317**, 44.
Chevalier, R. A. and Liang, E. P., 1989, *Ap. J.,* **344**, 332.
Chiosi, C. and Maeder, A., 1986, *Ann. Rev. Astr. Ap.,* **24**, 329.
Ensman, L. M. and Woosley, S. E., 1988, *Ap. J.,* **333**, 754.
Fabbiano, G., 1988, *Ap. J.,* **330**, 672.
Fesen, R. A., Becker, R. H., and Blair, W. P., 1987, *Ap. J.,* **313**, 378.
Filippenko, A., 1988, *A. J.,* **96**, 1941.
Fransson, C., 1986, in *Radiation Hydrodynamics in Stars and Compact Objects*, ed. D. Mihalas and K. H. A. Winkler (Berlin: Springer), p. 141.
Fransson, C., Cassatella, A., Gilmozzi, R., Kirshner, R. P., Panagia, N., Sonneborn, G., and Wamsteker, W., 1989, *Ap. J.,* **336**, 429.
Heckman, T. M., Armus, L., and Miley, G. K., 1987, *A. J.,* **93**, 276.
Huang, Z. P., Condon, J. J., Yin, Q. F., and Thuan, T. X., 1990, *IAU Circ.*, No. 4988.
Kronberg, P. P., 1988, in *Galactic and Extragalactic Star Formation*, ed. R. E. Pudritz and M. Fich (Dordrecht: Reidel), p. 391.
Kronberg, P. P., Biermann, P., and Schwab, F. R., 1985, *Ap. J.,* **291**, 693.
Kronberg, P. P. and Sramek, R. A., 1985, *Science,* **227**, 28.
Lacques, P., Nieto, J.-L., Vidal, J.-L., Augé, A., and Despiau, R., 1980, *Nature,* **288**, 145.
Lebofsky, M. J. and Rieke, G. H., 1979, *Ap. J.,* **229**, 111.
Lo, K. Y., Cheung, K. W., Masson, C. R., Phillips, T. G., Scott, S. L., and Woody, D. P., 1987, *Ap. J.,* **312**, 574.
Lugten, J. B., Watson, D. M., Crawford, M. K., and Genzel, R., 1986, *Ap. J. (Letters),* **311**, L51.
Mac Low, M.-M., McCray, R., and Norman, M. L., 1989, *Ap. J.,* **337**, 141.
McCarthy, P. J., Heckman, T., and van Breugel, W., 1987, *A. J.,* **93**, 264.
Nakai, N., Hayashi, M., Handa, T., Sogue, Y., and Hasegawa, T., 1987, *Pub. Astr. Soc. Japan,* **39**, 685.

Nomoto, K., Shigeyama, T., Yanagita, S., Hayakawa, S., and Yasuda, K., 1990, in *Chemical and Dynamical Evolution of Galaxies,* ed. F. Ferrini, J. Franco, and F. Matteucci (Pisa: Giondini), in press.
Noreau, L. and Kronberg, P. P., 1987, *A. J.,* **93**, 1045.
O'Connell, R. W. and Mangano, J. J., 1978, *Ap. J.,* **221**, 62.
Panagia, N., Sramek, R. A., and Weiler, K. W., 1986, *Ap. J. (Letters),* **300**, L55.
Parker, E. N., 1963, *Interplanetary Dynamical Processes* (New York: Interscience).
Rieke, G. H., Lebofsky, M. J., and Walker, C. E., 1988, *Ap. J.,* **325**, 679.
Rieke, G. H., Lebofsky, M. J., Thompson, R. I., Low, F. J., and Tokunaga, A. T., 1980, *Ap. J.,* **238**, 24.
Rupen, M. P., van Gorkom, J. H., Knapp, G. R., Gunn, J. E., and Schneider, D. P., 1987, *A. J.,* **94**, 61.
Sanders, D. B., Scoville, N. Z., Young, J. S., Soifer, B. T., Schloerb, F. P., Rice, W. L., and Danielson, G. E., 1986, *Ap. J. (Letters),* **305**, L45.
Schaaf, R., Pietsch, W., Biermann, P. L., Kronberg, P. P., and Schmutzler, T., 1989, *Ap. J.,* **336**, 722.
Sramek, R. A., Panagia, N., and Weiler, K. W., 1984, *Ap. J. (Letters),* **285**, L59.
Sramek, R. A., Weiler, K. W., and Panagia, N., 1990, *IAU Circ.,* No. 49.
Tenorio-Tagle, G. and Bodenheimer, P., 1988, *Ann. Rev. Astr. Ap.,* **26**, 145.
Tomisaka, K. and Ikeuchi, S., 1988, *Ap. J.,* **330**, 695.
Turner, J. L. and Ho, P. T. P., 1985, *Ap. J. (Letters),* **299**, L77.
Van Buren, D. and Norman, C. A., 1989, *Ap. J. (Letters),* **336**, L67.
Van den Bergh, S. and Tammann, G. A., 1991, *Ann. Rev. Astr. Ap.,* in preparation.
Wampler, E. J., Wang, L., Baade, D., Banse, K., D'Odorico, S., Gouiffes, C., and Tarenghi, M., 1990, *Ap. J. (Letters),* submitted.
Watson, M. G., Stanger, V., and Griffiths, R. E., 1984, *Ap. J.,* **286**, 144.
Weaver, R., McCray, R., Castor, J., Shapiro, P. R., and Moore, R. T., 1977, *Ap. J.,* **218**, 377.
Weiler, K. W., Panagia, N., and Sramek, R. A., 1990, *Ap. J.,* in press.
Weiler, K. W., Sramek, R. A., Panagia, N., van der Hulst, J. M., and Salvati, M., 1986, *Ap. J.,* **301**, 790.
Weliachew, L., Fomalont, E. B., and Greisen, E. W., 1984, *Astr. Ap.,* **137**, 335.
Wheeler, J. C. and Levreault, R., 1985, *Ap. J. (Letters),* **294**, L17.
Wood, R. and Andrews, P. J., 1974, *M.N.R.A.S.,* **167**, 13.
Woosley, S. E. and Weaver, T. A., 1986, *Ann. Rev. Astr. Ap.,* **24**, 205.
Young, T. R. and Branch, D., 1989, *Ap. J. (Letters),* **342**, L79.

DISCUSSION

N. Walborn: If the M83 supernova was a typical Type II, *i.e.,* from a red supergiant progenitor, then it is probably unrelated to the ionizing cluster of the giant HII region. A similar configuration will occur if one of the peripheral red supergiants in 30 Doradus explodes. As discussed by Roberta Humphreys, stars of masses as high as those ionizing giant HII regions do not become red supergiants.

Chevalier: The M83 supernova SN 1968L occurred in the nuclear starburst region of the galaxy. It is plausible that this region is producing supernovae as well as copious ionizing radiation, as is the nuclear region of M82.

P. Conti: I noticed that you very carefully avoided identifying SN Ib with Wolf-Rayet progenitors, even though you alluded to an origin as initially massive stars with, now, lower (and uniform) masses. I would suppose that a problem with such an interpretation is that the SN Ib radio emission is not consistent with a shock hitting a pre-existing Wolf-Rayet wind?

Chevalier: The problems of identifying SN Ib with Wolf-Rayet stars are the fact that most observed Wolf-Rayet stars may be too massive to give a sufficiently narrow light curve and have winds of too low density to give the observed turn-on time of the radio emission. It is possible that many Wolf-Rayet stars evolve to a final state where these problems are solved. Another possibility for the radio emission is that the winds are clumped so that the mass loss estimate based on free-free absorption is an overestimate.

A. Moffat: Concerning the question of clumpiness in Wolf-Rayet winds: there is growing evidence of "blobs" in the winds of Wolf-Rayet stars (cf. Moffat et al. 1988, Ap. J.). We only see a few percent of the emission line flux in the form of subpeaks, but this may be only the "tip of the iceberg" with a much larger number of smaller mass blobs, i.e., the whole wind could be granular, or fractal, in structure.

R. Kennicutt: In starburst nuclei, the ratio of nonthermal radio emission to thermal emission is much higher than in disk star forming regions. Is there any physical mechanism which would make the supernova remnants in the starburst regions more efficient radio emitters, or is some other explanation required?

Chevalier: The thermal emission simply depends on the number of ionizing photons. The nonthermal emissivity depends on the energy densities of the relativistic electrons and the magnetic field, both of which are likely to be enhanced in a region of high thermal pressure. This nonlinear dependence can increase the ratio of nonthermal to thermal emission in a high pressure region.

G. Rieke: Given the wide range in supernova intrinsic radio luminosities and the importance of the interstellar medium in the integrated radio luminosity of a supernova in a starburst, how can one explain the tight observed correlation between far-infrared luminosity and radio luminosity in starbursts?

Chevalier: While a general correlation between far-infrared and radio luminosity is expected, the linear relation between the two is much tighter than I would have anticipated. Tim Heckman noted that the pressures in starbursts do not vary much over a range in luminosity. It may be that both the infrared and the radio luminosity are proportional to the area of the disk covered by the starburst.

Z. Wang: Is there a clear example of supernova remnant triggering of star formation in the Milky Way Galaxy or the Magellanic Clouds?

Chevalier: While some cases of supernova remnant triggered star formation in our Galaxy have been proposed, I do not believe that they have held up under close scrutiny. The 30 Doradus nebula in the LMC, discussed by Nolan Walborn at this meeting, appears to show regions of propagating star formation. It may be that high pressures over a large volume are needed to trigger star formation and that a single supernova remnant is not sufficient.

G. Miley: I would just like to point out that if your supernova winds do produce star formation that much more star formation might be expected if a powerful radio source goes off in the center of a gas rich galaxy. There is in fact now considerable evidence that such processes are going on in distant radio galaxies.

J. Shields: Will ionizing photons escape in directions perpendicular to the disk?

Chevalier: The hot, low density flow perpendicular to the disk of M82 should allow ionizing photons to escape. Support for this hypothesis comes from the fact that dust grains along the minor axis are observed to scatter blue light coming from the center of the galaxy.

OBSERVATIONS AND MODELS OF BLUE COMPACT DWARF GALAXIES

Trinh X. Thuan
Astronomy Department
University of Virginia
Box 3818
Charlottesville, VA 22903
USA

1. INTRODUCTION

Blue compact dwarf (BCD) galaxies are low luminosity ($M_B \gtrsim -18$) extragalactic objects undergoing intense star formation, as evidenced by their blue UBV colors, their optical spectra which show strong narrow emission lines, and their ultraviolet spectra showing a steeply rising continuum toward the blue, characteristic of OB stars (figure 1). In contrast to the optical spectra, the UV spectra do not exhibit strong nebular emission lines because the stellar continuum is much more important in the UV than in the optical relative to the nebular emission. There are strong Si IV $\lambda 1405$ and C IV $\lambda 1550$ absorption lines with P-Cygni profiles, implying the presence of radiatively driven stellar winds from the OB stars, with terminal velocities between 2000 and 4000 km s^{-1}.

BCDs can be found by searching for very compact objects (Zwicky 1971, Binggeli, Sandage and Tammann 1985), by looking for objects which appear unusually blue in multicolor photographic surveys of galaxies like the one carried out with the Kiso Schmidt in Japan (*e.g.*, Takase and Miyanchi-Isobe 1988) or by searching for objects with a large UV excess or with emission line spectra in objective prism surveys with Schmidt telescopes. The latter include the Markarian survey with the Byurakan Schmidt (*e.g.*, Markarian, Lipovetskii and Stepanian 1981, Markarian, Stepanian and Erastova 1986), the Tololo survey (Smith, Aguirre and Zemelman 1976, Bohuski, Fairall and Weedman 1978), and the University of Michigan survey (*e.g.*, MacAlpine and Williams 1981) with the Curtis Schmidt telescope at the Cerro Tololo Interamerican Observatory, the Case (*e.g.*, Pesch and Sanduleak 1988) and Wasilewski (1983) surveys with the Burrell Schmidt at Kitt Peak National Observatory, and the survey by Kunth and Sargent (1986a) with the Palomar and UK Schmidt telescopes. The selection criteria described above pick out, besides BCDs, a wide variety of other objects such as Seyfert galaxies, quasars or starburst galaxies. Follow-up spectroscopic work is required

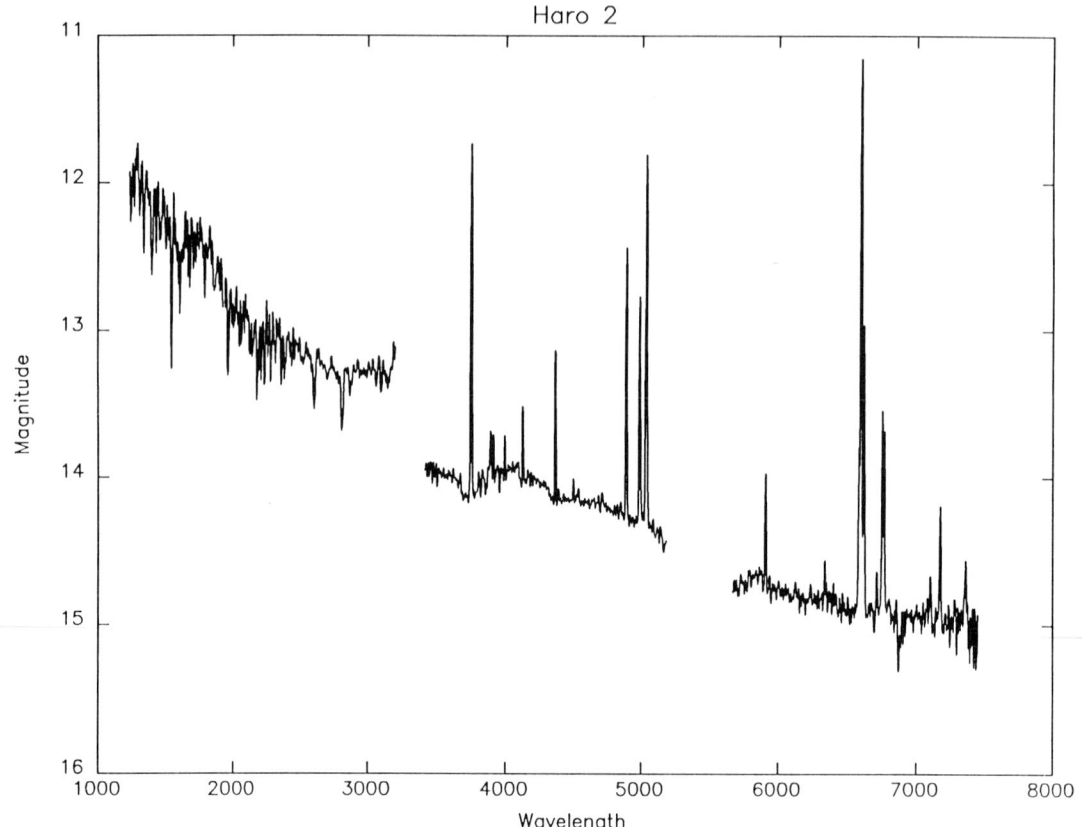

Figure 1. *IUE far and near-ultraviolet and IIDS blue and red spectra of the BCD Haro 2. There is a shift in the continuum level between the UV and optical spectra because they were taken through different apertures. The IUE aperture is a $10'' \times 20''$ rectangle while the IIDS aperture is a $6''\!.1$ diameter circular aperture.*

to disentangle the different categories of objects. Table 1 taken from Thuan (1985) gives the mean statistical optical and HI properties of BCDs. Section 2 discusses the evidence for starbursts in BCDs. Section 3 discusses the star formation history in BCDs and Section 4 compares the burst phenomenon in BCDs with that in larger, more massive starburst galaxies. We shall define here a starburst to be an event in which, during a time short compared to the Hubble time, the star formation rate (SFR) increases to be several orders of magnitudes greater than the mean SFR, averaged over the lifetime of the galaxy.

2. THE EVIDENCE FOR STARBURSTS IN BCDS

Ever since their discovery by Sargent and Searle (1970), it has been known that BCDs make massive stars in bursts. This is because of several observational constraints (Thuan 1987):

TABLE 1. *Optical and HI properties of BCDs*[a]

a_{25}	$39''.3 \pm 28''.0$
r_{25}	0.75 ± 0.18
d_{25} (kpc)	1.62 ± 0.96
S_B (mag arcsec^{-2})	22.8 ± 0.9
M_B	-14.8 ± 0.76
ΔV_{20}(km s^{-1})	97 ± 36
M_{HI} ($10^8 \, M_\odot$)	0.96 ± 1.2
M_T ($10^8 \, M_\odot$)[b]	6.8 ± 6.8
L_B ($10^8 L_{B_\odot}$)	1.7 ± 1.1
M_{HI}/M_T	0.15 ± 0.09
M_{HI}/L_B	0.62 ± 0.65
M_T/L_B	4.72 ± 4.36

[a] Based on a sample which consists of all blue compacts with $M_B \geq -16$ observed by Thuan and Martin (1981), a total of 19 galaxies. A Hubble constant of 75 km s^{-1} Mpc^{-1} has been adopted.

[b] Total masses M_T are calculated by taking the HI diameter to be 4 times d_{25}.

2.1. The Metallicity Constraint

BCDs are metal-deficient with respect to the solar neighborhood, with a metallicity ranging from $\sim Z_\odot/3$ to $\sim Z_\odot/30$. The abundance distribution peaks at $\sim Z_\odot/6$, with a sharp drop-off beyond $Z_\odot/10$ (Kunth and Sargent 1986b). For a long time, IZw18, one of the first BCDs discovered (Sargent and Searle 1970) held the record as being the most metal deficient known galaxy ($Z \sim Z_\odot/30$). Recently, the Byurakan group (Izotov et al. 1989) has found a BCD, SBS 0335–052, which is even more metal-deficient, with the oxygen and neon abundances being respectively 1.3 and 1.9 times lower than those in IZw18.

If metals are not removed from the star-forming region by galactic winds, or if the gas metallicity is not diluted by infalling pristine HI gas, then star formation at the rate presently observed in BCDs would produce too much metals ($Z \gtrsim Z_\odot/10$) after $\sim 10^7$ years (Lequeux et al. 1981, Viallefond and Thuan 1983, Kruger, Fricke and Loose, 1990). Kunth and Sargent (1968b) found that the observed oxygen abundance in IZw18 is reached after $\sim 4 \times 10^6$ years.

The metallicity contraint on the burst time scale can be relaxed somewhat if BCDs have galactic winds which blow the metal-enriched gas out of the star-forming region and mix it with the more pristine HI gas which surrounds it. Meurer (1989) has discussed examples of dwarf galaxies with a probable galactic wind: their Hα morphology suggests a gas outflow along the minor axis and reveals at least one cluster of young stars responsible for that wind. Moreover, BCDs are known to be embedded in large HI haloes, whose size can be 2 to 5 times larger than that of the star-forming region (Viallefond and Thuan 1983). The mixing of this gas with the metal-enriched gas can decrease the observed metallicity, although the HI interferometric maps do not show clear signs of infall. The presence of a galactic wind blowing out the metal-enriched gas or of infalling gas diluting the gas metallicity can also explain why many BCDs

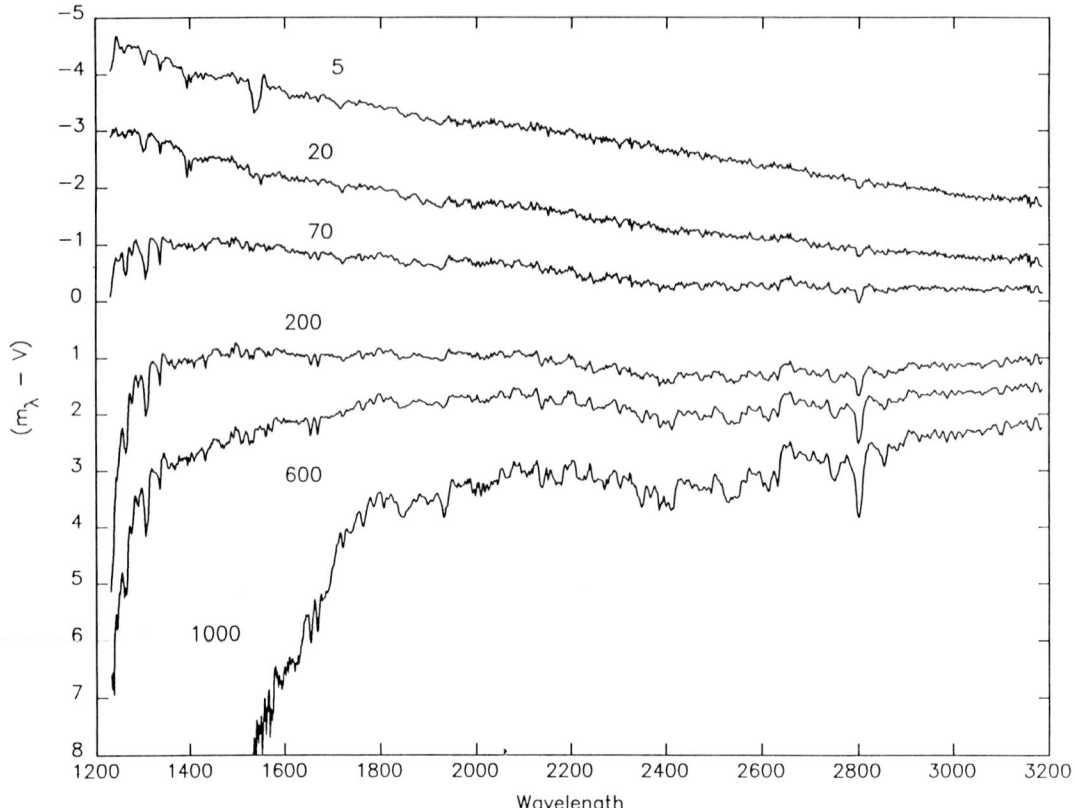

Figure 2. *Time evolution of the ultraviolet spectral energy distribution of a pure burst model (Fanelli 1990), based on synthetic cluster models by O'Connell (1983). Each curve is labelled by the time elapsed after the end of the burst, in units of 10^6 years.*

have a metallicity lower than that predicted by closed-box chemical evolution models (Matteucci and Chiosi 1983).

2.2. The Gas Constraint

BCDs have $\sim 10^8$ $M_\odot$ in neutral hydrogen (Thuan and Martin 1981 and Table 1). Star formation rates derived, for example, from spectral synthesis of BCD ultraviolet spectra (Fanelli, O'Connell and Thuan 1988) are in the range 0.01 to 10 $M_\odot$ yr^{-1}, giving a gas depletion time scale between 10^7 and 10^{10} years.

2.3. The Spectral Energy Distribution (SED) Constraint

The shape of the SED from the ultraviolet to the infrared (such as that shown for the BCD Haro 2 in figure 1 for the optical and UV wavelength ranges) puts strong constraints on the star formation history of the galaxy. Fanelli, O'Connell and Thuan (1988) have applied the optimizing population synthesis technique to the far-UV SEDs of several BCDs. They find that the solutions yield stellar luminosity functions which invariably display large discontinuities, indicative of discrete star formation episodes or bursts. Continuous star formation cannot account for the observed UV spectrum. The current burst of star formation in most objects appears to have lasted for less than $\sim 10^7$ years. This result can easily be seen by comparing the observed SED (such as the one shown in figure 1) with model SEDs in the ultraviolet. Figure 2 shows the time evolution of the integrated UV SED of a young star cluster model, as calculated by Fanelli (1990), based on synthetic cluster models by O'Connell (1983) and the ultraviolet stellar library of Fanelli, O'Connell and Thuan (1987). This represents the pure burst model. Figures 3a and 3b display the time evolution of the integrated UV SED of a burst of star formation of duration 10^7 years, superposed on a 9×10^9 years old underlying population, typical of an elliptical galaxy population (Fanelli 1990). The mass involved in the starburst is 10% of the total mass. The mass-to-light ratio of the underlying population is taken to be 5, while the temporal evolution of the mass-to-light ratio of the bursting component is taken from Larson and Tinsley (1978).

It is clear that, in order to reproduce the steeply rising UV continuum of BCDs, the burst age (the time elapsed after the end of the burst) cannot exceed $\sim 2 \times 10^7$ years. The pure burst and the burst + elliptical model are very similar at early times ($t \lesssim 2 \times 10^7$ years), because the UV light comes mostly from the burst component. The UV SED flattens at $\sim 2 \times 10^8$ years and the far UV fluxes drop off drastically after 10^9 years. Thus the shape of the continuum is a very sensitive age indicator for bursts in galaxies.

Some spectral features are also very sensitive to the age of the burst. For example, in both types of models, the CIV absorption feature at $\lambda 1550$ is very prominent at 5×10^6 years after the end of the burst, but has already weakened considerably after 2×10^7 years. On the other hand, other features such as the MgII feature at $\lambda 2800$ and the MgI feature of $\lambda 2852$ increase considerably in strength as the burst ages. Post-starburst galaxies like the ones discussed by Dressler and Gunn (1983), containing an old population mixed with A stars resulting from a burst 1 billion years ago, should have UV spectra similar to the SED labeled 1000 Myrs in figure 3b. Huchra (1987) has also computed evolutionary models for bursts, similar to the ones discussed above. Both Huchra's model and Fanelli's models differ appreciably from those of Struck-Marcell and Tinsley (1978) in that they both take into account the light contribution from red supergiants. Huchra's models also include the effects of gaseous emission. These models show that the slope of the Initial Mass Function (IMF) has little or no effect on the SED of a burst, although objects with steep slopes, *i.e.*, which are enriched in low mass stars, redden faster after the burst turns off.

Color-color plots are often used to detect and quantify subtle differences in SEDs. Of these, the optical-infrared (V-K, U-B) or (V-K, U-V) plots are most useful for deriving burst ages (Thuan 1983, 1985, Huchra 1987). Infrared magnitudes, when combined with optical magnitudes, give information on the relative importance of the young and old stellar populations, and hence on the star formation history of BCDs. Again burst ages of $\sim 10^7$ years are derived.

3. THE STAR FORMATION HISTORY IN BCDS

The observations can be used to ask the following questions on the star formation history of BCDs:
1. What are the properties of the present burst of star formation? What is the initial function (IMF) (*i.e.*, its slope and its upper and lower mass limits), the star

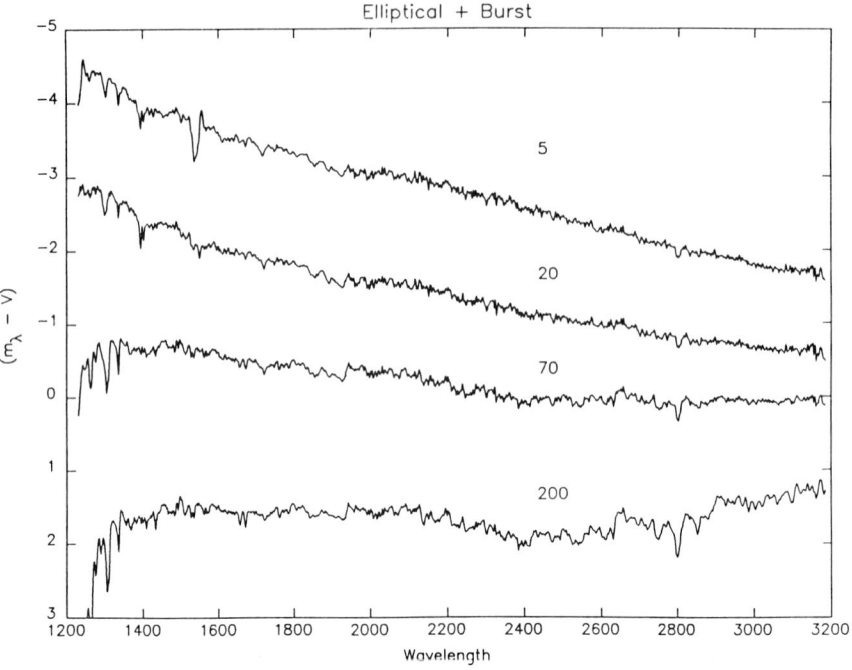
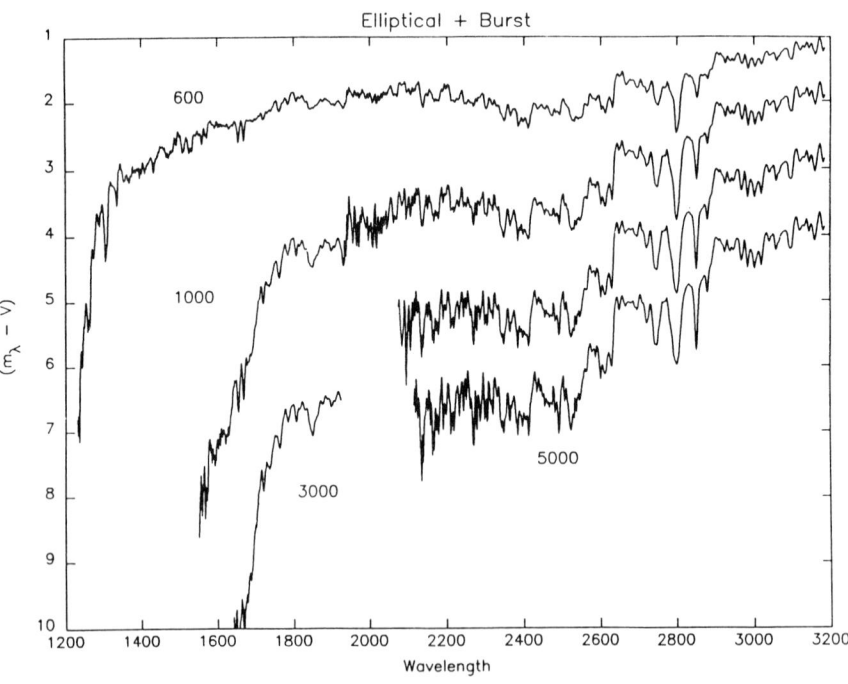

Figures 3a and 3b. *Time evolution of the ultraviolet spectral energy distribution of a burst superposed on an old elliptical galaxy stellar population (Fanelli 1990). Each curve is labelled by the time elapsed after the end of the burst, in units of 10^6 years.*

formation rate (SFR), the supernova rate, and the age or duration of the burst?
2. Is the present burst the first one in the history of the galaxy? If not, how many bursts were there previous to the present one and what is the time interval between bursts?

Multifrequency studies are needed to answer these questions. The ultraviolet probes the young O and B stellar populations, the optical probes the intermediate age B and A stellar populations, and the near-infrared probes the evolved giant stellar populations. The information is derived by using optimizing population synthesis techniques to reconstruct the observed SEDs (*e.g.*, Fanelli, O'Connell and Thuan 1988) or by comparing the observed SEDs with evolutionary models (Lequeux *et al.* 1981, Viallefond and Thuan 1983, Kruger, Fricke and Loose 1990). The optimizing synthesis approach attempts, by using a stellar library (such as the one constructed by Fanelli, O'Connell and Thuan 1987), to derive the distribution of stars in different evolutionary stages (main-sequence, giant, supergiant, etc.), subject to the constraint of astrophysical plausibility (for example, the main-sequence luminosity function must be a non-decreasing function of absolute magnitude), which best fits an observed SED. This approach makes it possible to explicitly evaluate the goodness of fit and the uncertainties in the derived parameters.

The evolutionary synthesis approach consists in calculating the time evolution of a composite SED, assuming a particular evolutionary scenario (for example, continuous star formation or bursts) and an IMF, and using the best available stellar evolutionary tracks and stellar atmosphere models. The resulting SED depends on a small number of parameters (age, metallicity, IMF slope, upper and lower mass limits, e-folding time for star formation), and constraints on these parameters are obtained by comparing the calculated to the observed SED, with no attempt to adjust the population parameters to obtain optimal fits.

Besides the constraint of the SED shape, other observations can be used to constrain the IMF parameters (see Scalo 1986 for a review). One frequently used constraint is the Hβ equivalent width, which measures the ratio of the number of O stars (through the ionizing flux required to reproduce the Hβ strength) to the number of B and A stars (through the continuum flux at Hβ). Other constraints include the effective temperature of the galaxy, defined as the effective temperature of a star with the same ratio of hydrogen Lyman continuum photons emitted above and below the helium Lyman limit at 504 Å (Lequeux *et al.* 1981, Viallefond and Thuan 1983). This quantity depends strongly on the slope of the IMF and its upper mass cut-off. Also very sensitive to these two parameters is the high-excitation nebular line [O III]λ5007 which is highly dependent on the number of massive stars (Campbell, Terlevich and Melnick 1986, Olofsson 1988). Sekiguchi and Anderson (1987) have proposed using the ratio of the equivalent widths of the Si IV λ1400 and C IV λ1550 ultraviolet absorption lines to put constraints on the IMF slope. Checks on the plausibility of the models can also be made by comparing the derived and observed radio continuum fluxes, the Hα fluxes, and the ionizing photon rates (Fanelli, O'Connell and Thuan 1988).

We first discuss the results concerning the IMF. The slope x of the IMF is defined here to be such that the number of stars born per unit $\log M$ at mass M is given by $dn(M)/d \ln M \propto M^{-x}$. Viallefond and Thuan (1983) using evolutionary models found $x = 1.5$ for 2 BCDs: IZw36, $(Z \sim Z_\odot/10)$ and IZw18 $(Z \sim Z_\odot/30)$, similar to the Salpeter slope of 1.35 but flatter than the solar neighborhood slope of 2. They obtained $x \sim 2$ for the HII complex NGC 5471 in M101, and noticed that there appears to be a tendency for slightly flatter IMFs in metal-deficient BCDs as compared with more metal-rich HII complexes in late-type galaxies. While a flattening of the IMF with decreasing metallicity is not theoretically implausible (less metals implies less cooling, and hence more massive star formation at the expense of low mass star formation), it is by no means clear that such a relationship exists. Bergvall (1985) found $x \sim 2$ in his study of the blue compact galaxy ESO 338–IG04 ($Z \sim Z_\odot/3$). IMF studies of a large sample of HII regions in the Large Magellanic Cloud ($Z \sim Z_\odot/3$) and the Small Magellanic Cloud ($Z \sim Z_\odot/8$) also yield a solar neighborhood slope ($x \sim 2$) (Massey, Parker, and Garmany 1990), although Melnick (1985) finds a considerably flatter slope ($x \sim 1.0$) for the 30 Doradus complex in the LMC. Both Terlevich (1985) and Melnick (1985) have argued for a metallicity dependence of the IMF slope. A major problem with their proposed correlation is that it depends strongly on the adopted values for the masses for the individual HII regions. These masses were derived from the observed linewidths, assuming that the HII regions and BCDs are in virial equilibrium. The latter assumption is highly questionable since there is direct observational evidence from spatially resolved velocity maps of HII regions (Kennicutt, this conference) and BCDs (Thuan, Williams and Malumuth 1987) that there are large-scale stellar wind and supernova-driven flows (*e.g.*, Gallagher and Hunter 1983). The Hα intensity map of the dominant HII region in the BCD IZw49 shows loops, filaments and large (50–300 pc) holes, suggestive of explosive events. In summary, the evidence for a metallicity dependence of the IMF slope is not very strong. A solar neighborhood IMF slope ($x \sim 2$) is probably adequate to fit the observations of most BCDs ($Z \gtrsim Z_\odot/10$) although very metal-deficient BCDs ($Z \lesssim Z_\odot/10$) may have a slightly flatter slope ($x \sim 1.5$).

Models with an upper mass cut-off $m_u \sim 100\ M_\odot$ appear to fit the BCDs IZw36 and IZw18 well (Viallefond and Thuan (VT) 1983). There is suggestive evidence that the lower mass cut-off m_L, must be relatively high in some BCDs. VT found that, if the SFR has been constant in time, $m_L \sim 4\ M_\odot$ in IZw36. This relatively higher value is needed to reproduce the equivalent width of Hβ. If the SFR was a spike in the past, the large lower mass cut-off is not needed. Similarly, Bergvall (1985) found that ESO 338–IG04 has $m_L \sim 10\ M_\odot$ if $x \sim 2$. High values of m_L are not needed in all BCDs. In IZw18, VT found $m_L \lesssim 2\ M_\odot$, essentially because it has a weaker Hβ equivalent width than IZw36. Evidently, low mass star formation is not necessarily suppressed by either the lower metallicity in BCDs or their enhanced gas stirring by the massive star formation.

Ever since their discovery, there has been controversy on whether BCDs are truly young objects. Is the present burst the first one in the lifetime of the galaxy, or are they old galaxies, with episodic bursts of star formation superposed on an older stellar population? Thuan (1983, 1985), on the basis of small aperture near-infrared photometry centered on the star-forming regions, concluded that the latter hypothesis was correct, although the question was not settled because near-infrared colors cannot unambiguously distinguish between old giant and young red supergiant stars (Gondhalekar *et al.* 1984, Campbell and Terlevich 1984). The best way to check for an older stellar population in a BCD is by obtaining deep optical and near-infrared images. Loose and Thuan (1985, 1986) and Kunth *et al.* (1988) found that, in more than 90% of the BCDs they investigated, there is an extended low-surface-brightness regularly shaped component underlying the compact high-surface-brightness irregular star-forming region, supporting the old galaxy hypothesis. Figure 4 shows a CCD frame in the B band of the BCD MK86 (Loose and Thuan 1990). While there are four main morphological types of BCDs (Loose and Thuan 1985), MK86 is representative of the iE type which is by far the most common ($\sim 70\%$). The iE galaxies show a complex structure with several

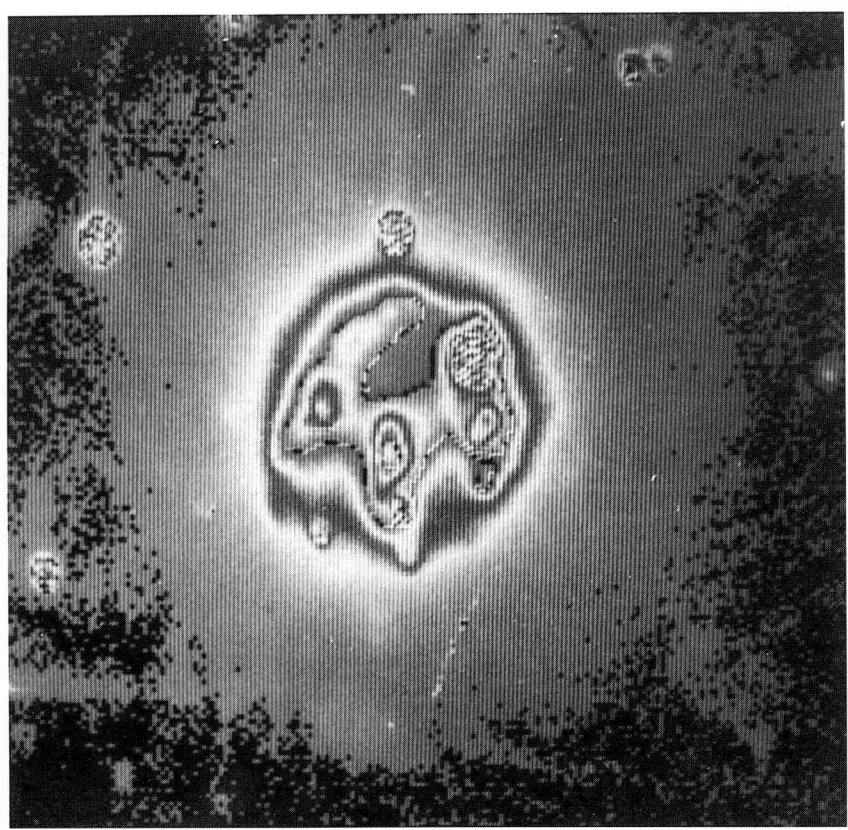

Figure 4. *CCD B picture of the BCD MK86 (Loose and Thuan 1990).*

centers of star formation and irregular (i) isophotes in the central regions while the isophotes of the underlying low-surface-brightness component are elliptical (E). Figure 5 shows an infrared K picture of MK 86 (Freitag, Loose and Thuan 1990). The two brightest star-forming regions in the B picture are also clearly visible in the K picture: evidently, the K band also picks up the light of the young red supergiants, in addition to the light from the older underlying population. The difference is more evident when color maps are compared. Figure 6 shows the B-R optical color map of MK86, while figure 7 shows the J-H near-infrared color map. While there are steep B-R color gradients in the vicinity of the star forming regions, the latter being bluer than the underlying stellar population, there are virtually *no* structures in the J-H (or H-K) color maps, confirming the hypothesis that the light from the underlying component comes from a smoothly distributed older stellar population. Even IZw18, one of the most metal-deficient ($Z \sim Z_\odot/30$) BCDs known, appears to have an underlying low-surface-brightness component (see a CCD picture of IZw18 in Thuan 1986). If IZw18 has not had previous formation of massive stars, before the present burst, the presence of a low-surface-brightness outer component (not coexistent spatially with the star-forming region, and arising presumably from lower mass stars) may imply a bimodal mode of star formation, where high and low mass stars are made separately.

In BCDs less extreme than IZw18, the stellar populations derived from spectral

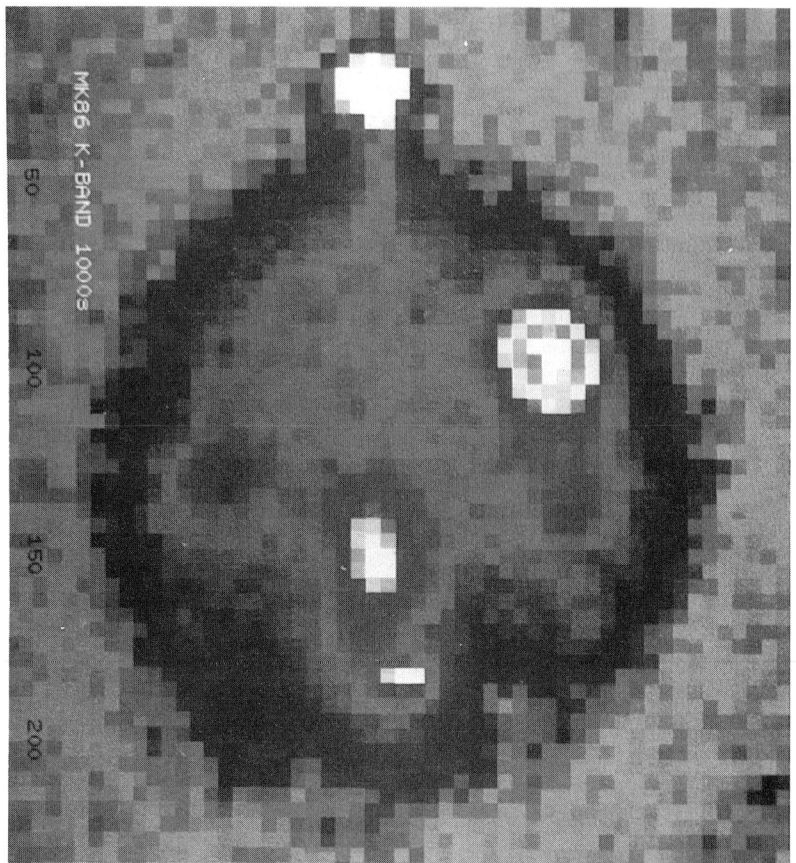

Figure 5. $2.2\mu m$ K picture of the BCD MK86 (Freitag, Loose and Thuan 1990).

synthesis of the far-UV spectra (Fanelli, O'Connell and Thuan 1988) generally show evidence of earlier star formation episodes. These manifest themselves as large discontinuities in the derived stellar luminosity functions. These discontinuities are like fossil records of the past history of the BCD: their number is equal to the number of previous bursts and their location in the luminosity function gives the dates of the bursts. There is generally an important A star component in the synthesis solutions, indicating significant star formation activity $\sim 5 \times 10^8$ years ago. The number of previous bursts is a few. For the BCD Haro 2 for example, there were at least two earlier bursts, the most recent of which ended not more than $\sim 2 \times 10^7$ years ago. Davidge and Maillard (1990) have obtained $2\mu m$ spectroscopic observations of Haro 2 and found CO absorption, possibly indicative of an older stellar population.

The ratio of the mass of stars made during the burst to the mass in the old galaxy is $\gtrsim 0.1$ (Lequeux et al. 1981, Thuan 1983, 1985). Far-ultraviolet synthesis yields directly the star formation rate for stars with $M \geq 10\ M_\odot$, without any assumption regarding the IMF. Fanelli, O'Connell and Thuan (1988) found the star formation rate of BCDs in the mass range superior to 10 $M_\odot$ to vary between 4 and $4 \times 10^{-4}\ M_\odot$ yr^{-1}, which would give a supernovae rate between 0.2 and 2×10^{-5} SN yr^{-1}. With a lower mass cut-off of 2 $M_\odot$ and an IMF slope $x = 1.8$, the SFR is between 17 and $2 \times 10^{-3}\ M_\odot$ yr^{-1}.

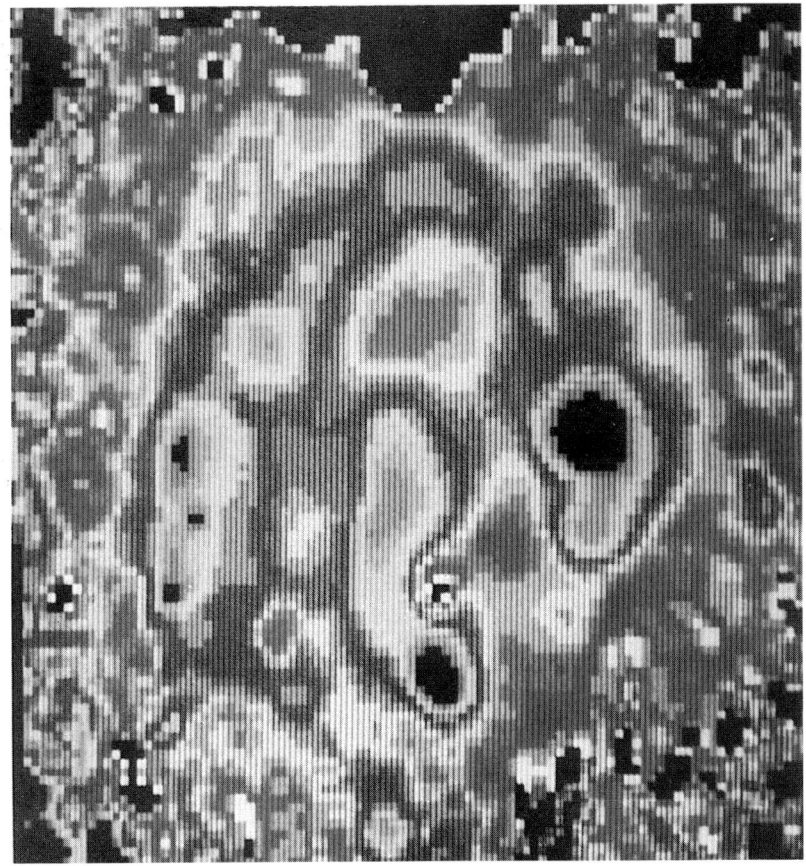

Figure 6. *The (B-R) optical color map of the BCD MK86 (Loose and Thuan 1990). The darkest zones are the bluest ones.*

4. COMPARISON WITH MASSIVE STARBURST GALAXIES

We now compare the starburst phenomenon in BCDs with that in more massive galaxies. The latter are more metal-rich ($Z \sim Z_\odot$), more luminous and dustier (as shown by their large IRAS far-infrared fluxes).

4.1. The Triggering Mechanism

The starburst triggering mechanism in the two types of galaxies appears to be different. The majority of the massive starburst galaxies show clear signs of galaxy mergers (disturbed morphology, multiple nuclei, tidal tails, etc.). For example, one of the prototype massive starburst galaxies, Arp 220, shows an apparently single nucleus in a low resolution ($\sim 1''\!.3$ FWHM) 2.2μm image (Carico et al. 1990). But when observed with the VLA at 3.8 cm, with a resolution of $0''\!.25$ FWHM, the single nucleus is resolved into two distinct nuclei separated by $\sim 1''$ and in the process of merging (figure 8, Huang et al. 1990). The starburst in massive galaxies is probably triggered by

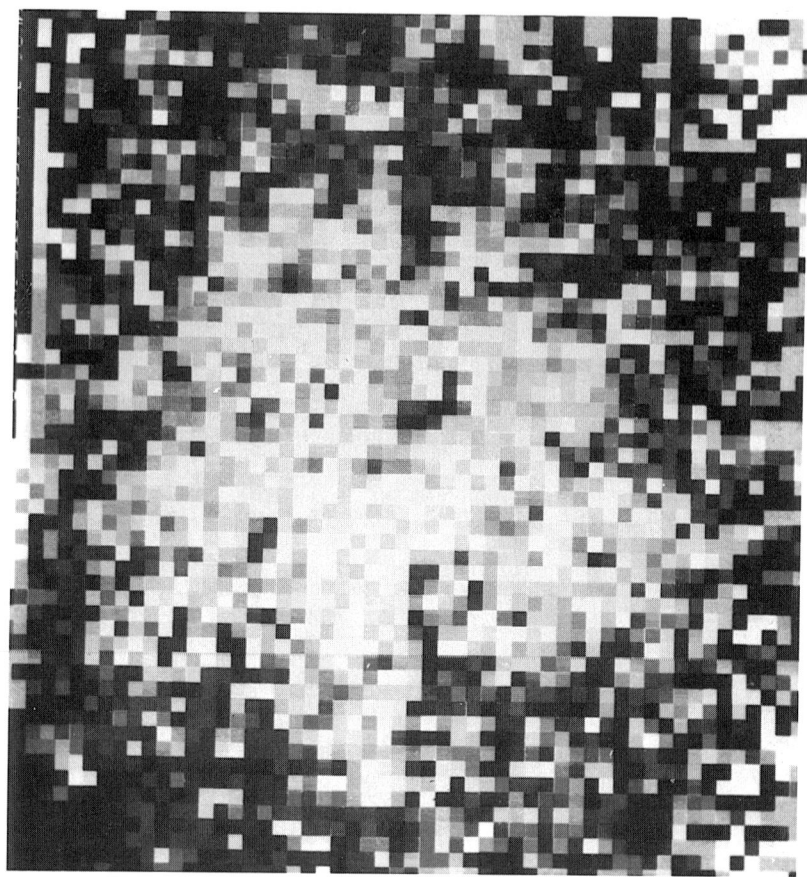

Figure 7. *The (J-H) near-infrared color map of the BCD MK86 (Freitag, Loose and Thuan 1990).*

the shock waves induced in the gas by the merging process. The situation is different for BCDs. They are mostly isolated (in at least 50% of the cases), and gravitational interaction with other galaxies cannot be invoked to explain starbursts in the majority of the BCDs. These are probably triggered by cloud collisions. High resolution HI maps of BCDs show a clumpy structure, with several distinct HI cloud complexes, moving in a disorderly way with respect to each other (see the HI map of IZw18 by Lequeux and Viallefond 1980 and the HI map of IZw36 by Viallefond and Thuan 1983). These cloud collisions trigger shocks which compress the HI gas to high densities. These high densities can be seen in high resolution HI maps such as those obtained by Viallefond, Lequeux and Comte (1987). These authors find peak HI densities of 18 H atoms cm^{-3} in IZw18 and 28 H atoms cm^{-3} in IZw36. This is to be compared with the mean densities of HI complexes in nearby galaxies which range from 1 to 8 H atoms cm^{-3}. The starburst is triggered once a critical threshold density is reached (Larson 1987).

The presence of turbulent random motions is also evident in the Hα Fabry-Perot interferometric maps of 2 BCDs obtained by Thuan, Williams and Malumuth (1987). Velocity gradients of $\sim$ 30 km s^{-1} on scales of $\sim$ 10 pc in the BCD VIIZw403 and $\sim$ 50 pc in the BCD IZw49 are seen. The corresponding dynamical time scale is only

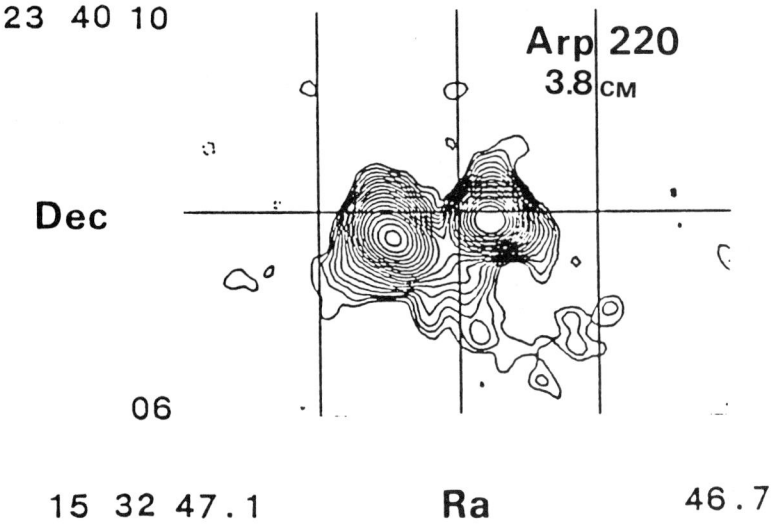

Figure 8. *A very high resolution (0″.25 FWHM) VLA map of the starburst galaxy Arp 220 at $\lambda = 3.8$ cm (Huang et al. 1990).*

$\lesssim 10^6$ years, one order of magnitude smaller than the burst duration of $\sim 10^7$ years. Complex velocity fields such as those seen in VIIZw403 and IZw49 cannot be maintained much longer than several dynamical times. Thus the ionized gas in BCDs cannot be in equilibrium and must have been stirred by the ongoing starburst.

4.2. The Stellar Populations

The ratio of the mass of stars made during the burst to the mass in the old galaxy is 1 to 2 orders of magnitude smaller in massive galaxies than in BCDs (ranging from ~ 0.001 to ~ 0.01 vs. $\gtrsim 0.1$; Thuan 1985). As in BCDs, the light from the older stellar populations in the massive starburst galaxies forms a smooth component underlying the star-forming regions. Figure 9 shows a $2.2\mu m$ picture of the prototype starburst galaxy M82 (Freitag, Loose and Thuan 1990). All the disturbed morphology seen in the optical (which drew attention to M82) has vanished, and M82 is now just an ordinary spiral galaxy seen edge-on. The view of M82 in the radio (at 3.8cm) shows yet another perspective (figure 10, Huang et al. 1990). Because of the very high angular resolution (0″.25 FWHM), individual supernova remnants (SN) resulting from the death of massive stars formed in previous starbursts can be seen. The derived relation between the surface brightnesses and diameters of the M82 supernova remnants has the same slope as the relations established for the SN remnants in the Magellanic Clouds and the Milky Way (Huang et al. 1990). According to Rieke et al. (1980), the supernova rate in M82 is 0.3 SN yr^{-1}. Given that the mass of the central part of M82 is $\sim 5 \times 10^8$ $M_\odot$, the SN rate per unit mass in M82 is comparable to the highest SN rates found in BCDs, and is 50 to 100 times the rate in the solar neighborhood and in late-type galaxies.

How do the IMFs in BCDs and starburst galaxies compare? Scalo (1989) has recently reviewed the evidence for top-heavy IMFs in massive starburst galaxies (*i.e.*, for IMFs with large lower mass cut-offs). With the IMF slope having the solar neighborhood value ($x \sim 2$), Scalo (1989) found that there is suggestive but not conclusive evidence

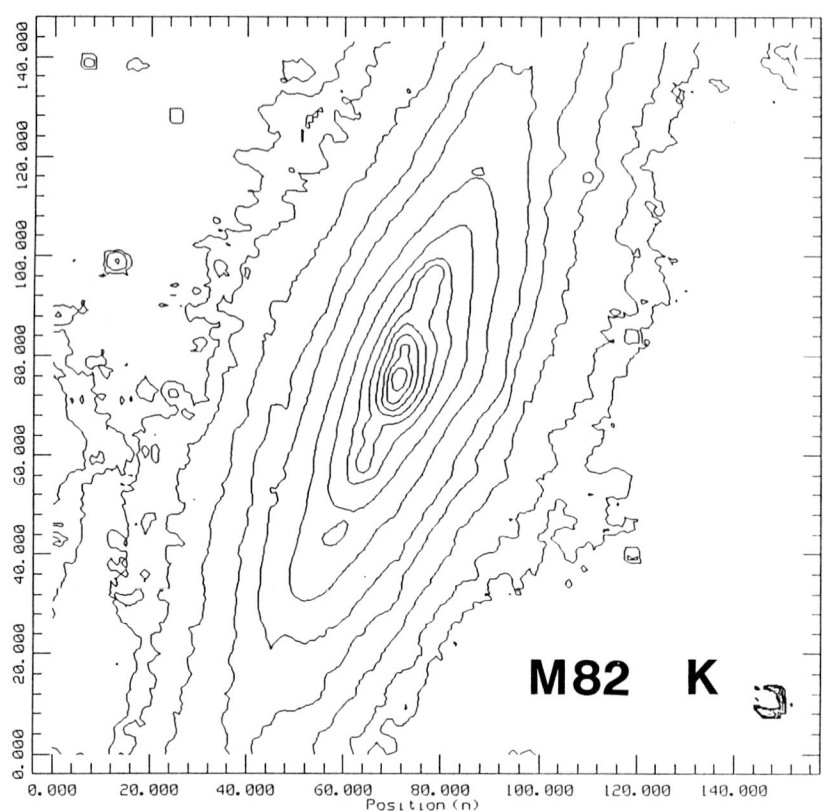

Figure 9. *A 2.2μm K picture of M82 (Freitag, Loose and Thuan 1990).*

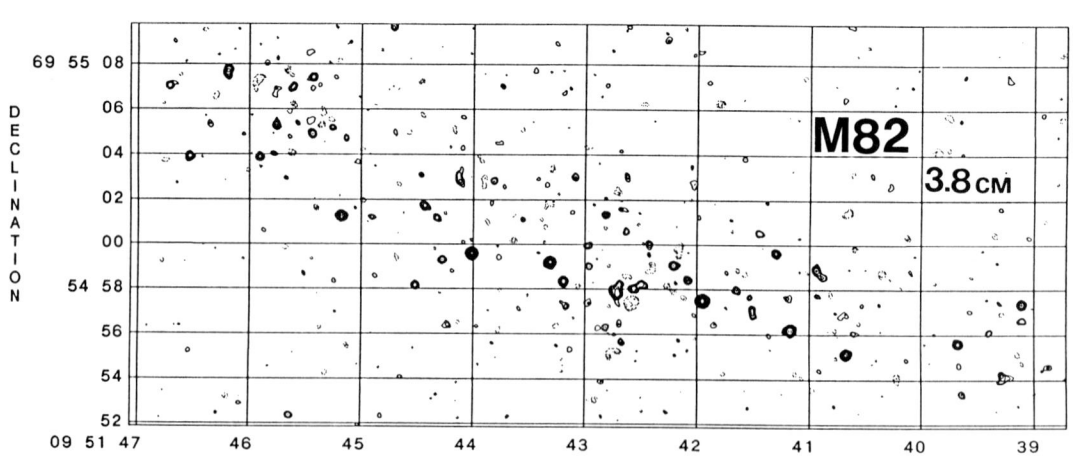

Figure 10. *A very high resolution (0″.25 FWHM) VLA map of M82 at $\lambda = 3.8$ cm, showing the radio supernova remnants (Huang et al. 1990).*

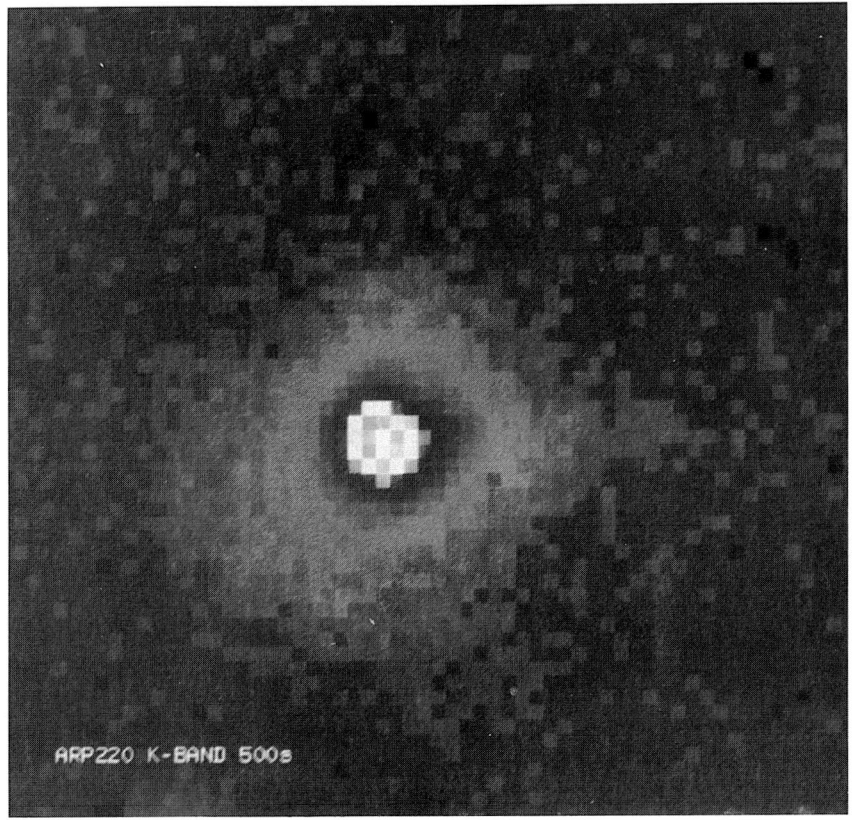

E

Figure 11. *A 2.2µm K picture of Arp 220 (Freitag, Loose and Thuan 1990).*

that $m_L \sim 3$–10 $M_\odot$, if the upper mass limit m_u is not significantly larger than ~ 80–100 $M_\odot$. Other investigators also find a high value of m_L: Rieke's (this conference) latest model for M82 yields $m_u = 25\ M_\odot$ and $m_L = 10\ M_\odot$, while Joseph (this conference) finds $m_u = 25\ M_\odot$ and $m_L = 3$–6 $M_\odot$ in his modelling of NGC 3256. There appears to be a significant fraction (~ 10–50 percent) of massive starburst galaxies with an excess of massive stars.

We examine finally the relationship between thermal and non-thermal activity in BCDs and massive starburst galaxies. Sanders et al. (1988) have presented a model in which 'ultraluminous' infrared starburst galaxies with infrared luminosities $L_{IR} \geq 10^{12}\ L_\odot$ represent a phase in the formation of quasars. In this scenario, a massive starburst event results in the formation of massive stars which die and form stellar mass black holes which coalesce to form a 'monster' black hole of $\sim 10^8\ M_\odot$ which then powers the infrared luminosity of the starburst galaxy. The quasar becomes visible when the ultraluminous starburst galaxy sheds its dust cocoon. Figure 11 shows a 2.2µm picture of the ultraluminous infrared starburst galaxy Arp 220 (Freitag, Loose and Thuan 1990). The high-surface-brightness star-forming region is clearly visible on top of an underlying low-surface-brightness older stellar population. In contrast to the

Figure 12a. *The (J-H) near-infrared color maps of Arp 220 (Freitag, Loose and Thuan 1990).*

situation in most BCDs (the iE type), the star-forming region in massive starburst galaxies is located at the center of the underlying component (at the bottom of the potential well). The star formation is probably triggered by shocks caused by the infall and collision of the gas of the 2 merging galaxies in that well.

Figures 12a and 12b show the near-infrared J-H and H-K color maps of Arp 220 (Freitag, Loose and Thuan 1990). While in BCDs there is *no* near-infrared color gradient (figure 7), there exists a steep color gradient in Arp 220. The central parts of ultraluminous IRAS starburst galaxies have very red infrared colors, suggesting a significant amount of non-thermal infrared emission powered by the 'monster' in the center. As the distance from the center increases, the colors get bluer, approaching those of normal galaxies at large distances. Arp 220 has J-H = 1.34 ± 0.10, H-K = 0.90 ± 0.10 in an aperture of $2''.5$ diameter, and J-H = 0.88 ± 0.10, H-K = 0.48 ± 0.10 in a $27''$ aperture. The fact that a 'monster' may be lurking in the heart of the most luminous starburst galaxies has also been confirmed by Huang et al. (1990) who have obtained very high resolution ($0''.25$ FWHM) VLA maps at $\lambda = 3.8$ cm for all IRAS galaxies with $L_{FIR} > 10^{12}$ $L_\odot$. Nearly 40% of the galaxies in this sample have unresolved radio sources, presumably associated with very compact objects.

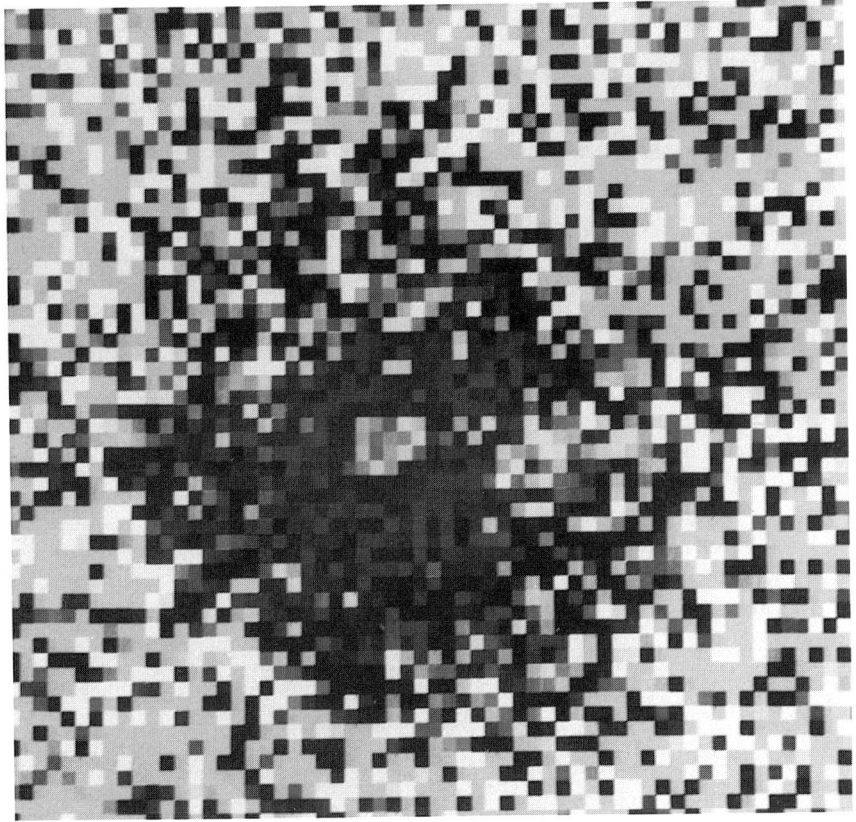

N

E

Figure 12b. *The (H-K) near-infrared color maps of Arp 220 (Freitag, Loose and Thuan 1990).*

BCDs, because of their small mass, do not appear to go through this 'active galactic nuclei' phase.

I thank J. Condon, M. Fanelli, V. Freitag, Z. P. Huang, H. Loose and Q. F. Yin for allowing me to discuss some of our joint work in advance of publication. This work was supported in part by Air Force Office of Scientific Research grant 89-0467 and NASA grant NAG 5-1389.

REFERENCES

Bergvall, N. 1985, *Astr. Ap.*, **146**, 269.
Binggeli, B., Sandage, A. and Tammann, G. A. 1985, *A. J.*, **90**, 1681.
Bohuski, T. J., Fairall, A. P., and Weedman, D. W. 1978, *Ap. J.*, **221**, 776.
Campbell, A. W. and Terlevich, R. J. 1984, *M.N.R.A.S.*, **211**, 15.
Campbell, A. W, Terlevich, R. J., and Melnick, J. 1986, *M.N.R.A.S.*, **223**, 811.
Carico, D. P., Graham, J. R., Matthews, K., Wilson, T. D., Soifer, B. T., Neugebauer, G. and Sanders, D. B. 1990, *Ap. J. (Letters)*, **349**, L39.

Davidge, T. and Maillard, J. P. 1990, *Ap. J.*, **351**, 432.
Dressler, A. and Gunn, J. E. 1983, *Ap. J.*, **270**, 7.
Fanelli, M. N. 1990, Ph.D. Thesis, University of Virginia.
Fanelli, M. N., O'Connell, R. W. and Thuan, T. X. 1987, *Ap. J.*, **321**, 768.
_____. 1988, *Ap. J.*, **334**, 665.
Freitag, V., Loose, H. H. and Thuan, T. X. 1990, in preparation.
Gallagher, J. S. and Hunter, D. A. 1983, *Ap. J.*, **274**, 141.
Gondhalekar, P. M., Morgan, D. H., Dopita, M., and Phillips, A. P. 1984, *M.N.R.A.S.*, **209**, 59.
Huang, Z. P., Condon, J. J., Yin, Q. F. and Thuan, T. X. 1990, in preparation.
Huchra, J. P. 1987, in *Starbursts and Galaxy Evolution*, eds. T. X. Thuan, T. Montmerle and J. T. T. Van (Editions Frontieres: Paris), p. 199.
Izotov, Y. I., Lipovetsky, V. A., Guseva, N. G. and Stepanian, J. A. 1989, preprint.
Kruger, H., Fricke, K. J. and Loose, H. H. 1990, Poster Paper at the 'Massive Stars in Starbursts' meeting, Baltimore.
Kunth, D. and Sargent, W. L. W. 1986a, *A. J.*, **91**, 761.
_____. 1986b, *Ap. J.*, **300**, 496.
Kunth, D., Maurogordato, S. and Vigroux, L. 1988, *Astr. Ap.*, **204**, 10.
Larson, R. B. 1987, in *Starbursts and Galaxy Evolution*, eds. T. X. Thuan, T. Montmerle, and J. T. T. Van (Editions Frontieres: Paris), p. 467.
Larson, R. B. and Tinsley, B. M. 1978, *Ap. J.*, **219**, 46.
Lequeux, J. and Viallefond, F. 1980, *Astr. Ap.*, **91**, 269.
Lequeux, J., Maucherat-Joubert, M., Deharveng, J. M. and Kunth, D. 1981, *Astr. Ap.*, **103**, 305.
Loose, H. H. and Thuan, T. X. 1985, in *Star-Forming Dwarf Galaxies and Related Objects*, eds. D. Kunth, T. X. Thuan, and J. T. T. Van (Editions Frontieres: Paris), p. 73.
_____. 1986, *Ap. J.*, **309**, 59.
_____. 1990, in preparation.
MacAlpine, G. M. and Williams, G. A. 1981, *Ap. J. Suppl.*, **45**, 113.
Markarian, B. E., Lipovetskii, V. A. and Stepanian, D. A. 1981, *Astrofizika*, **17**, 619.
Markarian, B. E., Stepanian, D. A., and Erastova, L. K. 1986, *Astrofizika*, **25**, 551.
Massey, P., Parker, J. W., and Garmany, C. D. 1990, *A. J.*, **98**, 1305.
Matteucci, F. and Chiosi, C. 1983, *Astr. Ap.*, **123**, 121.
Melnick, J. 1985, in *Star Forming Dwarf Galaxies and Related Objects*, eds. D. Kunth, T. X. Thuan and J. T. T. Van, (Editions Frontieres: Paris), p. 171.
Meurer, G. R. 1989, Ph.D. Thesis, Australian National University.
O'Connell, R. W. 1983, *Ap. J.*, **267**, 80.
Olofsson, K. 1988, Ph.D. Thesis, Uppsala Astronomical Observatory.
Pesch, P. and Sanduleak, N. 1988, *Ap. J. Suppl.*, **66**, 297.
Rieke, G. H., Lebofsky, M. J., Thompson, R. I., Low, F. J., and Tokunaga, A. T. 1980, *Ap. J.*, **238**, 24.
Sanders, D. B., Soifer, B. T., Elias, J. H., Madore, B. F., Matthews, K., Neugebauer, G. and Scoville, N. Z. 1988, *Ap. J.*, **325**, 74.
Sargent, W. L. W. and Searle, L. 1970, *Ap. J. (Letters)*, **162**, L155.
Scalo, J. M. 1986, *Fund. Cosmic Phys.*, **11**, 1.
Scalo, J. M. 1989, in *Windows on Galaxies*, eds. A. Renzini, G. Fabbiano and J. S. Gallagher (Kluwer: North Holland).

Sekiguchi, K. and Anderson, K. S. 1987, *A. J.,* **94**, 644.
Smith, M. G., Aguirre, C. and Zemelman, M. 1976, *Ap. J. Suppl.,* **32**, 217.
Struck-Marcell, C. and Tinsley, B. M. 1978, *Ap. J.,* **221**, 562.
Takase, B. and Miyauchi-Isobe, N. 1988, *Ann. Tokyo Astr. Obs., Ser. 2,* **22**, 41.
Terlevich, R. 1985, in *Star Forming Dwarf Galaxies and Related Objects,* eds. D. Kunth, T. X. Thuan and J. T. T. Van, (Editions Frontieres: Paris), p. 395.
Thuan, T. X. 1983, *Ap. J.,* **268**, 667.
_____. 1985, *Ap. J.,* **299**, 881.
_____. 1986, in *Nearly Normal Galaxies,* ed. S. M. Faber (Springer-Verlag: New York).
_____. 1987, in *Starbursts and Galaxy Evolution,* eds. T. X. Thuan, T. Montmerle and J. T. T. Van (Editions Frontieres: Paris), p. 129.
Thuan, T. X. and Martin, G. E. 1981, *Ap. J.,* **247**, 823.
Thuan, T. X., Williams, T. B. and Malumuth, E. 1987, in *Starbursts and Galaxy Evolution,* eds. T. X. Thuan, T. Montmerle and J. T. T. Van (Editions Frontieres: Paris), p. 151.
Viallefond, F. and Thuan, T. X. 1983, *Ap. J.,* **269**, 444.
Viallefond, F., Lequeux, J. and Comte, G. 1987, in *Starbursts and Galaxy Evolution,* eds. T. X. Thuan, T. Montmerle and J. T. T. Van (Editions Frontieres: Paris), p. 139.
Wasilewski, A. J. 1983, *Ap. J.,* **272**, 68.
Zwicky, F. 1971, *Catalogue of Selected Compact Galaxies and of Post-Eruptive Galaxies* (Guemligen, Switzerland).

DISCUSSION

J. Bland Hawthorn: Given that metallicities in BCDs are about one tenth solar, do you find the dust content to be reduced by the same factor?

Thuan: Helou (1985, in Star-forming dwarf galaxies and related objects, eds. D. Kunth *et al.*, Editions Frontieres) has studied this question by using IRAS observations of dwarf galaxies. He found that the relation between 100 μm emission and neutral hydrogen content does not require a systematic deficiency in the dust to gas ratio in dwarf systems compared to spirals.

J. Bland Hawthorn: Are blue compact dwarfs detected in CO?

Thuan: To date, very few BCDs have been detected in CO (see F. Combes in Star forming dwarf galaxies and related objects, 1985 eds. D. Kunth, T. X. Thuan, and J. T. T. Van, Editions Frontieres, for a review). Their low metallicity does not favor dust and molecule formation. Arnault *et al.* (*Astr. Ap.,* **205**, 41, 1988) surveyed 12 BCDs with the IRAM 30m telescope and detected none. Their upper limits are consistent with the relation $L(CO)/M(HI) \propto [O/H]^{2.2}$ found for spiral galaxies and giant irregulars.

J. Young: Two questions. First, the gas consumption timescales you quoted assume an IMF forming high and low mass stars. If there is a low-mass cut-off, the present SFRs can be sustained considerably longer. Second, what fraction of dwarfs are BCDs? Can they be merger remnants, like IIZw40?

Thuan: If the lower mass cut-off is 10 $M_\odot$, the spectral synthesis results (Fanelli, O'Connell and Thuan 1988) give directly, without any assumption about the IMF, SFRs between 0.2 and 10^{-4} $M_\odot$ yr^{-1}, which give gas consumption time scales between 2×10^9 years and 10^{11} years.

Unfortunately, no complete statistical surveys of low-surface-brightness dwarfs and BCDs have been performed in the same volume of space, so that the fraction of dwarfs that are BCDs is not known precisely. If a burst lasts $\sim 2 \times 10^7$ years, and there have been ~ 5 bursts in the lifetime of the galaxy, one would expect 1 out of 100 dwarfs to be a BCD. Finally, I do not believe the majority of BCDs to be the product of mergers.

A. Campbell: I think it's a little misleading to conclude from the $(U-V)$ and $(B-K)$ diagram that the young stellar populations must be blue and old stellar populations must be red. At an age of 3×10^7 years, like IZw18, red supergiants should dominate the near-IR light. The near-IR offers a powerful tool for studying the *young* stellar population.

Thuan: Yes, I agree. It would be nice to obtain near-IR spectra of BCDs. Combining them with the UV and optical spectra would further constrain the star formation history of BCDs.

A. Campbell: Also, I don't think the metal content really constrains the duration of the burst, because the potential wells of BCDs are so shallow that they can easily throw out metals.

B. Rocca-Volmerange: Did you try to reproduce the *nebular* spectra of BCDs with your models. The emission-lines give you constraints on the Lyman continuum luminosity.

Thuan: Yes. Our models give consistent Hβ luminosities and equivalent widths.

R. Kennicutt: Some cautionary notes about lower mass limits derived from Balmer equivalent widths. First, my models show that the optical continuum is dominated by supergiants. Stars below $4M_\odot$ make only a negligible contribution to the integrated light. A small tweaking of the post-main-sequence evolutionary tracks for O and B stars with low metallicities could wipe out any information about a low mass cut-off. Furthermore, a low-mass cut-off is inconsistent with your assertion that the underlying light comes from old giants produced in previous bursts. To make old giants, you need low mass stars! Finally, if you had an IMF that made only stars above $4M_\odot$ the metal enrichment rate would probably violate the metallicity constraints unless you blow all the metals away.

Thuan: I agree that the lower mass limit is not a firm one. We need to look into the problem in more detail.

T. Heckman: The visual impression given by the images you showed is that the regions of current star-formation are preferentially located in the central part of the BCD (though not necessarily at the precise center). Have you tried to determine quantitatively whether these starbursting regions are more centrally concentrated than the underlying smooth light distribution?

Thuan: There are several morphological types of BCDs (Loose and Thuan, in Starforming dwarf galaxies and related objects, 1985, eds. D. Kunth et al., Editions Frontieres). There are BCDs with several centers of star formation near but not exactly at the center of the underlying component (type iE, MK86 in figure 4 is of that type). When there is only one center of star formation (type nE) (Haro 2, discussed by Loose and Thuan (1986) is of that type), the centers of the star-forming region and of the underlying older stellar population coincide.

K. Fricke: Is the metallicity in BCDs smoothly distributed? Is it due to the chemical evolution of the underlying component or to the cumulative effect of the starbursts?

Thuan: Because of the lack of spatial resolution, we do not have any information on the spatial distribution of the metallicity in BCDs. Maybe that is a job for HST! The metallicity results from the cumulating effect of successive starbursts. We do have observational evidence that stars have the same metallicity as the ionized gas, from the strengths of the stellar CIV and SiIV UV absorption lines (Fanelli, O'Connell and Thuan 1988).

R. Kudritzki: I am a bit puzzled that you try to model the UV spectrum of an extremely metal-poor galaxy like IZw18 by synthesizing observed spectra of massive stars in the Galaxy, with normal metallicities. Is this because you are only interested in the continuum?

Thuan: Ideally, it would have been nice to have metal-deficient stars in our library, but unfortunately UV observations and stellar atmosphere models of metal-deficient stars are not yet available. Thus, I am very encouraged by the progress you reported at this meeting on your stellar atmosphere modelling work. Clearly we need to use these types of models for our synthesis work in the future. The results I have reported here depend mainly on the fit to the continuum and should not be greatly affected by the mismatch between the metallicity of the galaxy and that of the library stars. In very metal-deficient galaxies like MK 209 $\equiv$ IZw36 ($Z \sim Z_\odot/10$), the synthesized spectrum predicts too strong Si IV and C IV absorption lines compared to what is observed (Fanelli, O'Connell and Thuan 1988).

THE INITIAL MASS FUNCTION IN M82

G. H. Rieke
Steward Observatory
University of Arizona
Tucson, AZ 85721
USA

Abstract. Recent observations provide a large number of tight constraints on starburst models of M82. The most critical of these constraints are now deduced independently from a variety of measurements often using differing techniques. Comparison with starburst models using modern stellar evolutionary tracks demonstrates that these constraints cannot be met with a conventional local initial mass function; formation of low mass stars must be less common in M82 than predicted by such an IMF.

1. INTRODUCTION

Theoretically, a variety of factors might influence the process of fragmentation of an interstellar cloud and hence the stellar initial mass function. Although it is clear that the IMF varies over relatively small regions of space, the observational search for large-scale variations in the IMF has produced few if any universally accepted examples. The problems arise from the need to identify a well defined region of space containing a young enough population of roughly coeval stars that the massive members still survive, but in a circumstance where a census can extend to low mass members. It has proved quite difficult even to determine the IMF for the solar neighborhood, due both to the fact that massive stars have evolved into obscurity and also to the problems in enumerating our faintest, lowest mass neighbors.

Under certain conditions, variations from the local IMF can be explored propitiously in starbursts in other galaxies. The requirements for a coeval, young stellar population are met by definition. Counts of the low mass members of a starburst are inaccessible. However, in sufficiently extreme cases the presence of these stars can be tested by searching for the mass they would contain. M82 is an optimum site for such studies because it is close and because it is a low mass galaxy whose properties are dominated by the powerful starburst occurring there.

2. STARBURST MODELING

Although the edge-on aspect of M82 and the consequent heavy extinction preclude easy access to commonly used optical indicators of stellar populations, we will show in the following section that many other observational constraints are available. It is possible to fit these constraints by a series of ad hoc assumptions about mixtures of stellar types whose emission is combined with that from other sources (*e.g.*, Lester *et al.* 1990). However, as pointed out originally by Tinsley (1968) in a different context, such modeling is likely inadvertently to make unrealistic assumptions about stellar evolution and/or to require bizarre forms for the IMF and time sequence of star formation. These difficulties arise because the conventional IMF and laws of stellar evolution tightly constrain the composite spectrum of a stellar population as functions of stellar formation rate and age.

Tinsley (1968) showed how these important constraints could be included in stellar population studies by building computer models in which stars form in accordance with specified initial mass functions and rates. After their formation, the evolution of the stars is followed in the computer. Information about stellar evolution and the correspondence of observable parameters with evolutionary phases is included in the models and they are used to predict properties of the composite population for comparison with observations. Modifications in the IMF or in the assumptions about other input conditions can be introduced in a systematic way.

Despite the improvements in population modeling through this technique, the predictions are only as good as the input information regarding the evolution and observable properties of the relevant types of stars. It is difficult to conduct formal error analyses on these issues so a conservative attitude is needed in drawing conclusions.

This type of modeling was originally applied to starbursts in the study of M82 by Rieke *et al.* (1980) and has been extended to a number of other galaxies by various workers since then. Updates to the models of M82 have been reported by Rieke (1988) and by Bernlöhr (1990a, b). The organization of these models is illustrated for the case of M82 in Figure 1. In addition to placing a sound foundation under the starburst hypothesis for M82, the models have lead to the interesting—and still controversial—conclusion that the IMF in this galaxy is biased against low mass stars compared with the IMF in the solar neighborhood.

3. OBSERVABLE PARAMETERS

The parameters that can be used to compare with starburst models of M82 are listed in Figure 1; below, we describe the status of each from an observational perspective. We will take M82 to be at a distance of 3.2 Mpc and the starburst to lie within a 30″ diameter region, based on the high resolution far infrared measurements by Telesco and Harper (1980) and Joy, Lester, and Harvey (1987).

3.1. Luminosity

Three components are important contributors to the total luminosity of the starburst in M82. The thermally reradiated far infrared flux produces 3×10^{10} $L_\odot$ (*e.g.*, Telesco and Harper 1980). The absolute K magnitude of the red stars implies that they emit a total of 3×10^{10} $L_\odot$ (see discussion below). Notni (1985) notes from the pattern

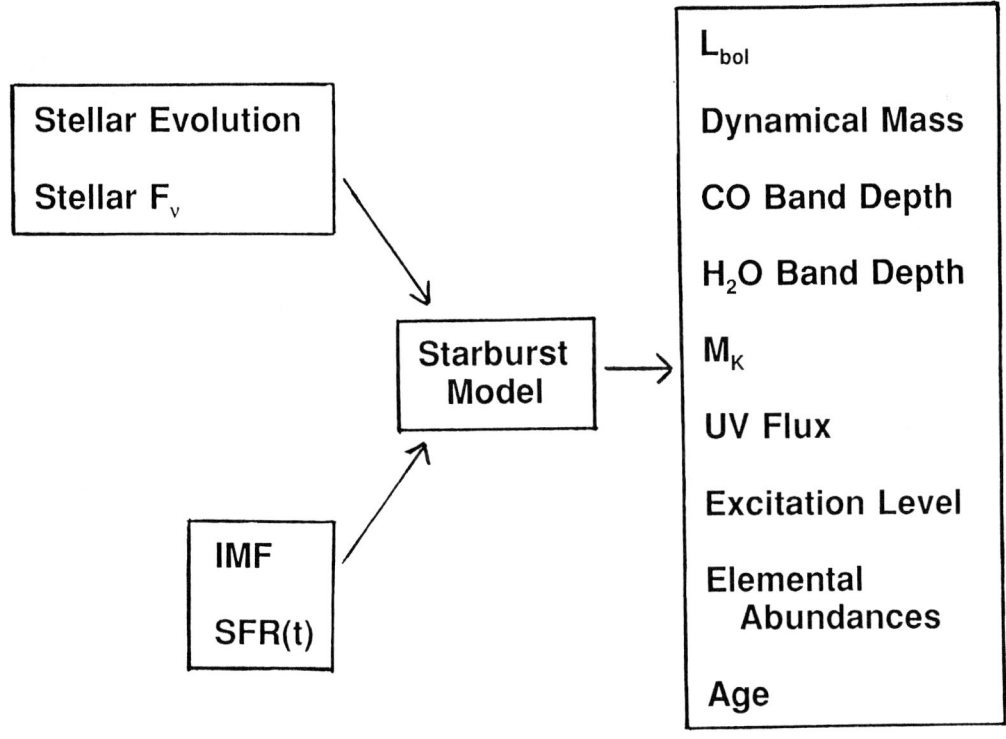

Figure 1. *Schematic diagram showing how starburst models can be used to predict observable quantities.*

of polarization vectors perpendicular to the galaxy plane that most of the scattered light comes from the nuclear region and is bluer than the rest of the galaxy. This result would imply that most of the luminosity of the red stars and some of that from the blue ones escapes; that is, it does not get absorbed and thermally reradiated. From these arguments, we place a lower limit of $L_{bol} > 7 \times 10^{10}\ L_\odot$.

3.2. Mass

The rotation curve of M82 has been determined in many different ways: 1.) optical data, *e.g.*, O'Connell and Mangano (1978); 2.) infrared fine structure lines at wavelengths that should penetrate the extinction, *e.g.*, Beck et al. (1978); 3.) mm-wave interferometry, *e.g.*, Lo et al. (1987); and 4.) high resolution single dish measurements in the mm-wave, *e.g.*, Nakai et al. (1987). Despite the range of angular resolutions employed and the susceptibility of the measurements to varying systematic errors (*e.g.*, effects of extinction), the results are remarkably consistent and indicate a total dynamical mass of $6 \times 10^8\ M_\odot$ for the starburst region if it has spherical symmetry. Because the estimated mass would be reduced for disk-like symmetry, this value can be taken as an upper limit.

It is improbable dynamically that a nuclear starburst could form in a galaxy where the total mass of the starburst exceeded that of the pre-existing nucleus. In addition, there is a substantial amount of gas in the central 30″ of M82. Estimates of the mass of gas are typically $1-2 \times 10^8$ $M_\odot$ (e.g., Lo et al. 1987), but with substantial uncertainties (Maloney and Black 1988; Stacey et al. 1990). We therefore adopt an upper limit to the stellar mass in the starburst of 2.5×10^8 $M_\odot$.

3.3. CO Index

The most prominent stellar features in the near infrared spectrum of M82 are the ^{12}CO absorptions extending longward from the 2-0 bandhead at 2.30μm. Conventionally, the strength of this absorption is quoted in terms of a photometric "CO index" measured through two narrowband filters and expressed in magnitudes. Measurements of CO indices for a broad variety of stars and stellar systems provide a context for interpreting the strength of the absorption in M82. However, the strong reddening of M82 makes the photometric prescription for determining the CO index invalid; it must be estimated from spectroscopy. The results of Walker, Lebofsky, and Rieke (1988) indicate a CO index of 0.20 magnitudes within a 7.8″ beam centered on the nucleus of M82, while Lester et al. (1990) find a value of 0.22 within a 3.8″ beam on the nucleus and a similar value 8″ to the southwest. These values are identical within the measurement errors, and an average of 0.21 is the best estimate of this parameter.

3.4. H$_2$O Index

A second prominent cool stellar feature in the near infrared is the absorption centered near 1.9μm due to steam, whose strength is also conventionally quoted in terms of a photometric index. Lester et al. (1990) report the H$_2$O feature to be very strong, corresponding to an index in magnitudes of 0.20. At least three previous observations (Willner et al. 1977; Rieke et al. 1980; Walker, Lebofsky, and Rieke 1988) would easily have detected a feature of this strength but did not. Lester et al. (1990) suggest that the feature may be strongly peaked in the nucleus of the galaxy and have escaped detection in previous observations with larger beams than they used (e.g., 7.8″ vs. 3.8″). This hypothesis seems unlikely on the basis of the new spectra in Figure 2 obtained with a 5.4″ beam both centered on the nucleus of the galaxy and on a position 8″ west and 2″ south (the second spectrum has been normalized to the first at 2.29μm). Neither spectrum shows any indication of a broad absorption centered at 1.9μm (Rieke and Rieke, 1990). We therefore reject the result of Lester et al. (1990); the four other measurements are all in agreement with an H$_2$O index < 0.07.

3.5. Absolute K magnitude

The H and K magnitudes of M82 in a 30″ diameter beam can be obtained by interpolation from Aaronson (1977) to be $m_K = 6.03$, $H-K = 0.61$. The H−K color is far redder than expected for a normal stellar population; both observations of stellar populations and theoretical starburst models predict an intrinsic H−K of 0.2. Conversion

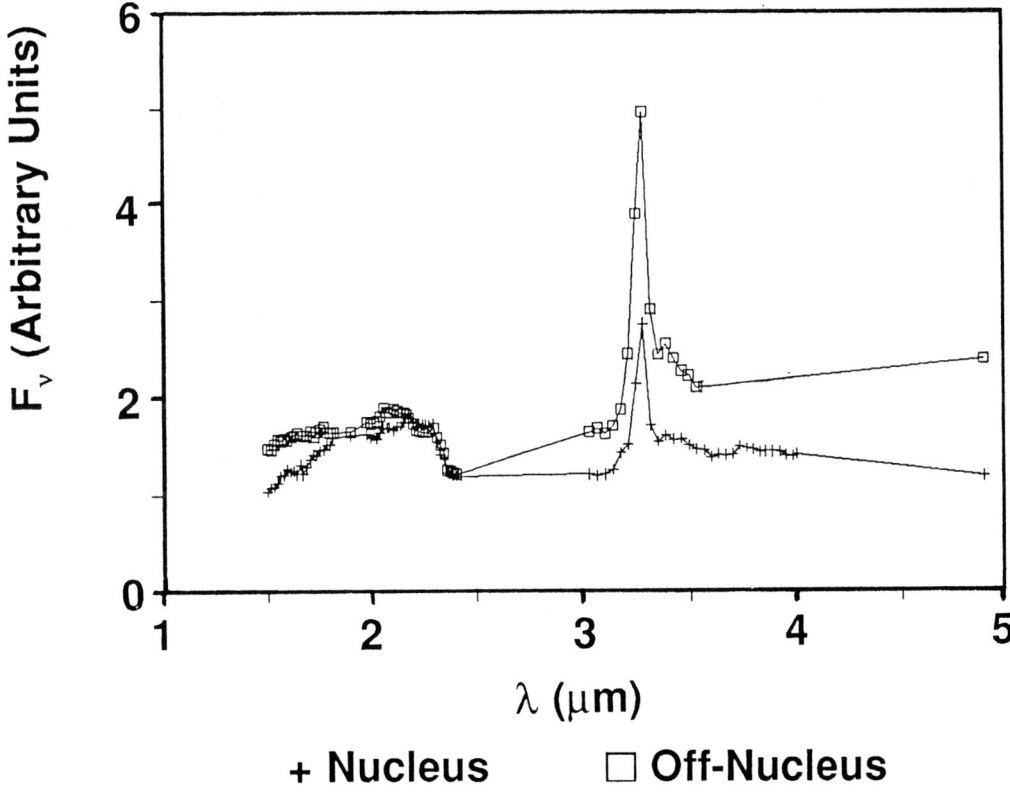

Figure 2. *Two near-IR spectra of M82 obtained with a 5.4″ beam. The crosses are the spectrum obtained on the nucleus of the galaxy and the boxes are the spectrum obtained at a position 8″ West and 2″ South of the nucleus. The two spectra have been normalized in flux at 2.29 μm.*

of the measurements to the intrinsic m_K requires that the origin of this red color be identified and removed. Interstellar extinction plays a role in these colors, but estimating its effects depends critically on the whether the K magnitude contains a significant contribution by warm dust and on how the obscuring dust is mixed with the stars in the galaxy.

Lester et al. (1990) argue that the nuclear K magnitude contains a 30% contribution (at 2.29μm) from warm dust that accounts for much of the red H−K. This dust would dilute the strength of the CO absorptions at 2.3μm; Lester et al. are forced to hypothesize a stellar population with a much larger CO index than 0.21. This requirement is a significant weakness in their argument, since starburst models show that 0.21 is about as large an index as can be plausibly produced (Bernlöhr 1990a and b). A second difficulty can be seen from Figure 2. This figure demonstrates that the hot dust component is stronger by a factor of 2 or more relative to the stellar flux at the position SW of the nucleus than on the nucleus, as indicated by the 3.3μm emission feature and by the shape of the continuum. However, the CO band strengths are identical (the similarity of the CO band strengths at these two positions is demonstrated in detail by Lester et al. (1990)). If 30% of the 2.29μm flux on the nucleus were due to hot dust, the CO bands would be very heavily filled in at the off-nuclear position. Finally,

Telesco et al. (1990) argue that the proportion of hot dust emission remains relatively low within the central 30″ of the galaxy. From these arguments, the H−K color from the starburst region appears not to be significantly affected by emission by hot dust.

Red stellar populations that could dominate the near infrared emission of M82 all have very similar colors at these wavelengths. Hence, the extinction can in principle be derived accurately if the extinction law and distribution of dust can be determined. The extinction is highly non-uniform over the face of the galaxy and it is likely that there are significant optical depth effects. Rieke et al. (1980) developed a crude model to account for these effects and derived an absolute K magnitude of −23. Unfortunately, this model has been misunderstood and misrepresented to make it appear seriously discordant with more naive models which assume only a foreground screen (e.g., Lester et al. 1990). The maximum extinction in the optically deep model was $A_V = 25$, but the equivalent extinction for a foreground screen would only be about $A_V = 12$. Recently, point-by-point extinction corrections have been obtained from infrared images which indicate both that the extinction is highly variable over the visible galaxy and that M_K is −22.5 to −23 (Telesco et al. 1990; Rieke and Rieke 1990), in good agreement with the earlier estimate.

3.6. UV Flux

Given the very heavy and patchy extinction, optical and near infrared line measurements are weighted toward regions of relatively light obscuration and tend to underestimate the total amount of ionized gas. This difficulty can be avoided by using mid-infrared or radio recombination lines. The recent measurement by Puxley et al. (1989) of H53α should be especially reliable because it is of good signal to noise and at high enough frequency that stimulated emission is not a problem.

The continuum flux at 3mm is 0.55 Jy (Jura, Hobbs, and Maran 1978; Carlstrom 1988). To be consistent with the recombination line strengths, it is necessary that the electron temperature be low, $T_e \sim 5000$K. With this value for the electron temperature, consistent estimates of the ionizing flux are obtained from both radio recombination lines and dereddened Brα as illustrated in Table 1. From these data, log (UV) ∼ 53.9. Because dust should compete with gas in absorbing ionizing photons, this value should be taken as a lower limit.

3.7. Excitation Level

Puxley et al. (1989) summarize arguments regarding the effective temperatures of the stars that excite the ionized gas in M82. Using the [OIII] and [NIII] line strengths from Duffy et al. (1987) and their own H53α observation, they show that the upper mass limit of the current-day mass function, m_u, is close to 25 $M_\odot$, with little dependence on various possible sources of uncertainty such as slope of the IMF or changes in metallicity or in the state of the interstellar medium. Similar conclusions can be reached from the mid-infrared fine structure lines of argon, sulphur, and neon (Willner et al. 1977) and from the optical forbidden lines of oxygen (Peimbert and Spinrad 1970).

TABLE 1. *UV flux in M82*

Reference	Line	Log (UV phot/sec)
Seaquist, Bell, and Bignell (1985)	Various Radio	53.91
Puxley et al. (1989)	H53α	54.04
Willner et al. (1977)	Brα	54.01
Simon, Simon, and Joyce (1979)	Brα	53.81

3.8. Elemental Abundances

Puxley et al. (1989) used previous measurements of fine structure lines of argon and neon (e.g., from Willner et al. 1977) to show that the abundances of these elements are within ±50% of solar. Using the results of Duffy et al. (1987), it follows that the abundance of oxygen is also within ±50% of solar. However, these authors find that N/O is anomalously low, consistent with the likely enrichment of the interstellar medium in M82 by massive supernovae. Assuming the ISM has been enriched, the stars in M82 would have lower metal abundances than the ISM; that is, it is unlikely that the stellar metallicity is greater than solar.

A dilemma arises from the apparently nearly solar elemental abundances in the ISM in M82, particularly of oxygen. Massive supernova explosions should be copious sources of oxygen (e.g., Arnett 1978); given the high efficiency of star formation, there should be relatively little unprocessed gas to dilute the enrichment products. Some possibilities are that the oxygen is thrown out of the galaxy in the supernova driven wind or that it is removed from the ionized gas into the neutral gas or onto dust grains. An intrinsically small oxygen enrichment would imply that few stars more massive than about 30 $M_\odot$ have ever formed in the starburst.

3.9. Age

X-ray observations show the hot, supernova-driven wind to extend to 8 kpc from the plane of the galaxy. Comparison with wind models (Tomisaka and Ikeuchi 1988) and free expansion timescales (Fabbiano 1988) suggests an age for this structure of 5×10^7 years. Allowing 1×10^7 years for the most massive stars to evolve to supernovae, the total age of the starburst should be about 6×10^7 years.

The starburst has a well developed red supergiant population and a large portion of its luminosity is produced by these stars. Depending only slightly on initial conditions, such a state is produced in starburst modeling only after about 3×10^7 years. Therefore, the age deduced by X-ray wind models seems to be confirmed by the state of the stellar population.

3.10. Summary

The most important constraints on starburst models for M82 are summarized in Table 2. Two other constraints enter indirectly. The argument that the stars in M82 should not be super metal-rich excludes models which invoke high metallicities to fit spectral features. The stellar-based age limits will be automatically honored by models that produce adequate near infrared flux, although adoption of the ages suggested by the X-ray wind models would provide a more stringent constraint.

The new observational data have generally confirmed the estimates by Rieke et al. (1980) or acted to make them more stringent, reflecting the conservatism assumed in that paper.

TABLE 2. *Observational constraints on starburst models for M82*

Constraint	Value	Local IMF
Luminosity ($L_\odot$)	$> 7 \times 10^{10}$	4.6×10^{10}
Mass ($M_\odot$)	$< 2.5 \times 10^{8}$	2.4×10^{8}
CO Index	0.21	0.21
H_2O Index	< 0.07	0.06
M_K	< -22.5	-22.7
Log UV (phot/sec)	> 53.9	53.1
m_u ($M_\odot$)	~ 25	31

4. COMPARISON WITH MODELS

Predictions for the observable parameters in Table 2 have been calculated by Bernlöhr (1990a and b), utilizing more modern stellar evolutionary tracks than in previous work. His results for the conventional local initial mass function are listed in Table 2. The primary difference with the earlier models is the greater brightness at K that results from the greater luminosity of red supergiants in the new evolutionary tracks. It can be seen that the local IMF model is seriously deficient in both total luminosity and ionizing flux, despite placing the upper mass limit somewhat too high.

Although this particular starburst model violates just three of the constraints, it is wrong to hold that the constraints that are met are not useful (as in Telesco et al. 1990). For example, reducing the age of the starburst from that in Table 2 can raise its luminosity sufficiently to meet the constraint on this parameter and its UV flux to within about a factor of two of the target. However, younger models would fall substantially short in absolute K magnitude and CO band strength because of the inadequate development of their red supergiant populations. Thus, deriving unambigous conclusions from starburst models usually depends on an understanding of how all the observational constraints are affected by variations in model parameters.

As argued for example by Rieke et al. (1980), Kronberg, Biermann and Schwab (1985), Rieke (1988), Puxley et al. (1989), and Bernlöhr (1990a), satisfying the full set of observational constraints requires that the formation of high mass stars be favored in the M82 starburst compared with the conventional local IMF. Future work, especially more detailed starburst modeling, may help determine how this bias appears—e.g., as a low mass cutoff or a flatter IMF.

In summary, recent work has strengthened the evidence already available to show M82 to be an extremely strong case for large scale departures in the IMF from its local form. The discrepancies between the predicted outputs of local IMF starbursts and the observed parameters are now established by a variety of observations. The changes that have occurred in starburst models over the last decade are much smaller than the existing discrepancies, making it implausible that further modifications in stellar evolution theory will remove these differences. Given this "existence proof", theoretical work can examine how the extreme conditions in a nuclear starburst have a strong influence on the process of star formation as reflected in the modifications to the initial mass function.

I thank D. Arnett, R. Kennicutt, and M. Rieke for helpful discussions and K. Bernlöhr for communication of unpublished results. K. Loken provided critical assistance in preparing this review. This work was partially supported by the National Science Foundation.

REFERENCES

Aaronson, M. 1977, Ph.D. Thesis, Harvard University
Arnett. W. D. 1978, *Ap. J.*, **219**, 1008.
Beck, S. C., Lacy, J. H., Baas, F., and Townes, C. H. 1978, *Ap. J.*, **231**, 28.
Bernlöhr, K. 1990a, this conference.
_____. 1990b, private communication.
Carlstrom, J. E. 1988, in *Galactic and Extragalactic Star Formation*, ed. Pudritz and Fich (Kluwer: Dordrecht), p. 571.
Duffy, P. B., Erickson, E. F., Haas, M. R., and Houck, J. R. 1987, *Ap. J.*, **315**, 68.
Fabbiano, G. 1988, *Ap. J.*, **330**, 672.
Joy, M., Lester, D. F., and Harvey, P. M. 1987, *Ap. J.*, **319**, 314.
Jura, M., Hobbs, R. W., and Maran S. P. 1978, *A. J.*, **83**, 153.
Kronberg, P. P., Biermann, P., and Schwab, F. R. 1985, *Ap. J.*, **291**, 693.
Lester, D. F., Carr, J. S., Joy, M., and Gaffney, N. 1990, *Ap. J.*, **352**, 544.
Lo, K. Y., Cheung, K. W., Masson, C. R., Phillips, T. G., Scott, S. L., and Woody, D. P. 1987, *Ap. J.*, **312**, 574.
Maloney, P., and Black, J. H. 1988, *Ap. J.*, **325**, 389.
Nakai, N., Hayashi, M., Handa, T., Sofue, Y., Hasegawa, T., and Sasaki, M. 1987, *Pub. Ast. Soc. Japan*, **39**, 685.
Notni, P. 1985, *Astron. Nach.*, **306**, 273.
O'Connell, R. W., and Mangano, J. J. 1978, *Ap. J.*, **221**, 62.
Peimbert, M., and Spinrad, H. 1970, *Ap. J.*, **160**, 429.
Puxley, P. J., Brand, P. W. J. L., Moore, T. J. T., Mountain, C. M., Nakai, N., and Yamashita, T. 1989, *Ap. J.*, **345**, 163.
Rieke, G. H., Lebofsky, M. J., Thompson, R. I., Low, F. J., and Tokunaga, A. T. 1980, *Ap. J.*, **238**, 24.
Rieke, G. H. 1988, in *Galactic and Extragalactic Star Formation*, ed. Pudritz and Fich (Kluwer: Dordrecht), p. 561.
Rieke, G. H., and Rieke, M. J. 1990, in preparation.
Seaquist, E. R., Bell, M. B., and Bignell, R. C. 1985, *Ap. J.*, **294**, 546.
Simon, M., Simon T., and Joyce R. R. 1979, *Ap. J.*, **227**, 64.
Stacey, G. J., Geis, N., Genzel, R., Lugten, J. B., Poglitsch, A., Sternberg, A., and Townes, C. H. 1990, preprint.
Telesco, C. M. and Harper, D. A. 1980. *Ap. J.*, **235**, 392.

Telesco, C. M., Campins, H., Joy, M., Dietz, K., and Decher, R. 1990, preprint.
Tinsley, B. M. 1968, *Ap. J.*, **151**, 547.
Tomisaka, K., and Ikeuchi, S. 1988, *Ap. J.*, **330**, 695.
Walker, C. E., Lebofsky, M. J., and Rieke, G. H. 1988, *Ap. J.*, **325**, 687.
Willner, S. P., Soifer, B. T., Russell, R. W., Joyce, R. R., and Gillett, F. C. 1977, *Ap. J. (Letters)*, **217**, L121.

DISCUSSION

N. Scoville: I am puzzled by your constraint on the upper mass limit on the IMF derived from the oxygen abundance. Doesn't this constraint apply to every galaxy?

Rieke: Maybe it means that the IMF in the Milky Way doesn't go all the way up to 100 $M_\odot$ either. I certainly don't have a universal solution for this problem.

A. Maeder: I want to follow up on this. The stellar yields from Arnett that you used to derive this constraint were computed when IUE had not even been launched and models for massive stars had no mass loss in them! Current models that incorporate mass loss decrease the oxygen yield by factors of 2 to 3.

Rieke: Even a factor of 2 or 3 doesn't change my argument much. An IMF extending past 40 $M_\odot$ gives something like 10 times solar oxygen abundances.

N. Walborn: I really think that you are taking the models too literally. We just don't know enough about real supernova explosions and very massive star evolution to derive strong constraints.

Rieke: We think we understand why the chemical abundances in the solar system are what they are. I'm sure a factor of 2 or 3 could have been missed in the calculations, but the effect in M82 is much stronger than that.

A. Campbell: M82 is observed to have an enormous bi-polar outflow that indicates that at least part of the metals made inside the starburst are expelled out into the inter-galactic medium. Did you account for the oxygen that must have been lost in that way when you derived a metallicity constraint on the upper mass cut-off to the IMF?

Rieke: No, not explicitly. My limit assumes that all the metals are retained and remain mixed with the ionized gas.

J. Bland Hawthorn: If you look at M81 you see individual supergiants. M82 is at the same distance, so why don't you see individual supergiants in M82?

Rieke: Well, maybe because there's an upper mass limit on the IMF ... (laughter).

J. Bland Hawthorn: Given that high mass stars drastically affect their environment, is it possible that the IMF is complicated by sporadic star formation at the high mass end and continuous star formation at the low mass end?

Rieke: Because the deficiency of low mass stars is deduced from the dynamical mass, the conclusion about a deficiency of low mass stars can just as well be stated that such stars will never form in this region in sufficient numbers to build a conventional local IMF.

I. Mirabel: Can you make any comments on physical reasons for the low mass cutoff of 3 $M_\odot$? Could it be related to the high degree of turbulence in the molecular gas in M82?

Rieke: I could speculate about the role of the turbulence, the large gravitational field, the high temperature of the ISM. All these things ought to make it more difficult for low mass stars to form.

A. Campbell: It seems that the result that the IMF must be truncated at low mass depends quite strongly on what you assume for the luminosity of red supergiants. Roberta Humphreys pointed out yesterday that these stars are extremely luminous, so only a few of them would give you a large K luminosity with a small amount of mass. They would also increase the depth of the CO index, as required.

Rieke: The evolutionary models in the most recent starburst models include stars equivalent to the most luminous red supergiants. Moreover, the most difficult constraint to meet with a conventional IMF is the ionizing flux, given the minimum age implied by the presence of red supergiants. The details of the red supergiant outputs are not critical.

B. Rocca-Volmerange: In our Galaxy, carbon is essentially produced by low and intermediate mass stars. If the M82 IMF cuts off at 3 $M_\odot$, the carbon abundance should be low. Do you see any evidence for this effect?

Rieke: As we've heard, abundances are uncertain in any case, and we have no data on carbon.

T. Heckman: I'm worried that your age constraint comes from the estimated dynamical age of the X-ray nebula. This may be uncertain by an order-of-magnitude. If you relax your age constraint entirely, what happens?

Rieke: It doesn't help much. To get an adequate CO band strength and 2μm flux you must have a relatively old stellar population. The CO band is the best indicator because it grows monotonically with time. In fact it is hard for the models to give a CO band as strong as that observed. Using only the CO band strength, the ages are consistent with the X-ray ages to within a factor of two. Remember too that this is an over constrained problem and other constraints come to your rescue.

F. Bruhweiler: In your model you have assumed that the IR hydrogen recombination lines are completely due to the Lyman continuum of OB stars and hence derived the UV flux distribution. However, recombination can also arise in shocks in this apparently violent galaxy. Do you have any indication how shocks affect your deduced UV fluxes? How would they affect the final model for M82?

Rieke: Given the consistency in the amount of UV flux and the temperature of the exciting source obtained by a variety of indicators under the assumption of radiative excitation, it seems unlikely that shocks play a dominant role in M82.

C. Norman: If M82 were viewed face-on, your model predicts a nuclear magnitude in the visual of -21 to -22. Since such galaxies are rarely observed, does this present a problem for your model? In other words, where are the face-on M82's?

Rieke: The predicted absolute visual magnitude for M82 is about -20, in the absence of reddening. A dwarf galaxy with such a bright nucleus would indeed be uncommon; perhaps reddening plays a role in extinguishing the visible outputs of all these objects.

N. Devereux: I've studied a sample of 20 starburst galaxies with similar nuclear star-formation rates to M82. On the basis of the 2 μm, 10 μm, and far-IR luminosities, there is no need to invoke a non-standard IMF. While I don't have as many constraints as George does, I wonder whether M82 might be a very strange fish!

Rieke: It may be that since M82 is a bright starburst in a low mass galaxy, the starburst completely dominates the old underlying stellar population. That may not be the case in the more massive galaxies you've studied.

N. Devereux: Well, I don't understand what's going on! (Laughter)

POPULATION SYNTHESIS MODELS OF STARBURSTS

Brigitte Rocca–Volmerange
Institut d'Astrophysique de Paris, CNRS
F–75014 Paris
and
Université de Paris XI, Bat 205
F–91405 Orsay Cedex
France

Abstract. Ultraluminous starbursts have been discovered in the far-IR while spectral signatures of intense star forming processes are detected from the far–UV to the near-IR, inside blue compact dwarf galaxies, faint galaxies in deep surveys and most distant radio galaxies. Are such bursts of star formation related to infrared starbursts? This is possible if a merging scenario drives the evolution of galaxies. Typical signatures are analyzed and the interpretation of various samples, more specially of high-z galaxies, makes interactions of galaxies a common scenario of galaxy evolution. This simultaneously implies luminosity and number evolution.

1. INTRODUCTION

From the pioneering works on bursts of star formation in blue galaxies (compact dwarfs or HII regions) analyzed in the far–UV light (Searle *et al.* 1973, Larson and Tinsley, 1978, Huchra, 1977), an extensive advance on the concept of a starburst galaxy resulted from multispectral data. The important results from the IRAS satellite on ultraluminous galaxies (Soifer *et al.* 1984, Lonsdale *et al.* 1985) and their optical counterparts showed that starbursts are triggered by galaxy interactions. On the other hand, the present samples of powerful distant galaxies (radio galaxies, bright cluster members) also show evidences of an intense star forming process and of interactions which makes an identification with starbursts tantalizing (Djorgovski, 1987). However distant radio galaxies do not look like interacting IRAS galaxies: the main difference is the large amount of dust in IRAS galaxies observed in the far-infrared while radio galaxies show signs of activity in the far-UV/optical/near-infrared and huge nebular emission lines. These spectral signatures of star formation are amplified features of nearby HII regions. At a lower level, these blue signs were detected in most distant galaxy clusters (Butcher and Oemler, 1984), in spectra of elliptical galaxies with an A-star population (Dressler and Gunn, 1983), and more recently, in a large sample of distant clusters (Gunn, Hoessel and Oke, 1986). Are these fainter bursts related to the intense starbursts (triggering, dynamical evolution, star formation parameters, ...)? Besides analyses of individual galaxies, a more statistical approach was undertaken. Deep surveys show a population of extremely blue galaxies which was analyzed with starburst models: in Broadhurst et al. 1988, the blue excess of faint counts is interpreted as an strong luminosity evolution of the present low-luminosity objects while the best fits of the present data including Tyson's (1988) observations need luminosity and number evolutions simulating a merging effect (Guiderdoni and Rocca-Volmerange, 1990, Rocca-Volmerange and Guiderdoni, 1990, Koo, 1990). Differing from the distant radio galaxies which possibly have a non–thermal emission, these faint galaxy populations are likely "normal" field galaxies and then excellent witnesses of evolution models. What is the role of the starburst phenomenon in these populations? If we admit that a starburst is an interacting galaxy, how frequent is the interaction? What is the mass distribution of intervening galaxies? Is the merging the normal evolution scenario of galaxies? Is the dust formed during, after or before the interaction? These questions are largely debated. But the two simultaneous approaches, statistical and individual, will be the best way to understand the role of a starburst in the general scenario of the formation and evolution of a galaxy.

2. TYPICAL SPECTRAL FEATURES

The current concept of a starburst as a globally star forming galaxy needs a more precise definition. A burst of star formation is an episodic phenomenon which induces an extremely high star-formation rate on a short timescale. The relative strength of a burst is often estimated by $b = M_*/M_{tot}$, the mass of stars formed during the burst relative to the stellar mass present in the galaxy (Larson and Tinsley, 1978). M_*, depending on the star formation rate (SFR) and its time duration, can be estimated by an indirect way from the far-infrared and from the direct far-UV emission of the stellar population. In the two cases, the b parameter is measured by calibrating relative to the visible: L_{FIR}/L_B for IRAS galaxies and L_{UV}/L_V for redshifted radio galaxies.

What values of b correspond to a starburst? Massive stars are so bright (a factor $\geq 10^3$ in luminosity relatively to the solar luminosity) that a small number is sufficient to dominate the total emission, screening for a short time the underlying population. Values $b =0.1$-0.05 are adopted for starbursts from blue and red analyses (Larson and Tinsley, 1978, Struck-Marcell and Tinsley, 1978, Belfort, 1987) In the extremely distant radio galaxy 0902+34 (Lilly, 1988), an estimate of the current star-formation rate from the far-UV relative to the visible gives $b \simeq 0.01$ (Rocca-Volmerange and Guiderdoni, 1990). Higher values of b are expected during the epoch of galaxy formation. They imply a conversion of most of the total mass $\simeq 10^{12} M_\odot$ on a short timescale ($\simeq 1$ Gyr). However the search for the Lyα1215Å emission predicted by such models (Partridge and Peebles, 1967, Meier, 1976) is up to now unfruitful (Pritchet and Hartwick, 1989). Several timescales are used for starbursts: 10^8yrs corresponds to dynamical timescales of BCDG and distant galaxies of a few arcsec diameter, 10^7yrs or 10^6yrs to the duration of a radio jet if it is assumed to trigger the star formation process and 1 Gyr is mostly used for models of massive ($10^{12} M_\odot$) galaxies.

To model the stellar population of starbursts, we here limit our analysis to the far-UV/visible/near-IR wavelength range. The best examples of starbursts are detected in the far-UV light (HII regions or BCDG, visible interacting and distant radio galaxies). UV and optical counterparts of the ultraluminous IRAS galaxies (Soifer et al. 1984, Mirabel, this conference) as well as the typical galaxy M82 (Lester et al. 1990, Rieke, this conference) are strongly biased by a dominating obscuration effect. This may be interpreted as an evolutionary effect: the direct stellar energy distribution and strong emission lines can only appear when the dust in which forming stars are embedded has been swept off by winds and mass losses. The main UV to IR spectral features of a galaxy during a burst are typical of massive stars and possibly of evolved stars of the underlying population:

i) The **stellar continuum** which depends on the current star formation activity as well as the emission from older stars, witnesses of the past activity. Following observations through visible UBVRI and infrared JHKLMN broad bands, low resolution spectra in the far-UV with the IUE satellite (see stellar atlases from Wu et al. 1983, Heck et al. 1984 and the atlas of HII regions, Rosa et al. 1984) to the near-IR (Lester et al. 1990, Davidge and Maillard, 1990) considerably improved our knowledge of stellar signatures in galaxies. The question is to separate both old/new populations. In the UV, massive stars are dominant but evolved hot stars (horizontal branch or post asymptotic giant branch stars) can contribute to a post-burst phase while red supergiants or stars on the asymptotic giant branch can dominate in the infrared even if a large population of giants is present.

ii) **Stellar absorption and emission lines** from high-resolution spectra which are good indicators of the metal enrichment and, in some cases, of the age. Hydrogen absorption lines from Lymanα (1215Å) to Balmer and Brackett lines, HeI and HeII lines, MgII 2797Å, CaII H and K lines, NaI 5892Å, TiO bands, the CaII triplet at 8498, 8542 and 8662Å (Jones et al. 1984), as well as CO bands (2.35μm) are among the best indicators of star formation. Also the Wolf-Rayet spectral features HeII4686Å and NIII4640Å are found in blue compact galaxies (Kunth and Schild, 1986) and can be used to estimate star-formation rates of bursting galaxies. In the far-UV light, detailed absorption lines resulting from a better analysis of the IUE stellar spectra, (Walborn and Fitzpatrick, this conference) can considerably help for understanding most spectra of star forming galaxies. Spectra of distant radio galaxies also show stellar absorption lines (Chambers and McCarthy, 1990) even if the signal-to-noise ratio needs to be improved. These lines can be used as a metallicity indicator. Metallicity, estimated from equivalent widths or line ratios, is related to age, either from stellar models (Barbuy et al.), from a library of globular clusters of known ages (Bica and Alloin, 1986) or from chemical models (Rocca-Volmerange and Schaeffer, 1990). All of them are more or less directly related to models of internal structure.

iii) The **nebular emission lines** which are indicators of the ionizing photon number and then of the massive stellar population. UV and optical line intensity ratios are tabulated in Spitzer, 1978, Aller, 1984, Baldwin et al. 1981. The fluxes and equivalent widths of the Hα line (Kennicutt, 1983) as well as the Hβ line (Gallagher et al. 1984) are used to estimate the star-formation rates. A difficulty comes from the relative extinction and the contribution of a stellar absorption. Some predictions for synthetic spectra of galaxies have been calculated (Rocca-Volmerange and Guiderdoni, 1988, Olofsson, 1989).

All these typical features are found in starbursts which are identified in BCDG/HII galaxies, interacting galaxies and distant galaxies like radio galaxies and possibly field galaxies detected in faint counts. Concerning the stellar energy distribution of BCDG/HII galaxies, most of these objects have a strong UV excess, a high surface brightness, P-Cygni profiles for SiIV(1403Å) and CIV(1548Å) due to the mass loss of OB and WR stars, and high-excitation emission lines (see review by Kunth, 1989, Thuan, 1987 and this conference). The Lyα (1215Å) emission is surprisingly faint or nonexistent for such highly bursting galaxies (Deharveng et al. 1985). As reported by Huchra (1985) these objects differ from Markarian and Seyfert 2 galaxies which could be well represented by the bursting galaxy Mk171 (Augarde and Lequeux, 1985). P Cygni profiles of C IV and Si IV lines are found and the extinction E_{B-V} is estimated to $\simeq 0.28$ to 0.7 according to models. Most of these galaxies have strong emission lines and a decreasing or flat stellar energy distribution.

Other examples are given by interacting galaxies. Starburst nuclei (Balzano, 1983), spirals with close companions and the "Arp" sample of disturbed pairs (Keel et al. 1983) have been compared to a control sample. The authors conclude that very luminous starburst HII regions are only seen in the massive interacting spirals. They also present some relations with LINERS and Seyfert galaxies. Other samples of dwarf galaxies ESO400-IG43 (Bergvall and Jörsater, 1988) and more recently the elliptical galaxy ESO341-IG04 (Bergvall et al. 1989) and the merging galaxy ESO286-IG19 with optical, infrared and nebular emission (Johansson, 1990) extend the interaction range.

Spectrophotometry of the optical counterparts of 3CR radio galaxies up to $z \simeq 3$ shows strong emission lines and extreme blue colors, interpreted as an intense burst. As a consequence, Lyα equivalent widths ($\geq 1000\text{\AA}$) of the so-called Lyα galaxies (Spinrad, 1987, Djorgovski, 1987 and references therein) are mainly attributed to a photoionization process due to the massive stellar population (Valls-Gabaud et al. 1990). Moreover bursts can be analyzed through the optical and infrared counterparts of distant radio galaxies observed in the VIJK bands (Laing et al. 1983, Lilly and Longair, 1984, Allington-Smith et al. 1982, Downes et al. 1986 and Dunlop et al. 1989). The best example is the extremely distant ($z = 3.395$) radio galaxy discovered by Lilly, 1988. The galaxy 4C41.17 discovered at a redshift $z = 3.8$ (Chambers et al. 1990) seems to show similar properties. A surprising fact is that in spite of an active research, no other $z \geq 3$ galaxy has been discovered for two years. Is it due to the difficulty of detection or to the poorness of such a distant sample? The question is important for galaxy formation models which favor a bulk of galaxies at a redshift $z \simeq 2-3$. The most striking feature of these distant galaxies is the gap between the apparent flux in the K band and the plateau observed in the $U'BVRI$ bands. After redshift correction, this corresponds to a high visible emission in the galaxy frame, compared to the far-UV. The interpretation in terms of starburst will be discussed below.

At last, our galaxy count models (Guiderdoni and Rocca-Volmerange, 1990) fit the number, color and redshift distributions revealed by the deepest surveys in visible photometric bands, including Tyson (1988) down to $J = 27$. A direct consequence of the luminosity evolution is a low value of the deceleration parameter q_0. Number evolution is needed to save $\Omega_0=1$ (Rocca-Volmerange and Guiderdoni, 1990) which would imply some merging or interacting effect. According to their mass distribution, interaction events can be rare for massive galaxies but much more frequent for small components. Another interpretation of such blue galaxies in the Durham Faint Survey (Broadhurst et al. 1988, Colless et al. 1989) are strong bursts formed at low redshift ($z \leq 0.5$), principally affecting the faint end of the luminosity function. They also interpret their histogram of equivalent widths of the [OII]3727Å emission lines as due to burst effects. Can models significantly separate these various interpretations?

3. MODELS

Direct stellar and nebular emission is observable in the far–UV light till the near–infrared. It gives the respective contributions of young and old stars while indirect emission through the far-infrared reveals a huge amount of energy dissipated by heated grain reemission. The analysis of the gas and the stellar population (mass distribution, metallicity, star formation) in such starbursts is the first objective to estimate the role of starbursts in the large processes of galaxy evolution. This approach needs to know typical features of ionized gas and stars as well as the physical conditions of the starburst phenomenon (density, pressure, temperature) which eventually may regulate the star-formation parameters. A population synthesis needs a parameterization of the star formation laws, compatible with large scale observations of star forming regions. Concerning stars, detailed observed spectra of normal or/and typical subpopulations such as Wolf–Rayet stars, AGB- and carbon-stars are preferred and are used in a stellar library. The location of such stars in the HR diagram is also needed for fitting the various stellar evolution models of massive stars (Maeder and Meynet, 1988, Renzini and Voli, 1981, see also reviews in this conference). Galaxy models with or without evolution were proposed to reproduce observations of starbursts in the far–UV to near–infrared with metallicity effect. Most of them conclude that emission from young massive stars is dominant, often occulting any feature of a possible old giant population.

The classical way to model the evolution of a stellar burst is to create a coeval population according to an initial mass function (IMF) and a star formation rate (SFR) and then to follow the evolution of stars essentially from theoretical tracks in the HR diagram. The evolutionary phases with an important released energy have to be represented even if the phase duration is short (Renzini, 1986). A library of stellar spectra, or possibly of globular and open cluster spectra, is needed to reproduce details of galaxy spectra. While IMF and SFR are free parameters, the set of evolutionary tracks, the stellar library, extinction and ionized gas emission models are input data which give the quality of the model. Models based on this scheme are, in historical order: Tinsley, 1972, Searle et al. 1973, Rocca-Volmerange et al. 1981, Bruzual, 1983, Guiderdoni and Rocca-Volmerange, 1987, Olofsson, 1989. Another approach is based on a library of star cluster integrated spectra, observed in the visible and the near-infrared (Bica and Alloin, 1986). Metallicity calibrated from equivalent widths is statistically related to the ages of star clusters. This method which is similar to the stationary methods of population synthesis (Faber, 1972, Pickles, 1985) implicitly assumes a constant IMF and star formation history which are not explicited. Moreover they use a metallicity-age relation consistent with evolutionary tracks. Any change in stellar tracks corresponds to a change of the estimated age and then of the relation with the metallicity. Another question is to know if the stellar population of spiral bulges and elliptical galaxies is well represented from spectra of globular clusters (see discussion in Baum, 1990).

We can summarize the main input data for evolutionary models (Bruzual, 1983, Guiderdoni and Rocca-Volmerange, 1987, Rocca-Volmerange and Guiderdoni, 1990):

- Stellar evolutionary tracks are from internal structure models giving the luminosity, the effective temperature and the lifetime duration of a given mass star along evolutionary tracks in the Hertzsprung-Russell diagram. Theoretical Yale tracks are used in most first models. Recent improvements of stellar models including overshooting, new opacities and helium abundances, mass loss (VandenBerg, 1985, Maeder and Meynet, 1988, Bertelli et al. 1989) better fit the color-magnitude diagrams of the globular and open clusters.
- Stellar spectra have been extracted from spectral libraries based on the recent UV atlases observed by the IUE satellite (Wu et al. 1983) and connected to the

optical atlases (Gunn and Stryker, 1983). Several codes use theoretical models (Kurucz, 1979, Mihalas, 1972) which often present defaults in reproducing the energy emission and the absorption lines.

- Extinction is estimated by assuming that the dominant emitters through UV and visible bands are uniformly mixed with the gas. The effective optical thickness is essentially depending on the gas content.
- Nebular lines emitted from the gas ionized by massive stars are deduced from the Lyman continuum photon number. The emission-line ratio is an output of the PHOTO code (Stasińska, 1984). The energy emitted in the H lines, for instance in the Hβ line, is proportional to the number of Lyman–continuum photons: $F(H\beta,t) = E(H\beta)N_{Lyc}f$. This number N_{Lyc} is very sensitive to the slope of the IMF for $m_* > 15 M_\odot$ and to the upper mass limit, and can easily vary by a factor of 5 for a given SFR, as shown in Kennicutt, 1983. The adopted energy is $E(H\beta) = 4.757\ 10^{-13}$ erg photon^{-1}. The factor $(1-f)$ accounts for the amount of ionizing photons absorbed by dust grains or escaping from the HII regions.
- Data specific to chemical models (Rocca-Volmerange and Schaeffer, 1990) are nucleosynthesis products from supernovae of type I and II and from low and intermediate mass stars (which do not end as supernovae). Theoretical estimates have been calculated by Nomoto et al. 1984, Woosley, 1986, Renzini and Voli, 1981.

Synthetic spectra (Rocca-Volmerange and Guiderdoni, 1988) of a starburst are modeled with nebular emission lines at various ages. The IMF $\phi(m) \propto m^{-x}$ is limited between two mass limits (M_{inf} and M_{sup}). Normalized to 1, it is in some cases divided into a fraction ζ corresponding to the only visible component of forming stars, the other component being invisible stars, planets or VMO's (Bahcall, 1985). To respect the observed M/L ratio in nearby galaxies, we adopt a fraction $\zeta=0.5$. Figure 1 presents fluxes and lines for two different IMFs: a *massive* IMF with $x=0.9$, $M_{inf}=2\ M_\odot$ and $M_{sup}=120\ M_\odot$ and the so-called standard IMF with $x = 0.25$ ($0.1 \leq m/M_\odot \leq 1$), 1.35 ($1 \leq m/M_\odot \leq 2$) and 1.7 ($2 \leq m/M_\odot \leq 80$) from Scalo, 1986, instead of 2.3 from Miller and Scalo, 1979, or 2 from Lequeux, 1979. M_{inf} and M_{sup} are taken as 0.1 $M_\odot$ and 80 $M_\odot$, respectively. No significant difference appears between fluxes and lines emitted during the burst whatever the adopted timescale. Differences only appear in the far-UV range during the post-burst phase, essentially due to the stellar populations of 0.1 to 1 Gyr lifetimes, or $\simeq 2$ to $5 M_\odot$ initial masses.

4. COMPARISON WITH OBSERVATIONS AND DISCUSSION

Interacting or merging galaxies are the best examples of bursting galaxies to be compared with models. Results on IMF, age and the old population underlying the young massive population can be shortly summarized. In a sample of blue galaxies, the IMF appears flatter and with a lower M_{inf} than in the solar neighborhood (Olofsson, 1989). A similar result was deduced from different data on starbursts (Rieke and independently Joseph, this conference). The question is not clear about M_{sup} since these last studies conclude to $M_{sup} \simeq 30 M_\odot$ while the UV light and nebular emission lines in BCDG or in distant radio galaxies favor $M_{sup} \geq 60 M_\odot$. The weight of the old population can be estimated from models (Struck-Marcell and Tinsley, 1984, Bica and Alloin, 1986). A conclusion of the previous authors is that an old population is screened behind a massive starburst during 10^7 years, so long as 10 per cent of the galaxy mass is involved by the burst. This result assumes that the burst and the old

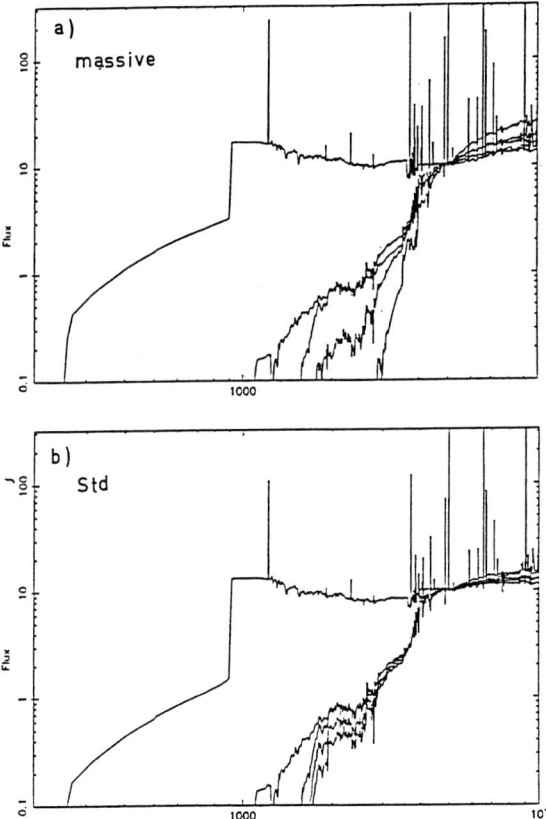

Figure 1ab. *Fluxes and nebular lines during and after a star formation burst consuming all the initial content of gas through 0.1 Gyr. A massive (a) and a standard (b) IMF as described in text is proposed. Fluxes in arbitrary units normalized at 4990Å are shown from 0.025 Gyr by a 0.5 Gyr step. The wavelength range is 200Å to 10^4Å and the spectral resolution is 10Å.*

population are well modeled by integrated spectra of star clusters and more specifically that the IMF and the age-metallicity relation are constant. Concerning the age, we shall mention two studies regarding BCDG and extremely distant galaxies. The galaxies Pox 186 and Tol 65 are proposed as young galaxies from detailed spectral analysis of their IUE and visible spectra (Olofsson, 1989). This essentially means a lack of evolved underlying population while an old population is found in the bursting elliptical galaxy ESO341-IG04 (Bergvall et al. 1989) and in the merging galaxy ESO 286-IG19 (Johansson, 1990). The metal-deficient galaxy IZw18 was also analyzed as a young galaxy (Lequeux and Viallefond, 1980).

The sample of nearby interacting galaxies analyzed by Keel et al. 1985, Bushouse, 1987, Kennicutt et al. 1987, have similar conclusions about the IMF which probably ranges between 10–100 $M_\odot$. Abnormally high star formation rates were estimated (3–10 $M_\odot$ yr^{-1}) for this mass interval, and very short timescales ($\simeq 10^8$ yrs) which can reach 1 Gyr when most of the gas is consumed.

For extremely distant galaxies ($z \geq 3$), an estimate of the age is of high interest since, for so high redshifts, an age ≥ 2 Gyrs would imply limits on the age of the

Universe (Rocca-Volmerange, 1988). The first interpretation of the spectral energy distribution by Lilly (1988) by means of Bruzual's model (1983) attributed this emission to an evolved population at an age ≥ 1 Gyr. In order to reproduce the rest–frame far–UV emission as well as the intense Ly-α emission line, Rocca-Volmerange (1988) had to introduce a small amount of current star formation with a current burst and the age consequently increased to 2 Gyr. Then younger ages (from $\simeq 0.2$ to 0.5 Gyr) were suggested by updating stellar populations of supergiants, asymptotic giant branch stars, (Bithell and Rees, 1990, Rocca-Volmerange and Guiderdoni, 1990) and/or by increasing the characteristic timescale of the star formation with a continuous law (Chambers and Charlot, 1990) or with two short successive bursts (Rocca-Volmerange and Guiderdoni, 1990). As shown in Fig. 1, the minimum age of 0.5 Gyr can increase to 1.5 or even 2.5 Gyr. In fact, the current star formation seems to erase most of the spectral features which would give a significant age. When photometric uncertainties and the possible contamination by emission lines (CIV for the R band of the observer or Hβ and [OIII] for the K band of the observer) are taken into account and because the present models do not include dust, metallicity effects and non-thermal components, ages have not hitherto been significantly constrained by spectrophotometry. The mass of the old population derived from the visible emission in the rest–frame is weakly sensitive to age and amounts to $2.3 \times 10^{11} M_\odot$ (for an age 0.5 Gyr) or $5.9 \times 10^{11} M_\odot$ (for an age 1.5 Gyr); and the mass of the current burst, determined from the 1250Å (in the rest–frame) continuum emission, is only $5.7 \times 10^9 M_\odot$, on the assumption that the current burst will last for 0.1 Gyr. In our model, we adopt a dark mass fraction $\zeta = 0.5$ which is needed to reproduce the mass/light ratios of the inner parts of giant ellipticals and spirals. Then the present star formation rate is 60 $M_\odot$ yr^{-1} while, during the initial burst, it was either 2300 $M_\odot$ yr^{-1} or 5900 $M_\odot$ yr^{-1} (for ages 0.5 to 1.5 Gyr).

In Guiderdoni and Rocca-Volmerange, 1990, we tentatively modeled the emission lines, more specifically the $W_\lambda([OII])$ equivalent width distribution as a function of the star formation efficiency $\epsilon = f N_{Lyc}/SFR$. SFR is the star formation rate and we adopted the ratio $([OII]3727\text{Å}/H\beta) = 1.8$, taking into account a logarithmic scatter 0.25. With the value $\epsilon = 7 \times 10^{43}$ Lyc photons s^{-1}/($M_\odot$ Gyr^{-1}) fitted on nearby galaxies, the model galaxies have the following $W_\lambda([OII])$ at age 16 Gyr: E/S0 (burst/UV–hot): 0 to 2.8Å; Sa–Sd: 8.0 to 26.2Å; Im: 31.4Å. These values are weakly sensitive to the age since the typical variation from 12 Gyr to 18 Gyr is less than a factor 2.

Figure 3 shows the distributions for the two Durham bright (DARS, Peterson et al. 1986) and faint (DFS, Broadhurst et al. 1988) deep surveys compared to the models computed with the values 0.5ϵ and 1.5ϵ. The apparent increase of efficiency with redshift by a factor $\sim 2 - 3$ could be due to a shallower IMF slope for very massive stars, or a larger upper mass limit, or dust destruction/stripping because of the stronger UV flux or SN rate. Such a trend is suggested by the strong Lyman α emission line in high–redshift radio galaxies. Values of 3ϵ - 5ϵ are required to fit the emitted number of Lyman α photons (up to 10^{56} Ly α photons s^{-1}) and the rest–frame equivalent widths (200–300Å), with SFRs as high as $\sim 600 M_\odot$ yr^{-1} predicted by the model for early epochs of $M_V = -23$ ellipticals (Rocca–Volmerange, 1989). So, because of the large uncertainty on this efficiency, it seems to us that there is no unambiguous evidence of strong bursts of star formation in the DFS sample. Anyway, this question has to be explored further.

Magnitude and color distributions given by published faint galaxy counts in various photometric bands (from $\simeq 3600$Å to $\simeq 7900$Å), are interpreted (Guiderdoni and

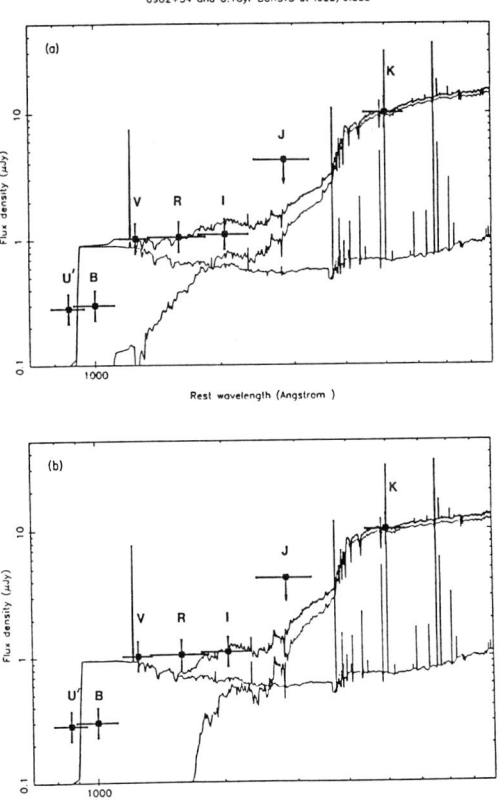

Figure 2. *Two fits of the U'BVRIJK photometry of the radio galaxy 0902+34 (Lilly, 1988, 1989) with different histories of star formation. (a) 0.1 Gyr initial burst followed by passive stellar evolution for 0.4 Gyr and new burst. (b) 1 Gyr initial burst followed by passive stellar evolution for 0.5 Gyr and new burst. The relevant timescale for the fit is the duration of the passive phase (0.4–0.5 Gyr). IMF is standard.*

Rocca-Volmerange, 1990) by analyzing the intrinsic evolution with our model of spectrophotometric evolution. A good fit of such blue counts is needed in order to give a cosmological significance to starbursts (see also Weedman, this conference). We show that, under the assumption of pure luminosity evolution, the magnitude and color distributions are consistent only with scenarios of evolution in which a significant fraction of the galaxies formed at high redshift ($z_{for} \simeq 10$ or more) in a universe with low q_0 (below $\simeq 0.15$) if $H_0 = 50$ km s^{-1} Mpc^{-1} and $\Lambda = 0$. Moreover among the plausible scenarios of evolution depending on the spectral type, and fitting the properties and distribution of nearby galaxies, scenarios with "active" past SFR, for instance with a large number of UV–hot early–type galaxies and SFR of late types proportional to the gas content, are to be preferred. More recently we proposed a unifying model of galaxy formation (Rocca-Volmerange and Guiderdoni, 1990) to save a flat universe with $q_0 = 0.5$ in interpreting the faint galaxy counts. This model needs number evolution in addition to pure luminosity evolution: present-day galaxies result from the merging of a large number of building blocks. Models of basic luminosity and color evolution fit the current Hubble sequence and the comoving number of these blocks

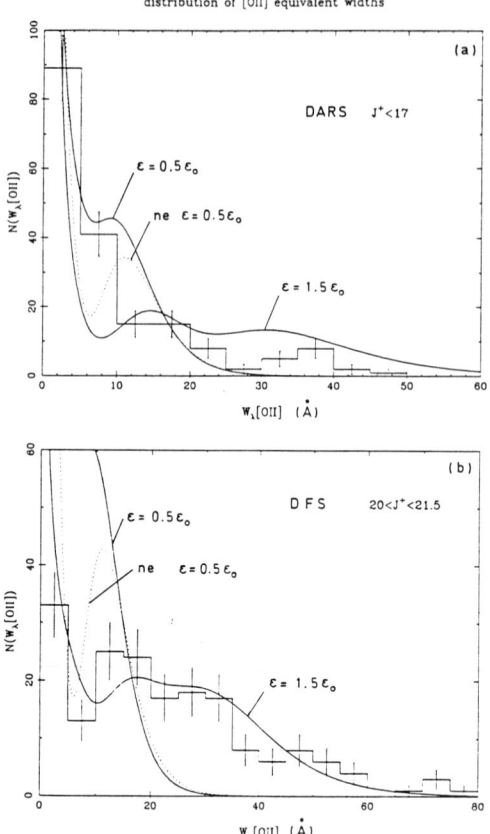

Figure 3a–b. *Distribution of equivalent widths of [OII] 3727Å. (a): Durham/Anglo-Australian Redshift Survey (Peterson et al. 1986). (b): Durham Faint Survey (Broadhurst et al. 1988). Two models are proposed with efficiencies 0.5ϵ or 1.5ϵ, ϵ is a standard efficiency fitting nearby galaxies, for the production of Lyc photons which actually ionize the gas.*

evolves as $(1+z)^\eta$. Field galaxies observed in faint counts are fitted with an evolutionary factor $\eta \simeq 1.5$ in the observable redshift range ($z \leq 4$), while the most distant radio galaxies would be objects accreting the available building blocks more rapidly than this average rate. Star formation can begin early in the blocks (at $z \geq 5$), while radio jets would only be responsible for current star formation. This would explain the alignment of the optical images with the radio axes, and the more compact IR images. The dispersion of the IR Hubble diagram (Rocca-Volmerange and Guiderdoni, 1990) is consistent with such a model. Improvements of calibration and homogeneity factors are needed from bright galaxies as in the APM survey (Maddox et al. 1990). The deepest surveys, in particular in the K band (Glazebrook et al. 1990, Lilly et al. 1990) will possibly bring better evidences of cosmological evolution.

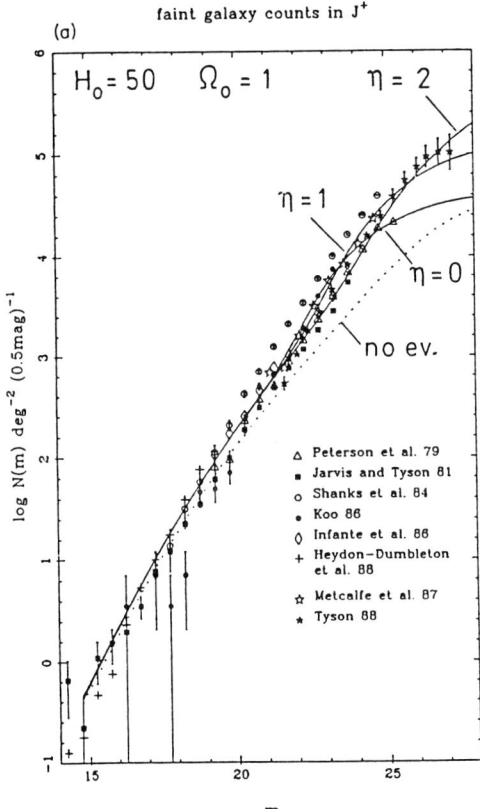

Figure 4. *Faint galaxy counts for a mix of z_{for} depending on type: 30 for the burst and UV-hot E/SO, 5 for the Sa, Sb and Sc and 2 for Sd and Im. $H_0 = 50$ and $q_0 = 0.05$. Solid line: with evolution. Dotted line: no evolution. The counts are not sensitive to the precise values of z_{for} provided that a significant fraction of galaxies formed at high z_{for}.*

5. CONCLUSION

An attempt to associate the starburst event, observed in most interacting galaxies, to a scenario of normal evolution of galaxies is possible by introducing a merging process to interpret the samples of high-z galaxies. Analyzing the spectrophotometric details of nearby and distant galaxy populations with consistent models is the only way to understand the role of a starburst in the evolution of galaxies, and possibly the detection of primeval galaxies. Important progress in this field will require an extension of the deepest surveys.

I am grateful to my collaborator B. Guiderdoni for reading the manuscript and making useful comments.

REFERENCES

Aller, L. H., 1984, Physics of Thermal Gaseous Nebulae, D. Reidel.
Allington-Smith, J. R., 1982, *M.N.R.A.S.*, **199**, 611.
Augarde, R., and Lequeux, J., 1985, *Astr. Ap.*, **147**, 273.
Bahcall, J. N., 1985, *Ap. J.*, **287**, 926.
Bahcall, S. R., and Tremaine, S., 1988, *Ap. J. (Letters)*, **326**, L1.
Baldwin, J. A., Phillips, M. M., and Terlevich, R., 1981, *Pub. A.S.P.*, **93**, 5.
Balzano, V. A., 1983, *Ap. J.*, **868**, 602.
Barbuy, B., 1983, *Astr. Ap.*, **123**, 1.
Baum, W. A., 1990, *Evolution of the Universe of Galaxies*, ed. R. Kron, A.S.P. Conference Series, 10, p. 119.
Belfort, P., 1987, Thèse de Doctorat, Universitè de Paris VII.
Bergvall, N., and Jörsater, 1988, *Nature*, **331**, 589.
Bergvall, N., Rönnback, J., and Johansson, L., 1989, *Astr. Ap.*, **222**, 49.
Bertelli et al. 1989, preprint.
Bica, E., and Alloin, D., 1986, *Astr. Ap.*, **162**, 21.
Bithell, M., and Rees, M. J., 1990, 1988, *M.N.R.A.S.*, **242**, 570.
Broadhurst, T. J., Ellis, R. S., and Shanks, T., 1988, *M.N.R.A.S.*, **235**, 827.
Bruzual, G., 1983, *Rev. Mex. Astron. Astrof.*, **8**, 63.
Bushouse, H. A., 1987, *Ap. J.*, **320**, 49.
Butcher, H. R., and Oemler, A., 1984, *Ap. J.*, **285**, 426.
Chambers, K., and Charlot, S., 1990, *Ap. J.*, **348**, L1.
Chambers, K. C., Miley, G. K., and Van Breugel, W., 1990, preprint.
Chambers, K. C., and McCarthy, P., 1990, preprint.
Colless, M., Ellis, R. S., Taylor, K., and Hook, R. N., 1990, *M.N.R.A.S.*, **244**, 408.
Deharveng, J. M., Joubert, M., and Kunth, D., 1985, *Star Forming Dwarf Galaxies*, Kunth, D., Thuan, T. X., Tran Thanh Van, eds, Editions Frontières, p. 431.
Djorgovski, S. G., 1987, *Starbursts and Galaxy Evolution* T. X. Thuan, T. Montmerle, Tran Thanh Van, eds., Editions Frontières, p. 401.
Davidge, T., and Maillard, J. P., 1990, preprint.
Downes, A. J. B., Peacock, J. A., Savage, A., and Carris, D. R., 1986, *M.N.R.A.S.*, **218**, 31.
Dressler, A., and Gunn, J. E., 1983, *Ap. J.*, **270**, 7.
Dunlop, J. S., Peacock, J. A., Savage, A., Lilly, S. J., Heasley, J. N., and Simon, A. J. B., 1989, *M.N.R.A.S.*, **238**, 1171.
Faber, S., 1973, *Ap. J.*, **179**, 731.
Gallagher, J., Hunter, D. A., and Tutukov, A. V., 1984, *Ap. J.*, **284**, 544.
Glazebrook, D., preprint.
Guiderdoni, B., and Rocca-Volmerange, B., 1987, *Astr. Ap.*, **186**, 1.
_____. 1990, *Astr. Ap.*, **227**, 362.
Gunn, J. E., Hoessel, J., and Oke, J. B., 1986, *Ap. J.*, **306**, 30.
Gunn, J., and Stryker, L., 1983, *Ap. J. Suppl.*, **52**, 121.
Heck, A., Egret, D., Jaschek, C., and Battrick, B., 1984, ESA SP-1052.
Huchra, J. P., 1977, *Ap. J.*, **217**, 928.
_____. 1985, *Starbursts and Galaxy Evolution*, eds T. X. Thuan, T. Montmerle, Tran Thanh Van, Editions Frontières, p. 199.

Johansson, L., 1990, preprint.
Jones, J. E., Alloin, D., and Jones, B. J. T., 1984, *Ap. J.*, **283**, 457.
Keel, W. C., Kennicutt, R. C., Hummel, E., and van der Hulst, J. M., 1985, *A. J.*, **90**, 708.
Kennicutt, R. C., 1983, *Ap. J.*, **272**, 54.
——————. 1990, *Paired and Interacting Galaxies, IAU Colloquium 124*, ed. Sulentic and W. Keel.
Kennicutt, R. C., Keel, W. C., Van der Hulst, J. M., Hummel, E., and Roettiger, K. A., 1987, *A. J.*, **93**, 1011.
Koo, D., 1990, *Evolution of the Universe of Galaxies*, ed. R. Kron, A.S.P. Conference Series, 10, p. 268.
Kunth, D., 1989, *Evolutionary Phenomena in Galaxies*, eds J. E. Beckman and B. E. J. Pagel, Cambridge Univ. Press, p. 22.
Kunth, D., and Schild, H., 1986, *Astr. Ap.*, **169**, 71.
Kurucz, R. L., 1979, *Ap. J. Suppl.*, **40**, 1.
Laing, R. A., Riley, J. M., and Longair, M. S., 1983, *M.N.R.A.S.*, **204**, 151.
Larson, R. B., and Tinsley, B. M., 1978, *Ap. J.*, **219**, 46.
Lequeux, J., 1979, *Astr. Ap.*, **80**, 35.
Lequeux, J., and Viallefond, F., 1980, *Astr. Ap.*, **91**, 269.
Lester, D. F., Carr, J. S., Joy, M., and Gaffey, N., 1990, *Ap. J.*, **352**, 544.
Lilly, S. J., and Longair, M. S., 1984, *M.N.R.A.S.*, **211**, 833.
Lilly, S., 1988, *Ap. J.*, **333**, 161.
——————. 1989, *Ap. J.*, **340**, 77.
——————. 1990, preprint.
Lonsdale, C. J., and Hacking, P. B., 1987, *High Redshift and Primeval Galaxies* Bergeron, J., Kunth, D., Rocca-Volmerange, B., Tran Thanh Van, eds, Editions Frontières, p. 141.
Maddox et al. 1990, preprint.
Maeder, A., and Meynet, G., 1988, *Astr. Ap. Suppl.*, **76**, 411.
Meier, D. L., 1976, *Ap. J.*, **207**, 343.
Mihalas, D., 1972, NCAR Technical Note STR76.
Miller, J., and Scalo, J. M., 1979, *Ap. J.*, **41**, 513.
Nomoto, K., Thieleman, F., and Yokoi, K., 1984, *Ap. J. (Letters)*, **286**, 644.
Olofsson, K., 1989, *Astr. Ap. Suppl.*, **80**, 317.
Partridge, R. B., and Peebles, P. J. E., 1967, *Ap. J.*, **147**, 868.
Peterson, B. A., Ellis, R. S., Efstathiou, G., Shanks, T., Bean, A. J., Fong, R., and Zen-Long, Z., 1986, *M.N.R.A.S.*, **221**, 233.
Pritchet, C. J., and Hartwick, F. D. A., 1990, *Evolution of the Universe of Galaxies*, ed. R. Kron, A.S.P. Conference Series, 10, p. 319.
Pickles, A. J., 1985, *Ap. J.*, **296**, 340.
Renzini, A., and Voli, M., 1981, *Astr. Ap.*, **94**, 175.
Renzini, A., 1986, *The Spectral Evolution of Galaxies*, Erice Workshop, eds C. Chiosi and A. Renzini (Dordrecht: Reidel).
Rocca–Volmerange, B., 1988, *ESO The Messenger*, **53**, 26.
——————. 1989, *M.N.R.A.S.*, **236**, 47.
Rocca–Volmerange, B., and Guiderdoni, B., 1988, *Astr. Ap. Suppl.*, **75**, 93.
——————. 1990, *M.N.R.A.S.*, (in press).

Rocca–Volmerange, B., and Schaeffer, R., 1990, *Astr. Ap.*, (in press).
Rocca-Volmerange, B., Lequeux, J., and Maucherat-Joubert, M., 1981, *Astr. Ap.*, **104**, 177.
Rosa, M., Joubert, M., and Benvenuti, P., 1984, *Astr. Ap. Suppl.*, **57**, 361.
Scalo, J. M., 1986, *Fundamentals of Cosmic Physics*, **11**, 1.
Searle, L., Sargent, W. L. W., and Bagnuolo, W., 1973, *Ap. J.*, **179**, 427.
Soifer, B. T., et al. 1984, *Ap. J.*, **283**, L1.
Spinrad, H., 1987, in *High–Redshift and Primeval Galaxies*, J. Bergeron, D. Kunth, B. Rocca–Volmerange and J. Tran Thanh Van eds., Editions Frontières.
Spitzer, L. Jr., 1978, Physical Processes in the Interstellar Medium.
Stasińska, G., 1984, *Astr. Ap. Suppl.*, **55**, 15.
Struck-Marcell, C. and Tinsley, B. M., 1978, *Ap. J.*, **221**, 562.
Thuan, T. X., 1987 *Starbursts and Galaxy Evolution*, eds T. X. Thuan, T. Montmerle, Tran Thanh Van, Editions Frontières, p. 129.
Tinsley, B. M., 1972, *Astr. Ap.*, **20**, 383.
Tyson, J. A., 1988, *A. J.*, **96**, 1.
Valls-Gabaud, D., Rocca-Volmerange, B., and Guiderdoni, B., in preparation.
Woosley, J. C., 1986, Sixteenth Advanced Course of the Swiss Society of Astronomy and Astrophysics, Saas-Fee.
Wu et al. 1983, IUE Ultraviolet Spectral Atlas, NASA No. 22.
VandenBerg, D. A., 1985, *Ap. J. Suppl.*, **58**, 711.

DISCUSSION

S. Charlot: You mentioned that Chambers and Charlot (1990) obtained young ages for the high-redshift radio galaxies by replacing red supergiants by AGB stars. I would like to make 3 comments: (1) The code we used already took into account the red supergiants. We simply added the AGBs empirically to complete the tracks. (2) The AGB correction changes our results by less than 20%. We are able to obtain young ages mainly by tuning up the shape of the SFR. In any case the stars which produce the "red bump" are the red supergiants as in your models. (3) At the time we wrote our paper, Lilly was arguing that the high-redshift radio galaxies *had to be* at least 1 to 2 Gyr old. We made the point that they *could* in fact be very young. We therefore produced a *lower-limit* to the ages of these objects, which does not contradict at all your results.

Rocca-Volmerange: The lowest limit is in fact obtained with a supergiant solution as proposed by Bithell and Rees, 1990 or by ourselves (in the Moriond, 1989 Conference).

G. Bruzual: Don't you think it is very dangerous to use spectral energy distributions of stars in the solar neighborhood to build spectral models of low metallicity galaxies? I think that the conclusions you derive may be misleading.

Rocca-Volmerange: I quite agree with this point. We did not apply our models to very low-metallicity galaxies. I did not try to fit Blue Compact Galaxies for this reason. I quite agree if we use the low-metallicity tracks, we should also use the low-metallicity stellar spectra.

G. Hensler: A question concerning the star formation rate. Your star formation rate is dependent on the total mass of the galaxies. How would you then distinguish between spiral and elliptical galaxies with the same initial mass because then I think the density plays a dominant role in the star formation rate.

Rocca-Volmerange: Our star formation rate does not only depend on the total mass of the galaxies. In fact, the star formation rates for spirals and ellipticals correspond to different gas consumption rates, which are given by different accretion rates in our models. The accretion rate is increasing for elliptical galaxies and lower and almost constant for spiral galaxies.

G. Miley: Birgitte, I have two points about the high-z radio galaxies. The first is that the Lyα maps are clearly different from the optical continuum maps, indicating that there is another ionization mechanism at work than ionization by stars.

Second, as I understand it, Chambers and Charlot were merely pointing out that a relatively young population of stars in these objects is *consistent* with the spectral energy distributions (SED), in conflict with the previously accepted folklore.

Finally, how many free parameters are there in your fits, particularly the counts? As an outsider I find it very difficult to judge the physical significance.

Rocca-Volmerange: On the first point: I agree with the fact that another mechanism of Lyα emission can exist. The question is what is its relative contribution compared to ionization by stars.

Concerning the ages of the radio galaxies, I only pointed out that, from the SED, many solutions are possible—from the youngest ages with supergiants to the oldest ones with giants.

As I underlined in my talk, the main parameters to fit the counts are the redshift of formation, the q_o value, and our modeling of galaxy evolution. It is not sensitive to the IMF but it may vary with the fraction of ellipticals which form at high redshift ($z \gtrsim 5$).

A. Maeder: Would you comment a bit more on the redshifts for the beginning of star formation in elliptical galaxies. Why did you choose a value of $z = 30$ for your models?

Rocca-Volmerange: There is no real difference between 5 and 30, because the evolution in time is so rapid in our models. We have one simple model with as few parameters as possible. More sophisticated models are currently being investigated.

H_2 AND INFRARED IN GLOBAL STARBURST GALAXIES

Nick Scoville and B.T. Soifer
Division of Physics, Mathematics and Astronomy
California Institute of Technology
Pasadena, CA 91125
USA

Abstract. Far infrared measurements from the IRAS survey combined with estimates of the molecular gas content provide a fundamental basis for the analysis of the starburst phenomena in galaxies. When the ratio of far infrared luminosity to molecular gas mass significantly exceeds that in normal galaxies like the Milky Way (4 $L_\odot$ $M_\odot^{-1}$), star formation is occurring on a shorter timescale, possibly with an initial mass function biased towards high mass stars. In the highest luminosity IRAS galaxies ($L_{IR} \geq 10^{11}$ $L_\odot$), the luminosity to H_2 mass ratio is typically 40 $L_\odot$ $M_\odot^{-1}$, indicating star formation rates of 10–100 $M_\odot$ yr^{-1} and cycling times for the ISM much less than 10^9 yr.

In the very luminous infrared galaxies, the optical morphology almost invariably shows evidence of a strong galactic interaction and a substantial fraction of the total molecular gas content is seen at radii ≤ 1 kpc. Dense molecular gas probably plays a pivotal role in the evolution of such dynamically disturbed systems: being dissipative, the gas can readily sink to the center of the interacting system where it may fuel a nuclear starburst and/or build up and fuel a central active nucleus. We show that the shape of the high luminosity end of the infrared galaxy luminosity function can be reproduced by a model in which normal spiral galaxies, represented in the Schecter function, undergo collision-induced starbursts. Statistics from the IRAS survey are consistent with the percentage of *all* spiral galaxies currently undergoing a *global* starburst being approximately 0.2% and the lifetime of the starburst being a dynamical time, approximately 2×10^8 years. The present epoch rate is therefore such that 2% of all galaxies participate in a merger every 10^9 years and with standard cosmological evolution, nearly all galaxies would be undergoing such merger-induced starbursts at z=1. Galactic merging and starburst activity must therefore play a central role in galactic evolution.

1. STARBURSTS

Starbursts come in two flavors—the formation of stars with a higher efficiency per unit mass of interstellar gas (resulting in rapid exhausion of the interstellar medium) or the formation of stars with an initial mass function more heavily weighted towards high mass stars than in normal galaxies like the Milky Way. Both forms of starbursts will result in a luminosity-to-interstellar gas mass ratio significantly exceeding that of normal galaxies. It is therefore useful to use the observed total luminosity-to-mass ratios as an operational signature (and definition) of *global* starbursts.

Inasmuch as the dust in star forming molecular clouds will significantly attenuate the visible and ultraviolet light from the young stars, re-radiating it in the infrared, the infrared luminosity is a more reliable probe of the star formation rate than optical indicators such as the short wavelength continuum or the emission lines from ionized gas. The IRAS survey has provided a fundamental database for the study of complete samples of starburst galaxies and the estimation of their total luminosities. Follow-up optical and near infrared imaging of the infrared galaxies has been most useful in provided sufficient spatial resolution to detail the morphology of the starburst and the host galaxy systems. In the most luminous infrared galaxies, the dominant neutral interstellar medium component is molecular rather than atomic gas and single dish CO data have been used for estimation of the total gas content. Approximately 25 of the luminous infrared galaxies have been mapped at high resolution using millimeter-wave interferometers in order to delineate the spatial distribution of the gas and its kinematics.

In this contribution, we use a 60μm flux limited galaxy sample from the IRAS survey as the basis for a discussion of starburst phenomena. In §2, the infrared luminosity function of galaxies is characterized, followed by a review of results for the global H_2 content and CO aperture synthesis for a sample of these galaxies. Arp 220 is then described in some detail as an example of the extreme merger-induced starburst galaxies (§3). In §4, we discuss the role of dynamical perturbations in initiating the starburst and present an empirical starburst model. The statistics of galactic interactions and the frequency of infrared starburst galaxies are analyzed in §5. A brief summary of our conclusions is provided in §6.

2. LUMINOUS INFRARED GALAXIES

A spectacular result of the IRAS survey was the discovery and recognition of a class of luminous galaxies emitting the bulk of their energy at far infrared wavelengths (Soifer et al. 1984). Soifer et al. (1987) showed that the infrared luminous galaxies are the dominant population in the local universe ($z \leq 0.1$) at luminosities greater than 3×10^{11} $L_\odot$ and at luminosities greater than 10^{12} $L_\odot$, the ultraluminous IRAS galaxies outnumber optically selected quasars 2:1. The luminous infrared galaxies are also extraordinarily rich in molecular gas with H_2 masses in the range $2-60 \times 10^9$ $M_\odot$ (Sanders et al. 1986, 1990)—1–20 times that of the Milky Way. Although the high luminosity ($10^{11}-10^{12}$ $L_\odot$) and ultraluminous ($\geq 10^{12}$ $L_\odot$) galaxies all have large masses of molecular gas, their gas contents are less elevated than the luminosities, relative to those of normal galaxies like the Milky Way. It is therefore clear that these objects are not simply scaled up, more massive versions, of normal spiral galaxies, but that they represent a transitory phase in the evolution of galaxies.

2.1. Infrared Luminosity Function of Galaxies

The IRAS Bright Galaxy sample (Soifer et al. 1987, 1989) is a statistically complete sample of 313 objects, designed to meet the following criteria: it is a complete flux-limited sample with 60 μm flux densities >5.2 Jy, galactic latitude $>30°$, and accessible to northern hemisphere telescopes. The galaxies range in distance from 0.6–300 Mpc and the median distance, excluding Virgo galaxies, is 32 Mpc ($H_o=75$ km s^{-1} Mpc^{-1}). The Virgo cluster provides a 10% contribution to the Bright Galaxy sample. The range of far infrared luminosities for galaxies in the sample is 10^8–2×10^{12} L$_\odot$ with the median at $\sim 2\times 10^{10}$ L$_\odot$. A similar distribution was found in a sample of galaxies studied by Lawrence et al. (1986).

In Figure 1, the luminosity function for IRAS galaxies in the bright galaxy survey is compared with those of normal optical galaxies, optically identified starburst galaxies, Seyfert galaxies, and quasars (Soifer et al. 1987). The infrared luminosity function* shown in Figure 1 is in good agreement with those determined independently by Lawrence et al. (1986), Rieke and Lebofsky (1986), Smith et al. (1987) and Yahil et al. (1990). At luminosities below 3×10^{11} L$_\odot$, the normal galaxy luminosity function represented by the Schechter function exceeds the infrared galaxy luminosity function but at higher luminosities, the space density of infrared galaxies is much greater than that of normal galaxies. It is also apparent from Figure 1 that the space density of infrared galaxies exceeds the space densities of the optically identified starburst galaxies, Seyfert galaxies, and quasars at luminosities above 10^9 L$_\odot$.

The normal galaxy luminosity function can be represented analytically by

$$N_o(L)dL = 2.95 \times 10^{-2} \left(\frac{L}{L_{*o}}\right)^{-1.25} e^{-L/L_{*o}} \frac{dL}{L_{*o}} \quad \text{Mpc}^{-3} \qquad (1)$$

where $L_{*o}=2.75\times 10^{10}$ L$_\odot$. The infrared galaxy luminosity function is fit by a 2-part power law:

$$N_I(L)dL = 1.1 \times 10^{-3} \left(\frac{L}{L_{*i}}\right)^{-1.6} \frac{dL}{L_{*i}} \quad \text{Mpc}^{-3} \quad for L < L_{*i} \qquad (2a)$$

$$= 1.1 \times 10^{-3} \left(\frac{L}{L_{*i}}\right)^{-3.3} \frac{dL}{L_{*i}} \quad \text{Mpc}^{-3} \quad for L > L_{*i} \qquad (2b)$$

where $L_{*i}=2.5\times 10^{10}$ L$_\odot$. [Equations (1) and (2) are the luminosity functions per linear interval in L; they differ from the logarithmic luminosity functions in Figure (1) by a factor L^{-1}.] These expressions will be used in §5.1 for a discussion of the frequency of starburst and interacting galaxies.

2.2. Molecular Gas Contents

We have recently completed a comprehensive survey of CO emission from luminous infrared galaxies [L_{IR} (8–1000 μm) $>2\times 10^{10}$ L$_\odot$] in order to determine the total masses of molecular gas. The majority of objects were selected from the IRAS bright

---

* In this discussion, the infrared luminosity is the λ=40-1000 μm luminosity in units of solar bolometric luminosity (3.83×10^{33} ergs sec^{-1}).

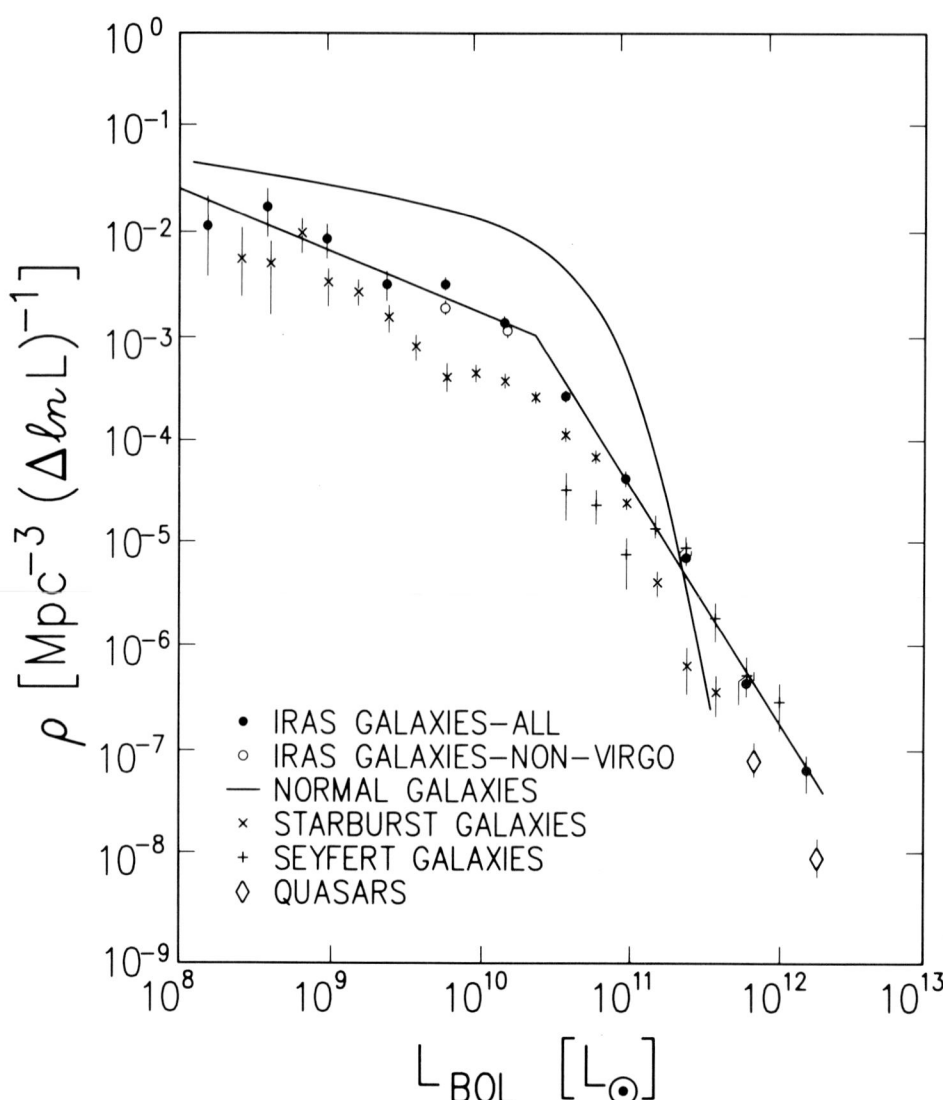

Figure 1. Luminosity function of a variety of classes of extragalactic sources, normalized to same Hubble constant ($H_o = 75$ km s^{-1} Mpc^{-1}) and plotted in units of bolometric luminosity (Soifer et al. 1987). Filled and open circles represent the far-infrared luminosity function derived for the IRAS Bright Galaxy sample, including and exluding the Virgo cluster respectively. The solid curve is an analytical fit (Equation 1) to normal galaxy luminosity function (Schechter 1976) that agrees with many observed luminosity functions (Felton 1977); crosses represent optically selected starburst galaxies, and plus signs optically selected Seyfert galaxies, both taken from Huchra (1977). Open diamonds represent optically selected quasars (Schmidt and Green 1983). Straight lines represent a fit of two power laws to bright galaxy luminosity function excluding Virgo galaxies (Equation 2).

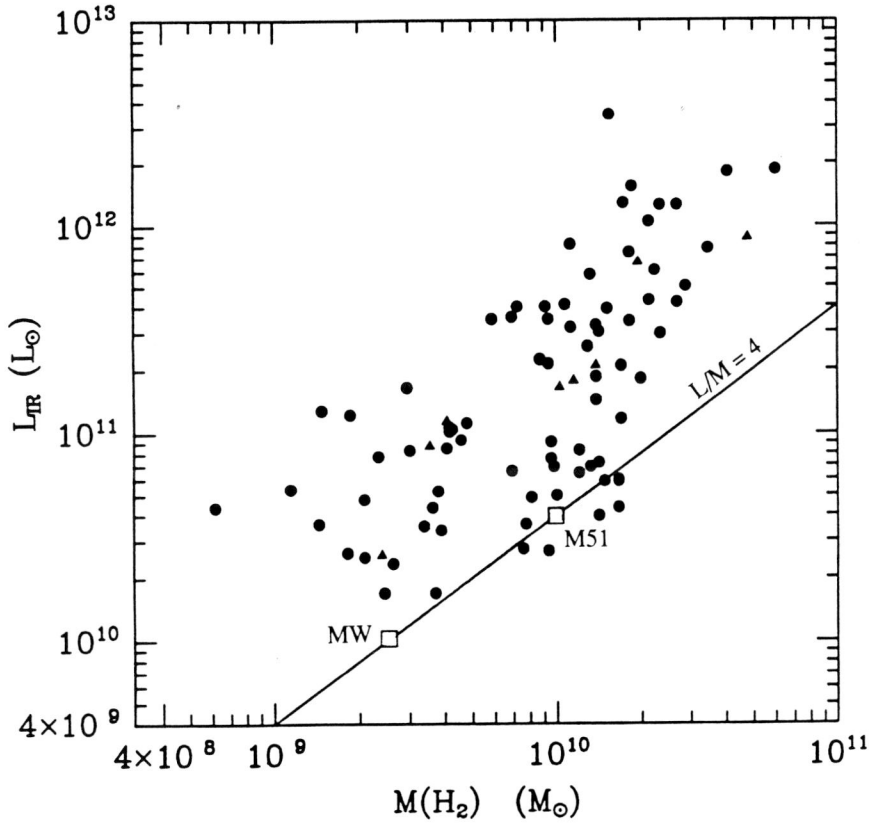

Figure 2. The infrared luminosity ($\lambda=8$–$1000~\mu m$) is shown as a function of the global molecular gas mass for a sample of 89 bright infrared galaxies (•; Sanders, Scoville, and Soifer 1990; Tinney et al. 1990). Comparable data are also shown for the Milky Way and M51.

galaxy sample (§2.1) with a few additional objects taken from an extension of the sample to lower galactic latitudes (Sanders, Scoville, and Soifer 1990, Tinney et al. 1990). CO emission line fluxes were converted into H_2 masses assuming a standard Galactic CO-to-H_2 conversion ratio $[\alpha=3\times10^{20}~\text{cm}^{-2}~(\text{K km s}^{-1})^{-1}$, cf. Scoville and Sanders 1987)]. The global H_2 masses are shown for these 89 galaxies (•) together with similar data for "normal" galaxies such as the Milky Way and M51. Virtually all of the luminous infrared galaxies are extremely rich in molecular gas with $M_{H_2}=10^9$–6×10^{10} $M_\odot$, corresponding to 0.4–20 times the mass of H_2 gas in the Milky Way. Figure 2 also shows a tendency for higher infrared luminosity-to-H_2 mass ratios in galaxies with higher infrared luminosities.

On the basis of their optical morphologies, the galaxies included in the CO survey have been differentiated with respect to the possibility of a recent galactic interaction (cf. Sanders, Scoville, and Soifer 1990). The galaxies were placed in one of four possible categories: isolated, a disturbed appearance, galactic pairs with non-overlapping disks, and advanced mergers. The latter category was distinguished from single galaxies on the basis of extended tidal tails or extremely close, double nuclei. In Table 1, the mean infrared luminosity-to-molecular mass ratios are given as a function of infrared

TABLE 1. *Luminosity-to-H_2 Mass Ratios*

L_{IR}/M_{H_2} (# Gal)	L_{IR} ($L_\odot$) 2×10^{10}–10^{11}	10^{11}–10^{12}	1–3×10^{12}	all
Morphology				
Isolated	6 (15)	—	—	6 (15)
Disturbed	12 (13)	22 (11)	—	17 (24)
Non-overlapping Disks	20 (4)	30 (7)	75 (1)	30 (12) ⎱ 39 (47)
Merger Pairs	32 (5)	34 (23)	74 (7)	42 (35) ⎰

L_{IR} is the infrared luminosity at $\lambda=8$-1000 μm and M_{H_2} is the H_2 mass assuming a Galactic CO-to-H_2 conversion ratio of 3×10^{20} cm^{-2} (K km s^{-1})$^{-1}$. The units are $L_\odot M_\odot^{-1}$ and the number of objects in each group sampled by Sanders, Scoville, and Soifer (1990) and Tinney et al. (1990) is given in parenthesis.

luminosity and interaction classification. The number of objects contributing in each group is given in parenthesis.

In addition to the trend for the highest luminosity galaxies to have higher luminosity-to-mass ratios, Table 1 clearly shows that the most intensely interacting system have the highest luminosity-to-mass ratios (at all luminosities). Combining the last two interaction classes (non-overlapping disks and merger pairs) and averaging over all luminosities, the mean luminosity-to-mass ratio is 39 $L_\odot M_\odot^{-1}$ which is similar to the ratio obtained in the most active Galactic GMCs. If the objects above 10^{12} $L_\odot$ are excluded (since they may be partially powered by an AGN), the ratio is 32 $L_\odot M_\odot^{-1}$. The global luminosity-to-mass ratio for the Galaxy is ~ 4 $L_\odot M_\odot^{-1}$ and the mean value for a sample of 40 Galactic GMCs containing giant radio HII regions is 7 $L_\odot M_\odot^{-1}$ (Scoville and Good 1989). The highest luminosity-to-mass ratio in any Galactic GMC is 40 $L_\odot M_\odot^{-1}$. If the infrared luminosity is taken to be a measure of the total mass of recently formed stars, the luminosity-to-mass ratio will be inversely proportional to the timescale over which interstellar matter is converted into young stars. Assuming that the initial mass function of the stars producing the luminosity in the infrared galaxies is similar to that in the Galaxy, the order of magnitude higher luminosity-to-mass ratios seen in the strongly interacting systems implies that the cycling time for the interstellar medium is approximately 10 times shorter than in normal galaxies. The normal galaxy cycling time is commonly estimated to be approximately 1–2×10^9 years; the interacting starburst systems must therefore have global star formation timescales $\sim 2\times10^8$ years. The gas exhaustion timescale will generally be a factor ~ 2 greater than the "cycling time" since a significant fraction of the stellar material is cycled back into the interstellar medium by high mass stars on timescales $<2\times10^8$ years. [The fraction of the stellar mass which is recycled back to the ISM is a function of both time and the low mass cutoff for the IMF (see §4).]

In the sense that the observed luminosity-to-mass ratio is indicative of the rate at which interstellar matter is converted into stars, the infrared galaxies with high luminosity-to-mass ratios clearly qualify as objects in which *global* starbursts are taking place. The correlation of the luminosity-to-mass ratio with the incidence of significant large-scale dyanamical perturbations strongly suggests that galactic interactions are probably the major (and possibly the only) mechanism for triggering a *global* starburst. Localized starbursts, for example in spiral arm regions, may, on the other hand, be initiated by smaller scale dynamical perturbations such as a central bar potential, an active nucleus, or the spiral arm potential well.

2.3. Nuclear Gas Concentrations

In the Galaxy, it has been suggested that high mass, OB star formation is triggered by the expansion of HII regions or supernovae (eg. Elmegreen and Lada 1977) or by cloud-cloud collisions (Scoville, Sanders, and Clemens 1986). Both processes depend on the *concentration* of molecular gas, not merely the total H_2 mass. In the former model, the rate of stimulation of molecular gas depends on the number of adjacent clouds within the sphere of influence of the expanding HII region or supernova shells. In the latter scenario, the rate of cloud-cloud collisions will depend quadratically on the number density of clouds and on their velocity dispersion. Dynamical perturbations (§2.2) can concentrate the gas and increase the cloud-cloud velocity dispersion. Localized regions with such effects may be found in nearby spiral galaxies. For example, in the disk of M51, the arm/interarm contrast in the surface density of molecular gas is approximately 3–5, compared with an arm/interarm contrast of 10 in $H\alpha$, (Vogel, Kulkarni, and Scoville 1988). A similar, approximately quadratic dependence, of the OB star formation rate on the molecular gas surface density is seen in the centers of seven Virgo cluster spiral galaxies studied at high resolution by Canzian (1990).

Most of the high luminosity infrared galaxies discussed in the preceding section are at sufficiently great distances that detailed comparison of the luminosity and interstellar gas distributions is not possible. On the other hand, it is useful to compare the luminosity-to-mass ratios in these galaxies with CO aperture synthesis measurements of H_2 surface density. In Table 2, the observational results are summarized for eighteen systems observed with the Owens Valley interferometer (cf. Scoville et al. 1990). The spatial resolutions of these maps are 2–7″, a factor of 10 better than is possible with single dish observations.

The galaxies listed in Table 2 span the luminosity range 2×10^{10}–3×10^{12} $L_\odot$ (λ=8–1000 μm) and the masses of molecular gas in the nuclear sources are 10^9–4×10^{10} $M_\odot$ (assuming the Galactic CO-to-H_2 conversion factor). All have infrared luminosity-to-molecular mass ratios significantly exceeding that of the Milky Way (4 $L_\odot M_\odot^{-1}$); the highest (Mrk 231) has a ratio of 200 $L_\odot M_\odot^{-1}$. In most cases, the size of the CO emitting region is $\lesssim 1$ kpc in radius (see Table 2).

In Figure 3, the luminosity-to-H_2 mass ratios are shown as a function of the central gas surface density ($M_\odot$ pc^{-2}) for the galaxies listed in Table 2. Only those galaxies for which actual size measurements exist (rather than upper limits to the unresolved source size) are plotted. Figure 3 clearly shows a correlation between the luminosity-to-mass ratio and the central gas surface density, suggesting that the high "efficiencies" of energy generation (L_{IR}/M_{H_2}) are linked to the high gas concentrations. The observed correlation is consistent with scenarios for high mass star formation via stimulated or binary processes (such as cloud-cloud collisions) which occur more rapidly when the gas is concentrated.

Given the concentrated distribution of the interstellar gas and the high luminosity energy densities it is unlikely that the molecular gas is arranged in GMCs similar to those containing the bulk of the Galactic H_2. It is expected that both the density and temperature of the molecular gas will be elevated and these changes will affect the rate of emission of CO line photons per unit mass of molecular hydrogen. One

Table 2. OVRO observations of high luminosity IRAS galaxies

Object	$\langle cz \rangle$ km s^{-1}	D Mpc	Radius " (kpc)	L_{IR} $10^{11} L_\odot$	Nuclear M_{H_2} $10^9 M_\odot$	M_{H_2}/M_{dyn}	CO morphology
Mrk 231	12660	174	<3.5 (2.9)	34.7	36.0		100% nuclear source[a]
IRAS 17208 − 0014	12850	175	1.5 (1.3)	27.0	55.0		100% nuclear source[b]
Arp 220	5452	77	1 (0.3)	15.5	16.3	0.90	70% nuclear source[a]
VII Zw 31	16245	221	2.5 (2.7)	8.7	29.4		60% nuclear source[a]
IRAS 10173 + 0828	14680	196	3.5 (3.3)	6.0	9.0		40% nuclear source[b]
NGC 6240	7285	101	3.5 (1.7)	6.6	11.2	0.77	interacting pair[c]
IC 694 (Arp 299)	3060	42	1.2 (0.3)	4.9	3.9		triple source[d]
VV114	6028	78	2.5 (0.9)	4.2	10.0		double source[a]
NGC 1614	4847	62	2 (0.6)	4.0	6.0		30% nuclear source[a]
Arp 55	11957	163	4 (3.2)	3.9	17.3	0.6	double source[e]
NGC 1068	1137	18	1.5 (0.13)	1.5	4.5		ring+nuclear source[f]
NGC 7469 (Arp 298)	4963	66	2.5 (0.8)	2.6	7.4	0.7	30% nuclear source[e]
ZW 049.057	3900	52	1.4 (0.4)	1.7	4.6		40% nuclear source[b]
NGC 828	5359	72	2.5 (0.9)	2.1	11.8	0.31	interacting pair[e]
NGC 2146	838	21	4/13 (0.4)	1.2	4		[g]
NGC 3079	1137	24	3/7 (0.3)	0.7	5	0.2	[h]
NGC 520 (Arp 157)	2261	29	2.5 (0.4)	0.6	3.2		interacting pair[e]
NGC 4038/39 (Arp 244)	1550	21	3.5 (0.4)	0.2	0.8		triple source[i]

References: a-Scoville et. al. 1989; b-Planesas, Mirabel, and Sanders 1990; c-Wang, Scoville, and Sanders 1990; d-Sargent and Scoville 1990; e-Sanders et. al. (1988); f-Planesas, Scoville, and Myers 1990; g-Young, Claussen, and Scoville 1988; h-Young et. al. 1988; i-Stanford et. al. 1990.

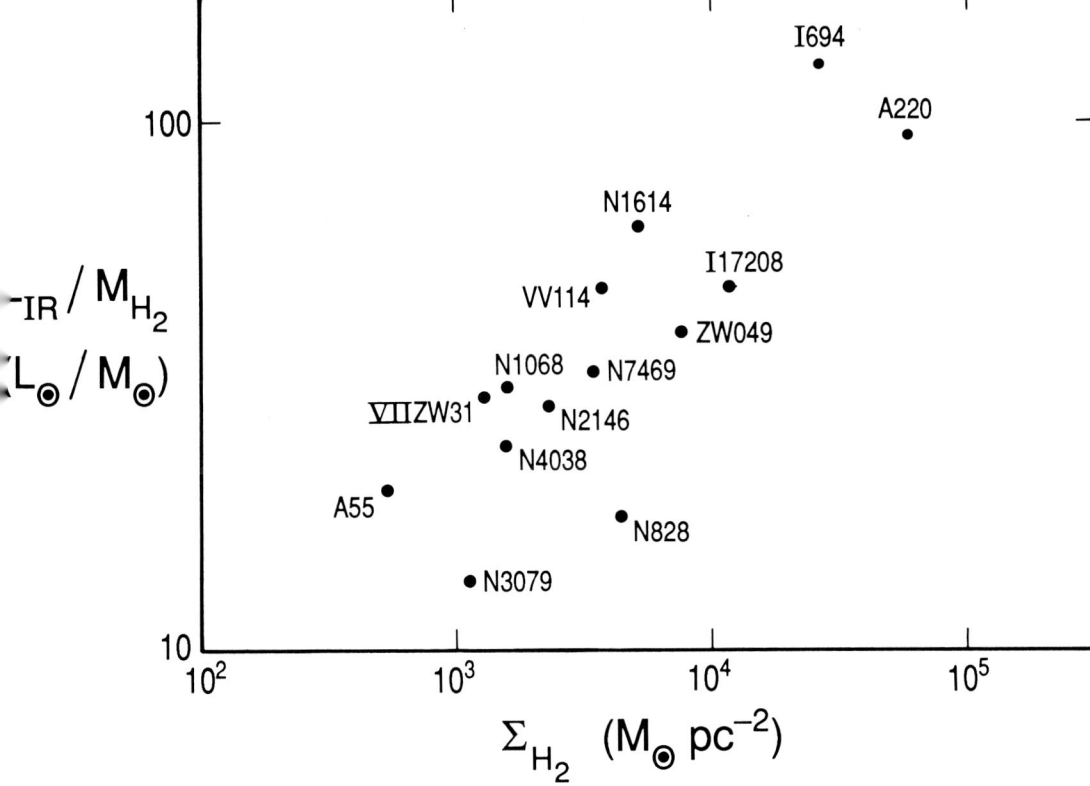

Figure 3. *The infrared luminosity-to-molecular mass ratios are shown as a function of the central gas surface density in 14 luminous infrared galaxies (Table 2). A general correlation is seen with increasing luminosity-to-mass ratios being found in the galaxies with higher gas surface densities. In computing the luminosity-to-mass ratios, the total infrared luminosity was assigned to the nuclear molecular gas concentration since the IRAS data have insufficient resolution to separate out any extended component. For NGC 1068, only the far-infrared and molecular gas components associated with the spiral arms at 10–15″ radius are used (cf. Planesas, Scoville, and Myers 1990).*

might therefore suspect that the molecular mass estimates are uncertain. However, since the CO-to-H_2 conversion ratio (α) scales approximately as $\rho^{1/2}/T_X$ (cf. Scoville and Sanders 1987), the variations in the gas density and temperature have cancelling effects on the CO-to-H_2 conversion ratio.

2.4. The Role of Galactic Interactions

Virtually all of the very luminous infrared galaxies show evidence of a significant galactic interaction (cf. Sanders 1990). The interstellar matter can play a central role in the dynamics of a galactic interaction since the gas is dissipative and hence responds irreversibly to the perturbation, sinking toward the center of the potential well. The H_2 masses for the luminous infrared galaxies are large but in most cases, not more than would be found in *two* galaxies such as M51 ($M_{H_2} \sim 10^{10}$ $M_\odot$, Scoville and Young

1983). [Notable exceptions are the highest luminosity systems, VII Zw 31, Arp 220 and Mrk 231 for which the H_2 masses are 2–5×10^{10} $M_\odot$.] In most cases, it is therefore plausible that the luminous infrared galaxies result from the merging of two gas-rich spiral galaxies and that the disturbed dynamics in the merging system results in dissipation of kinetic energy (and outward transport of angular momentum) in the ISM, leading to the deposition of a significant fraction of the original interstellar matter in the central region of the system. There binary processes such as cloud-cloud collisions or stimulated star formation can enhance the overall efficiency and rate of conversion of interstellar gas into young stars.

In Table 2, the gas mass fraction (M_{H_2}/M_{dyn}) is given for those galaxies in which the nuclear mass concentration is resolved and for which the velocity dispersion can be estimated from the CO line-width. For the six cases where this ratio has been evaluated, it is in the range 0.2–0.9, indicating that the interstellar gas constitutes a significant fraction of the total mass in the nucleus. Such high gas mass fractions may reflect the inappropriateness of using the Galactic CO-to-H_2 conversion factor. On the other hand, it is also possible that the gas does indeed constitute a large fraction of the central mass since numerical simulations of merging galaxies (Hernquist 1989) which include a gas component show that the gas sinks to the center of the merged system more readily than the stellar component. This is due to the fact that the gas is much more dissipative than the stars. Thus, it might be anticipated that there will be regions with very large gas mass fractions during the evolution of a merging galaxy system.

3. AN EXAMPLE–ARP 220

Arp 220 has an infrared luminosity at $\lambda=8$–1000 μm of 1.5×10^{12} $L_\odot$, exceeding that in the visual by nearly 2 orders of magnitude and placing it in the luminosity regime of quasars. Recent single dish measurements show the CO emission extending over a velocity range of 900 km s^{-1} and the derived H_2 mass is 3.5×10^{10} $M_\odot$, approximately a factor of 15 greater than that of the Galaxy (Solomon, Radford, and Downes 1990). At optical wavelengths, the galaxy appears approximately spherical with a central dust lane and tidal tails, both characteristics of galactic merging, extending up to 70 kpc away (Sanders et al. 1988).

In Figure 4, mm interferometric maps at $2''$ resolution (750 pc) are shown from a recent Owens Valley study (Scoville et al. 1990). The continuum (mostly dust emission) and the integrated CO emission are shown in the upper panels as a function of displacement coordinates from the cm-wave radio continuum peak (Norris 1988). Both the CO and 2.7 mm continuum fluxes peak within $0.5''$ of the western component of the near infrared nucleus (Graham et al. 1990). In addition to the compact nuclear source, an extended CO emission component can be seen in maps made with smaller velocity ranges. In the lower panels of Figure 4, the emission is shown for velocity windows ($\Delta V=104$ km s^{-1}), centered at 5330 and 5642 km s^{-1}. The low level contours of CO emission are clearly extended along a NE-SW direction. The full velocity range of CO emission is seen in the central component, while a more limited velocity range is seen in positions offset along the extended component.

The CO emission can be separated into two components: a core $1.4\times 1.9''$ and an extended component $7\times 15''$ (deconvolved sizes). The former contains 2/3 of the flux detected by the interferometer; the latter contains the remaining 1/3. Comparison of the total CO line flux detected in the interferometer maps with that seen in single dish measurements made with the IRAM 30 m telescope (Solomon, Radford, and Downes

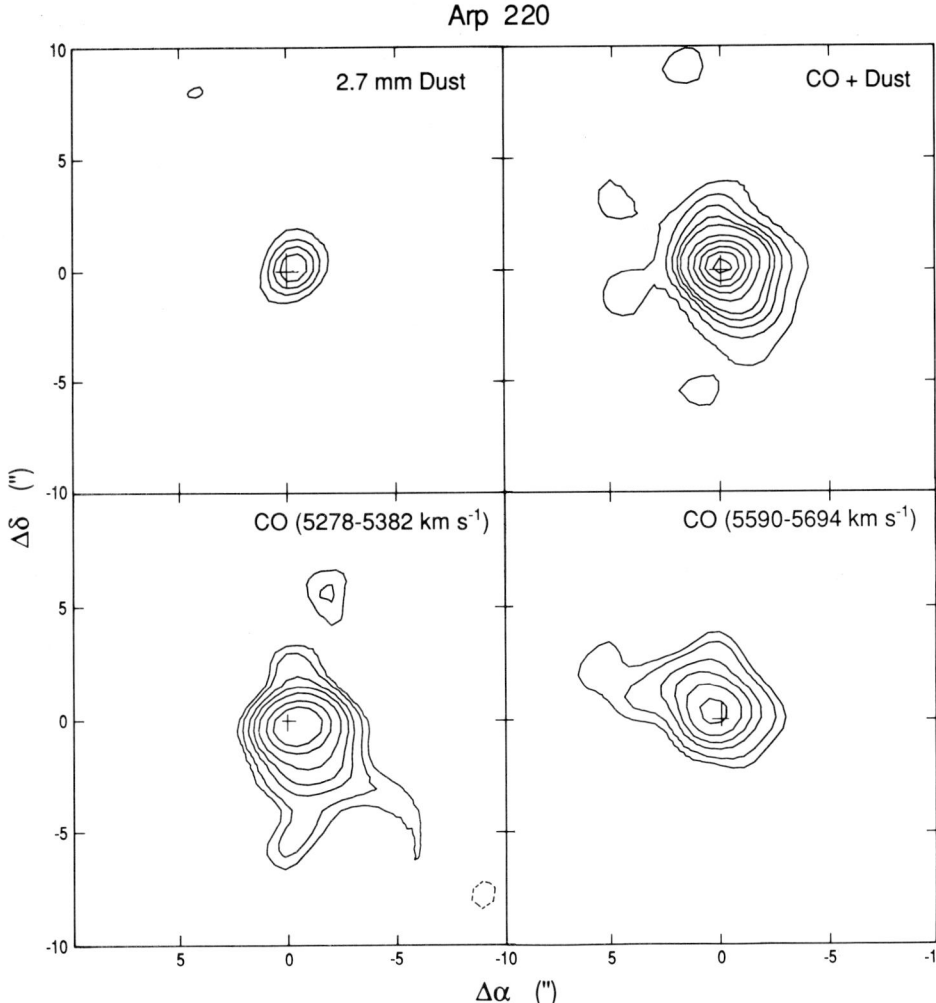

Figure 4. The $\lambda=2.7$ mm dust emission and integrated CO emission are shown in the upper panels as a function of displacement coordinates from the centimeter-wave non-thermal radio peak in Arp 220. In the lower panels, the CO emission at 5278–5382 and 5590–5694 km s^{-1} are shown. The synthesized beam is $1.9 \times 2.1''$ (PA=96°, Scoville et al. 1990).

1990) indicates that these two components account for 80% of the total CO emission from Arp 220. The highest velocity emission is probably all from the core source.

For the adopted distance of 77 Mpc, the H$_2$ masses in the core and extended components are 1.8×10^{10} M$_\odot$ and 9×10^9 M$_\odot$ respectively. The mean diameter of the core component ($\sim 1.7''$) corresponds to a radius of 315 pc and the H$_2$ density, smoothed out over the volume is 2900 cm^{-3}. If we assume that the full range of CO velocities ($\Delta V_{FWZI}=900$ km s^{-1}) exists within the core source of radius 315 pc, then the dynamical mass is 2.5×10^{10} M$_\odot$ (for a spherical distribution). This dynamical mass is almost precisely equal to the total gas mass (H$_2$+He) derived from the CO line flux for the core component.

In Figure 4 (lower panels), a velocity gradient in both the core and extended components may be seen—the lower velocity emission shifted to the Southwest of the nucleus, the higher velocity emission to the Northeast. The direction of this velocity gradient is parallel to the major axes (PA=52–63°) of the two CO emission components and to the dust lane seen prominently in optical photographs. Thus, the molecular gas has partially relaxed to a disk-like configuration with angular momentum playing a significant role in the gas distribution, probably inhibiting further collapse towards the center. The mean gas density (2900 cm^{-3}) derived for the central component from the CO measurements is somewhat lower than that (10^4–10^5 cm^{-3}) deduced by Solomon, Radford, and Downes (1990) from CS line measurements.

The λ=2.7 mm continuum flux (~30 mJy) provides a significant constraint on the emission measure of ionized gas and thus the O star luminosity in the nucleus of Arp 220 (Scoville et al. 1990). If the maximum residual 2.7 mm flux after subtraction of the non-thermal and dust components is taken to be $\lesssim 10$ mJy and it is assumed to be *entirely* free-free emission, then the upper limit for the emission rate of Lyman continuum photons is 7.5×10^{54} s^{-1}. For an O5 star (60 M$_\odot$, L=5×10^5 L$_\odot$), $Q_*=4\times10^{49}$ s^{-1} and the ratio of the bolometric luminosity to the ionizing photon production rate is 1.3×10^{-43} L$_\odot$ s^{+1} (Panagia 1973). The total luminosity corresponding to the derived upper limit to the ionizing photon production rate would therefore be 10^{11} L$_\odot$ for a population of O5 stars. Early O type stars by themselves would thus fail to provide the observed far infrared luminosity in Arp 220 by a factor of at least 15. In the next section, we present a series of starburst models in order to evaluate the luminosity, mass, and timescale constraints for a population of young stars with a range of masses and ages.

4. STARBURST MODELS

Constraints for starburst models derived from the observed luminosity, the inferred Lyman continuum production rate, and dynamical mass limits are most dependent on the assumed upper and lower mass cutoffs for the initial mass function (IMF) of the stellar population and the age of the starburst. To evaluate these constraints, we have computed starburst models in which the overall rate of star formation was assumed to be constant for a period of the order of 10^8 years. We feel this characteristic timescale is most relevant to the *global* starburst galaxies where the activity is triggered by an interaction. For such systems, it is difficult to imagine that the starburst could be synchronized on a timescale much less than the dynamical time ~2×10^8 yr.

The assumed initial mass function is a modified Miller-Scalo initial mass function

$$N(m) \propto m^{-1.4} \quad \text{for } m \leq 1 M_\odot \quad \text{and} \tag{3a}$$
$$m^{-2.5} \quad \text{for } m > 1 M_\odot. \tag{3b}$$

For stellar masses <10 M$_\odot$, the ZAMS mass-luminosity relation and main sequence lifetime were taken from Renzini and Buzzoni (1986) and for m$>$10 M$_\odot$, we adopt the relations from Maeder, (1987):

$$\log L = 2.55 \log m + 1.27$$
$$\log t = -0.963 \log m + 8.186 \quad \text{for } m = 10 - 40 M_\odot \quad \text{and} \tag{4a}$$

$$\log L = 1.84 \log m + 2.42$$
$$\log t = -0.430 \log m + 7.34 \quad \text{for } m > 40 M_\odot \tag{4b}$$

The total stellar lifetime including post-main sequence evolution was taken to be 1.2 times the main sequence lifetime and the post-main sequence luminosity was assumed to be constant. The production rate of Lyman continuum photons by high mass stars was adopted from Panagia (1973) with power law fits for each 5 $M_\odot$ interval for m>10 $M_\odot$. The amount of material recycled back into the interstellar medium as a result of stellar evolution was tracked assuming the remnant stellar masses specified by Renzini and Buzzoni (1986) and the ratio of the total stellar luminosity to that produced on the main sequence was taken from their Figure 8 which is appropriate to a constant rate starburst. The models were evolved with a constant rate of star formation; the lower mass cutoff was varied over the range 0.1–3 $M_\odot$ and the upper mass cutoff between 30–60 $M_\odot$.

4.1. Model Results

During the starburst, the total stellar luminosity (main sequence and post-main sequence) can be described quite accurately by the analytic expression

$$L = 1.18 \times 10^{10} \left(\frac{m_l}{1 M_\odot}\right)^\alpha \left(\frac{m_u}{45 M_\odot}\right)^{0.37} \left(\frac{\dot{M}}{1\, M_\odot\, yr^{-1}}\right) \left(\frac{t_B}{10^8 yr}\right)^{0.67} L_\odot \tag{5}$$

where t_B is the lifetime of the starburst and α=0.23 for m_l <1 $M_\odot$ and 0.55 for m_l >1 $M_\odot$. The rate of production of Lyman continuum photons is given by

$$Q = 9.85 \times 10^{52} \left(\frac{m_l}{1 M_\odot}\right)^\alpha \left(\frac{m_u}{45 M_\odot}\right)^\beta \left(\frac{\dot{M}}{1 M_\odot\, yr^{-1}}\right) sec^{-1} \tag{6}$$

where α=0.23 and 0.55 for m_l <1 $M_\odot$ and >1 $M_\odot$ and β=2.0 and 1.35 for m_u <45 $M_\odot$ and >45 $M_\odot$ respectively. Thus, the ratio of the total luminosity to the ionizing photon production rate is given by

$$\frac{L}{Q} = 1.20 \times 10^{-43} \left(\frac{m_u}{45 M_\odot}\right)^\alpha \left(\frac{t_B}{10^8 yr}\right)^{0.67} L_\odot\, sec \tag{7}$$

where $\alpha = -1.64$ and -1.0 for m_u <45 $M_\odot$ and m_u >45 $M_\odot$. Lastly, the ratio of the total luminosity to the mass of stars in the population is given by

$$\frac{L}{M_*} = 123 \left(\frac{m_l}{1 M_\odot}\right)^\alpha \left(\frac{m_u}{45 M_\odot}\right)^{0.37} \left(\frac{t_B}{10^8 yr}\right)^{-0.33} L_\odot\, M_\odot^{-1} \tag{8}$$

where α=0.23 and 0.55 for m_l <1 $M_\odot$ and >1 $M_\odot$ respectively. Relations (7) and (8) are accurate to ~20% for burst times in the range 10^7–5×10^8 years.

It is readily seen from Equation (5) that the infrared galaxies with luminosities in the range 10^{11}–10^{12} typically require star formation rates of 10–100 $M_\odot\, yr^{-1}$ for an IMF truncated at 1 $M_\odot$. For an IMF continuing down to 0.1 $M_\odot$, the required star formation rates ($\dot{M}$) are 1.7 times larger. Since the overall mass of interstellar gas in these galaxies is typically 10^{10} M$_\odot$, the duration of the starburst must be generally much less than 10^9 yr.

For Arp 220 where the emission rate of ionizing photons is limited by an observational upper limit on the free-free flux (see §3), $L/Q > 2 \times 10^{-44}$ L$_\odot$ sec. Thus, Equation (7) requires $m_u < 15$ M$_\odot$ for $t_B = 10^8$ years or the burst time must exceed 10^9 years for $m_u = 45$ M$_\odot$! On this basis, we suspect that significant fractions of the observed luminosity originate from both a nuclear starburst and a central non-thermal source.

In many of the ultraluminous galaxies, the gas mass fraction in the nucleus is large ($M_{H_2}/M_{DYN} = 0.2$–1, see Table 2). Thus, the ratio L/M_* must exceed the observed ratio L_{IR}/M_{H_2} which is typically 40 L$_\odot$ M$_\odot^{-1}$. Equation (8) then requires that the upper and lower mass cutoffs for the IMF be comparable with the normalizations given in Equation (8) in order to satisfy the dynamical mass constraint.

We have also computed the mean mass of stars contributing Lyman continuum photons as a function of the upper mass cutoff on the IMF. The models yield

$$m_Q = 31.8 \left(\frac{m_u}{45 M_\odot}\right)^{0.68} M_\odot \quad (9)$$

It is worthwhile noting that for a constant rate starburst, the ionizing photon production rate reaches a constant value within timescales $\sim 10^7$ yr and the spectrum of the UV radiation, characterized by m_Q, becomes independent of time for timescales $> 10^7$ yr (cf. Equation 9). The characteristic temperature of the stars emitting the Lyman continuum photons is thus approximately given by

$$T_Q \simeq 38000 \left(\frac{m_u}{45 M_\odot}\right)^{0.68} K. \quad (10)$$

The temperature of 38000 K corresponds to an O7 ZAMS star. The luminosity weighted stellar mass is given by

$$m_L = 22.3 \left(\frac{m_u}{45 M_\odot}\right)^{0.69} \left(\frac{t}{10^7}\right)^{-0.14} M_\odot \quad (11)$$

Thus, although the initial mass function extends to high masses, the majority of the ionizing photons and the luminosity are emitted by considerably lower mass stars. For this reason, one should be very cautious in evaluating the upper mass cutoff on the basis of the observed ionization state (eg. [OII]/[OIII]) of the gas in terms of a zero-age population model.

Lastly, it is of interest to evaluate that the fraction of the stellar matter put back into the ISM as a function of time and the mass cutoffs. For times $< 10^7$ yr, the percentage of recycled matter is $< 5\%$ for all reasonable parameters but increases to 10–50% for a burst of age 10^8 years and $m_l = 0.1$–3 M$_\odot$. During the starburst, the fraction of the time-integrated star formation rate ($\dot{M}t$) which is in stars (main sequence, post main sequence, and stellar remnants) is given approximately by

$$f(t) = 0.77 \left(\frac{t}{10^8 yr}\right)^{-0.13} \left(\frac{m_l}{1 M_\odot}\right)^\alpha \quad (12)$$

where $\alpha = -0.05$ to -0.08 for $m_l = 0.1$–3 M$_\odot$. After the starburst has stopped, the stellar mass fraction falls as

$$f(t) = f(t_B) \left(\frac{t}{t_B}\right)^{-0.25} \left(\frac{m_l}{1 M_\odot}\right)^{-0.14} \tag{13}$$

until it reaches assymptotic values corresponding to the mass fraction in stellar remnants (f=0.61, 0.32, and 0.12 for m_l=0.1, 1, and 3 $M_\odot$). At 10^9 years after the start of a burst lasting 10^8 years, the stellar mass fractions are 0.69, 0.44, and 0.12 for m_l=0.1, 1, and 3 $M_\odot$.

The temporal evolution of the luminosity for a constant rate burst may be divided into three phases. During the burst, the luminosity increases as $t^{0.67}$ (Equation 5). Immediately after star formation is halted, the luminosity falls precipitously, as $(t/t_B)^{-2.1}$, for $t=t_B \rightarrow 3t_B$ and then more gradually, as $t^{-1.2}$, until all the stars have evolved.

5. MERGER-INDUCED STARBURSTS

In §2.2, we showed that there exists a strong correlation between the starburst-infrared galaxies (ie. those with L_{IR}/M_{H_2} significantly higher than normal galaxies) and the occurrence of a recent galactic interaction. Virtually all of the strong *global* starbursts occur in such systems. The progenitors of such interacting starburst systems must have been normal optical galaxies represented in the optical galaxy luminosity function. To first order, one expects that the mass of the interstellar medium in the merging starburst galaxies is approximately the sum of those in the two progenitor galaxies and since *significant* dynamical perturbations are required to initiate a *global* starburst, comparable mass galaxies must be the progenitors. Based on these two simplified assumptions, we reason that the systems with infrared luminosity-to-molecular mass ratios enhanced by an order of magnitude might have originated from the interaction of two galaxies in the optical luminosity function each a factor ~20 lower in optical luminosity. The power law fall-off in the infrared luminosity function (which exceeds the optical luminosity function above 3×10^{11} $L_\odot$) may then be naturally understood since the highest luminosity infrared galaxies had optical galaxies with approximately 20 times lower luminosity as progenitors. At the very highest infrared luminosities ($L_{IR} > 10^{12}$ $L_\odot$), a significant fraction of the luminosity may also be contributed by an embedded active galactic nucleus (eg. Arp 220 and Mrk 231), rather than young stars.

5.1. The Frequency of Galactic Interactions

In Table 3, the space densities of optical galaxies, infrared galaxies, and interacting infrared galaxies are compared. The fraction of infrared galaxies which are interacting (adjacent pairs or merger remnants) is a strongly increasing function of luminosity, suggesting that the infrared active phase for the highest luminosity galaxies is comparable to the timescale for galactic merging (τ_{DYN} ~2×10^8 years). If we compare the number density of interacting galaxies at $L_{IR}=10^{10}-10^{12}$ $L_\odot$ with that of optical galaxies in a luminosity range displaced a factor of 20 lower, the number density ratio is 10^{-3} (integrating Equations 1 and 2 over the appropriate ranges). Since presumably the progenitors must be gas-rich, it is probably most appropriate to compare the IR galaxy density with that of optical *spiral* galaxies. We then obtain ~2×10^{-3} for the number density ratio. Since the duration of the infrared–starburst phase is probably approximately equal to the dynamical time (~2×10^8 years) and two galaxies must take part in the interaction, the present-day frequency of interaction-induced starbursts must

TABLE 3. *Optical and IR Galaxy Luminosities and Space Densities*

L_{BOL} ($L_\odot$)	N_{OPT} (Mpc^{-3})	$\mathcal{L}_{OPT}$ ($L_\odot$Mpc^{-3})	N_{IR} (Mpc^{-3})	$\mathcal{L}_{IR}$ ($L_\odot$Mpc^{-3})	Interating IR Gal. fract.	$\mathcal{L}_{Int}$ ($L_\odot$Mpc^{-3})
10^8–10^9	2.1×10^{-1}	7.3×10^7	3.8×10^{-2}	1.1×10^7	0	
10^9–10^{10}	1.0×10^{-1}	3.5×10^8	9.5×10^{-3}	2.9×10^7	~ 0.05	1.5×10^6
10^{10}–10^{11}	2.4×10^{-2}	5.4×10^8	1.8×10^{-3}	3.9×10^7	0.1	3.9×10^6
10^{11}–10^{12}	1.2×10^{-4}	1.5×10^7	2.0×10^{-5}	3.3×10^6	0.5	1.7×10^6
10^{12}–10^{13}	—		9.8×10^{-3}	1.7×10^5	1	1.7×10^5
10^8–10^{13}	3.4×10^{-1}	9.8×10^8	4.9×10^{-2}	8.2×10^7		5.7×10^6

Volume densities of optical and infrared galaxies obtained by integrating Equations (1) and (2). Column (6) lists the fraction of infrared galaxies classified as interacting.

be such that approximately 2% of the spiral galaxies would have passed through this phase in the last 10^9 years. Including the effects of cosmological evolution on galaxy interaction rates, we conclude that virtually all galaxies should have undergone such an interaction in the last 10^{10} years (ie. $z\simeq1$). Merger-induced starbursts must therefore be a significant factor in the past evolution of spiral galaxies, rejuvenating the stellar population and rearranging the galactic distribution of interstellar matter. Understanding how present day spirals have such cold and radially extended disks is an important theoretical issue.

5.2. The Infrared Starburst Luminosity Function

In order to test more explicitly the hypothesis that collision-induced global starbursts might account for the high luminosity tail on the infrared galaxy luminosity function, we have undertaken a Monte Carlo simulation. We assume that the progenitor galaxies are represented in the optical luminosity function and that in order to induce a global starburst, comparable mass galaxies must be involved. The luminosity of the merger product is then elevated by approximately a factor of 10 compared to the sum of the luminosities of the two progenitor galaxies. Thus, the rate of production of infrared galaxies with luminosity, L_{IR}, is given by

$$\dot{N}(L_{IR}) \propto N_o^2(L_{IR}/20) \qquad (14)$$

where N_o is given by Equation (1), and we have implicitly assumed, for simplicity, that variations in the galaxy-galaxy collision cross-sections and their relative velocities are insignificant compared to variations in the space density of galaxies as a function of luminosity. The duration of the elevated infrared-luminosity phase following an interaction was taken from the luminosity evolution found for the constant rate starburst models in §4. Specifically, we adopt

$$L_{IR} = 30 \left(\frac{\Delta t}{t_B}\right)^{0.67} L_o \qquad \Delta t < t_B$$

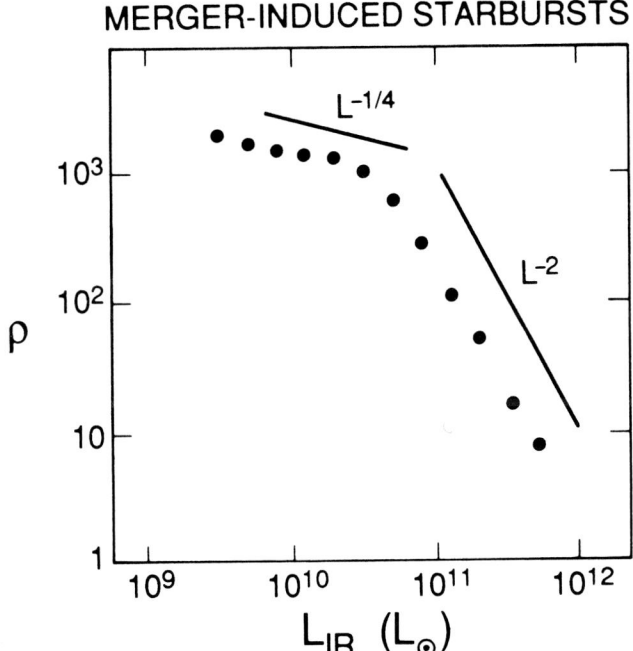

Figure 5. *The infrared luminosity function (•) resulting from merger-induced starbursts resulting from the Monte Carlo simulation described in the text. The derived distribution of galaxies per logarithmic luminosity interval may be represented by a 2-part power law ($L^{-0.25}$ and L^{-2}) with a break approximately $L = 3 \times 10^{10}$ $L_\odot$, in good agreement with the observed infrared luminosity function.*

$$= 30 \left(\frac{\Delta t}{t_B}\right)^{-2.1} L_o \quad t_B < \Delta t < 3 t_B \quad (15)$$

$$= 3 \left(\frac{\Delta t}{3 t_B}\right)^{-1.2} L_o \quad \Delta t > 3 t_B$$

where Δt is the elapsed time since the galactic collision, t_B is the duration of the constant rate starburst (taken to vary randomly between 10^8 and 3×10^8 years), and L_o is the luminosity of the progenitor optical galaxies. The time steps were taken to be $\Delta t = 2 \times 10^7$ years and at $\Delta t > 10^9$ years, the galaxies were removed from the infrared luminosity function.

In Figure 5, the distribution function for infrared merger galaxies is shown from a simulation in which optical galaxies were sampled over the range 10^9–10^{11} $L_\odot$. The derived distribution exhibits a 2-part power law spectrum with the number of galaxies (per logarithmic interval in luminosity) varying as $L_{IR}^{-0.25}$ at low luminosities and as L_{IR}^{-2} at higher luminosities. Given the simplistic and uniform prescription used to characterize the starburst activity, the agreement with the observed luminosity function (Figure 1) is impressive. In particular, the slope of the power law on the high luminosity tail and the presence of the break at 10^{10}–10^{11} $L_\odot$ are reproduced quite well. In this model, the high luminosity tail arises from merging of optical galaxies at $L \lesssim L_{*o}$ (cf. Equation 1). The merger product for these galaxies is elevated in luminosity by a factor ~ 10, based on the observed ratio L_{IR}/M_{H_2}.

In the context of the collision induced starburst scenario, the simulation described above may be used to predict systematic variations as a function of L_{IR}. For example, the very highest luminosity infrared galaxies should originate from merging of optical galaxies near the break in the Schecter function. In addition, since the starburst luminosity rises as $t^{0.67}$ during the burst, one would expect a high percentage of the most luminous IR galaxies to be aging, but still active starbursts. Such galaxies would have relatively high ratios for L/Q (see Equation 7) as found for Arp 220 (§3).

6. SUMMARY

We briefly summarize the results of infrared, optical, and molecular observations for high luminosity galaxies undergoing global starbursts:

1) The most noteworthy feature of the infrared galaxy luminosity function is the power law tail extending to $L \sim 10^{12}$ $L_\odot$. At the highest luminosities, the local space density of infrared galaxies significantly exceeds those of optical galaxies and quasars.

2) Extensive surveys of CO emission from the infrared galaxies have shown extraordinarily large abundances of molecular gas with H_2 masses 1–20 times that of the Milky Way.

3) The ratio L_{IR}/M_{H_2}, commonly taken as an indicator of the star formation efficiency, is typically 20–60 $L_\odot$ $M_\odot^{-1}$ in the luminous infrared galaxies. Interpreting this ratio as a signature of *global* starburst activity, we conclude that the overall star formation efficiencies in these galaxies are enhanced by an order of magnitude compared to normal spiral galaxies.

4) High resolution aperture synthesis measurements of the CO emission generally show a substantial fraction of the total ISM concentrated within the central ~ 1 kpc. Peak ISM surface densities reach 7×10^4 $M_\odot$ pc^{-2} (eg. Arp 220) and the gas mass fractions are 0.5–1.

5) Optical and near infrared images of the ultraluminous infrared galaxies reveal that virtually all are interacting/merging galaxy systems. Thus, the global starbursts are apparently initiated by galactic merging and the duration of the starburst must be comparable with the dynamical times $(1-2 \times 10^8$ years).

6) Starburst models with a standard IMF can account for the observed luminosities with star formation rates of 10–100 $M_\odot$ yr^{-1} for bursts extending over $\sim 10^8$ years. The total mass of young stars would then be $\gtrsim 10^9$ $M_\odot$.

7) The high luminosity power law and the break at $L_{IR} = 3-5 \times 10^{10}$ $L_\odot$ seen in the infrared galaxy luminosity function may be accounted for by global starbursts resulting from the merging of optical galaxies with comparable masses, sampled from the optical galaxy luminosity function.

8) The statistics of infrared interacting galaxies indicate that the percentage of galaxies undergoing merger-induced starbursts is approximately 2% every 10^9 years and $\sim 100\%$ at $z=1$.

It is a pleasure to thank our close collaborators in this work (C. Norman, D. Sanders, and A. Sargent) and Leona Kershaw's assistance in preparation of this contribution. This research was supported in part by NSF Grant AST 87-14405 and some of the ideas were developed while one of us (NS) was at the Aspen Institute for Physics. Some of the analysis presented in this contribution is to be contained in an article by the authors for *The Astrophysical Journal*.

REFERENCES

Canzian, B., 1990, Ph.D. Thesis, California Institute of Technology.
Chiosi, C. and Maeder, A. 1986, *Ann. Rev. Astr. Ap.*, **24**, 329.
Elmegreen, B. G. and Lada, C. J., 1977, *Ap. J.*, **214**, 725.
Felten, J.E. 1976, *A. J.*, **82**, 861.
Graham, J. R., Carico, D. P., Matthews, K., Neugebauer, G., Soifer, B. T., and Wilson, T. D., 1990, *Ap. J. (Letters)*, **354**, L5.
Hernquist, L., 1989, *Nature*, **340**, 687.
Huchra, J., 1977, *Ap. J. Suppl.*, **35**, 171.
Joseph, R. D., Wright, G. S., and Wade, R., 1984, *Nature*, **311**, 132.
Lawrence, A., Walker, D., Rowan-Robinson, M., Leech, K. J., and Penston, M. V., 1986, *M.N.R.A.S.*, **219**, 687.
Maeder, A., 1987, *Astr. Ap.*, **173**, 247.
Norman, C. A. and Scoville, N. Z. 1988, *Ap. J.*, **332**, 124.
Norris, R. P. 1988, *M.N.R.A.S.*, **230**, 345.
Panagia, N. 1973, *A. J.*, **78**, 929.
Planesas, P., Mirabel, I. F., and Sanders, D. B., 1990, *Ap. J.*, in press.
Planesas, P., Scoville, N. Z., and Myers, S. T., 1990, *Ap. J.*, (submitted).
Renzini, A. and Buzzoni, A. 1986, in *Spectral Evolution of Galaxies*, ed. C. Chiosi and A. Renzini (Dordrecht: Reidel), p. 195.
Rieke, G. H. and Lebofsky, M. 1986, *Ap. J.*, **304**, 326.
Sanders, D. B. 1990, *IAU Symposium No. 146 "Dynamics of Galaxies and Molecular Clouds Distribution"*, ed. F. Combes (Dordrecht: Kluwer).
Sanders, D. B., Scoville, N. Z., and Soifer, B. T. 1990, *Ap. J.*, in press.
Sanders, D. B., Scoville, N. Z., Sargent, A. I., and Soifer, B. T. 1988, *Ap. J. (Letters)*, **324**, L55.
Sanders, D. B., Scoville, N. Z., Young, J. S., Soifer, B. T., Schloerb, F. P., Rice, W. L., and Danielson, G. E., 1986, *Ap. J. (Letters)*, **305**, L45.
Sanders, D. B., Soifer, B. T., Elias, J. H., Madore, B. F., Matthews, K., Neugebauer, G., and Scoville, N. Z., 1988, *Ap. J.*, **325**, 74.
Sargent, A. I. and Scoville, N. Z., 1990, (in preparation).
Schechter, P. 1976, *Ap. J.*, **203**, 297.
Schmidt, M. and Green, R. 1983, *Ap. J.*, **269**, 352.
Scoville, N. Z. and Good, J. C. 1989, *Ap. J.*, **339**, 149.
Scoville, N. Z. and Sanders, D. B. 1987, in *Interstellar Processes*, ed. D. J. Hollenbach and H. A. Thronson (Dordrecht: Reidel), p. 21.
Scoville, N. Z. and Young, J. S. 1983, *Ap. J.*, **265**, 148.
Scoville, N. Z., Sanders, D. B., and Clemens, D. P. 1986, *Ap. J. (Letters)*, **310**, L77.
Scoville, N. Z., Sanders, D. B., Sargent, A. I., Soifer, B. T., and Tinney, C. G., 1989, *Ap. J. (Letters)*, **345**, L25.
Scoville, N. Z., Sargent, A. I., Sanders, D. B. and Soifer, B. T., 1990, *Ap. J. (Letters)*, in press.
Smith, B. J., Kleinmann, S. G., Huchra, J. P., and Low, F. J. 1987, in *Star Formation in Galaxies*, ed. C. J. Persson (Washington, DC: US Government Printing Office), in press.
Soifer, B. T., Boehmer, L., Neugebauer, G., and Sanders, D. B., 1989, *A. J.*, **98**, 766.

Soifer, B. T., Helou, G., Lonsdale, C. J., Neugebauer, G., Hacking, G., Houck, J. R., Low, F. J., Rice, W., and Rowan-Robinson, M. 1984, *Ap. J. (Letters)*, **283**, L1.

Soifer, B. T., Sanders, D. B., Madore, B. F., Neugebauer, G., Danielson, G. E., Elias, J. H., Lonsdale, C. J., and Rice, W. L., 1987, *Ap. J.*, **320**, 238.

Solomon, P. M., Radford, S. J. E., and Downes, D., 1990, *Ap. J. (Letters)*, **348**, L53.

Stanford, S. A., Sargent, A. I., Sanders, D. B., and Scoville, N. Z., 1990, *Ap. J.*, **349**, 492.

Tinney, C. G., Scoville, N. Z., Sanders, D. B., and Soifer, B. T. 1990, *Ap. J.*, in press.

Vogel, S. N., Kulkarni, S., and Scoville, N. Z. 1988, *Nature*, **334**, 402.

Wang, Z., Scoville, N. Z., and Sanders, D. B. 1990, *Ap. J.*, (submitted).

Yahil, A., Strauss, M. A., Davis, M., and Huchra, J. P. 1990, *Ap. J.*, (submitted).

Young, J. S., Claussen, M. J., and Scoville, N. Z. 1988, *Ap. J.*, **324**, 115.

Young, J. S., Claussen, M. J., Kleinmann, S. G., Rubin, V. C., and Scoville, N. Z., 1988, *Ap. J. (Letters)*, **331**, L81.

DISCUSSION

C. Lonsdale: I assume you've used the standard CO to H II conversion ratio from the Milky Way, yet the conditions in these nucleii are presumably quite different with higher gas temperatures and higher densities. My only uncertainty is on your derived H_2 masses.

Scoville: Well, obviously, the masses can't be much higher than they appear to be, because they would violate the dynamical mass constraints. But there are, well, for those who aren't familiar with the whole issue, basically the conversion between CO line flux and H_2 mass should vary as the temperature of the gas divided by the square root of the mean density within the clouds. And so there are two issues; one is, as Carol's pointed out, the clouds are very likely to be much hotter. The standard GMC in a disk in our galaxy might be 10 K kinetic temperature. Based upon the infrared color temperatures of the dust in these galactic nucleii, they are typically 40 or 50 K, so the gas temperature might be, say, 5 times higher on the mean than the gas temperature in standard GMC on which this conversion factor is based. The other problem, though, is that the mean density of the clouds is probably higher. The standard GMC in which the conversion factor is derived has a density of about 200 cm^{-3}, that is their smoothed out density within the cloud. And I've already showed that in the case of Arp 220, the mean smoothed-out density is a $\sim 2 \times 10^3$ cm^{-3}. And I don't really know how to balance those factors. My feeling is that they tend to be somewhat self-compensating. There's one independent method of deriving the gas masses, and that's based upon the dust continuum emission. Some people are happy about that, some aren't. I saw some wincing. In that case, the problem is that you're looking at long wavelengths in order to try to pick up even the cold dust, say, at 1 mm. One problem which arises, of course, is you don't really know what the emissivity law is in the far infrared, but that situation will probably be remedied. But you also really don't know what the opacity is per unit mass of gas and dust. And the people who've done the dust measurements, let's say at

1 mm, have tended to end up with higher mass estimates than is derived by CO, but my feeling is that they're unreasonably high because they violate the dynamical mass constraints. So, I'm sure that ours are correct to 5%.

T. Heckman: It seemed to me you were halfway to the answer to a really interesting question. In fact ...

Scoville: Maybe you can finish it then ...

T. Heckman: ... a most interesting question about starbursts, and that is, over the history of the universe, what fraction of the massive stars have ever formed, formed during a starburst as opposed to just sort of normal processes, say, in spiral density ways.

Scoville: OK, I was going to start to answer your question by saying that I actually think that essentially all high mass star formation may occur in this either collisional or stimulated mode, and in a sense, I consider that to be a starburst. On the other hand, if you say that, half or maybe all galaxies have undergone a severe interaction or have had a steep bar potential at one point in their past, and therefore had a nuclear starburst. I don't know the answer to the question.

C. Norman: I think that this is just a paradigm for virtually galaxy formation, exactly the process that's going on, so my answer to your question is nearly all the stars are formed in this way—massive stars.

Scoville: Well, but wait. How do you know that. (Laughter)

C. Norman: It's an ex cathedra statement.

Scoville: I mean that's my belief too, but .... In some nearby galaxies, in particular M33, Chris Wilson, who's a student also at Cal Tech, has been measuring individual molecular clouds and does find a ratio of CO estimated mass versus virial mass for those clouds that you can actually map at high resolution, which is consistent with the galactic ratio, but that's not the kind of environment you've been talking about.

J. Bland Hawthorn: On the issue of the triggering mechanism, you were raising the issue of NGC 1068. In the kinematics you see clear elliptical streaming in the Fabry-Perot data. It's very clear indeed and it's been quite well modeled. Now, to make the comparison with Arp 220, it's quite easy to show that in a disk system with or without halos when you start to merge or collide these systems, you set up elliptic streaming, very effectively indeed in the central regions at least. So I think the triggering mechanism in all cases may be just this point of elliptic streaming and this Harwitt idea of colliding pancakes of gas to trigger star formation may not be the correct one.

Scoville: No, I don't think that's the correct one. Harwitt's idea in fact was that you take the gross kinetic energy of the galactic collision and convert it directly into luminosity. And I think that the time scale that that can produce the luminosity is a few 10^5 years and it's clear that this phenomena's going on longer because you see these dynamical interactions take longer than that to actually occur. One thing I didn't quite point out as you're probably aware—sometimes one forgets to say the obvious—I was trying to actually build an evolutionary path here, and I would picture something like the antennae system as being fairly early in the overall evolution of a merger, Arp 220 being quite advanced, but having not really shed its gas. Something like 1068 might be a more evolved system, where the material in the inner disk, is perhaps a result of a past interaction that formed a strong bar in the center of the disk of the galaxy. Obviously bars and interactions can be related. But NGC 1068 obviously has a well-developed non-thermal central source and I think it's not obviously equivalent to say Arp 220. I didn't mean to imply that and, in fact, the triggering of present day star formation in NGC 1068 may not be the same as in Arp 220. In Arp 220 it may be just sort of random motions of clouds colliding. It looks like a sufficiently disordered system, I wouldn't want to say that there is elliptical streaming in Arp 220, but in 1068 that may well be the case.

J. Bland Hawthorn: The kinematics are very well-behaved, streaming is ...

Scoville: Well I think kinematics of the stars may be well-behaved, but the gas being fairly disordered I don't think it's going to have nice stream lines. I mean there's going to be shock fronts all through the system.

J. Bland Hawthorn: But those are only in two locii.

Scoville: Yes.

P. Stockman: You pointed out several times during the talk that in the IRAS sources they had greater amount of H_2 mass than typical galaxies, than other galaxies in the Schechter luminosity function. So that would suggest that unless, in fact, you create H_2 gas during this collision, which you started this talk by saying you didn't think that was true ...

Scoville: Yes.

P. Stockman: ... that some of the things that made an IRAS source are what the galaxies bring with them into the merger or interaction versus something that occurs in the interaction, in which case in a sense the IRAS sources are, you're selecting interacting systems in which the two galaxies are bringing an unusual amount of H_2 mass with them.

Scoville: Yeah, I think that's true to some extent. I think what you're saying is a good point, and I should have made it. Basically, the IRAS galaxies have to be galaxies which, in view of the fact that there's a higher luminosity to mass ratio, the galaxies which came from down here in, say, the optical luminosity function and got pushed up in luminosity. And so, one of the problems in, say, comparing the space density of infrared galaxies with optical galaxies is, I mean it's intuitively clear to me and obviously to you that one shouldn't really be comparing quite the same luminosity ranges. It's just very hard to know quite what you should do. I mean, how far down in the optical luminosity function you should be comparing the space density.

P. Stockman: That is, that if you're promoting low luminosity optical sources ...

Scoville: Well, these low luminosity optical sources may have a substantial H_2 mass. It may not be very active in terms of producing luminosity. But the real way the problem should be attacked, I think, is to take the H_2 mass as the indicator of what the galaxy's intrinsically like, and if somebody had the equivalent luminosity function or mass function for H_2 or interstellar gas content, then you could say go down to the same overall gas content in both the infrared active and nonactive galaxies and compare the two. But, I mean that's an approach which maybe could be tried. I'm not really—Judy Young and Nick Devereux have a large survey. Maybe they could do it.

G. Rieke: With regard to the active nucleus vs. starburst heating in ultraluminous galaxies, it seems to me there's a limit from the fact that the ultraluminous galaxies are not bright x-ray sources, as you might expect from the Type I Seyfert spectra. Can you comment on what that limit would be?

Scoville: Quick answer—No. There was a paper actually, which of course you wrote, which I think was comparing the x-ray flux from a sample of active galaxies with the x-ray flux from IRAS galaxies. I'm not that familiar with the paper but I got the impression that the x-ray active galaxies were picked from an x-ray selected sample, so there might be some uncertainty. The other thing is, at least in the case of Arp 220 the column densities are pretty huge, but if you can see it in the near infrared, you have to be able to see it in the x-ray. I can't answer your question, George.

G. Rieke: ... not x-ray active. It seems to me it pushes you toward saying that starbursts are probably responsible for a lot of luminosity in those galaxies.

Scoville: That's a good point.

S. Lamb: I just wanted to return to a point that was beginning to be made earlier. You talked about the actual dynamical triggering of the starbursts and I wanted to just mention that, of course, the galaxies will actually pass through one another, perhaps several times before this merger takes place.

Scoville: The stars will ...

S. Lamb: Yes. The stars of the whole. And that, in fact, one doesn't anticipate all of the gas to get left behind the very first time they pass by one another. So you might expect to see a series of starbursts leading up to some kind of final starburst perhaps, or in fact the very first passage through might produce the very largest effect on star formation. So that when one's comparing observations and theory, one should bear this in mind.

Scoville: My feeling would be that, you know, if you have two galaxies which pass through each other, most of the masses in the star systems and the gas which doesn't collide will continue on and I would have thought that the next near passage of those two galaxies would be many 10^8 years in the future. So that you'd end up basically the first starburst would have died away long before you see the second one, but that's just my intuition.

S. Lamb: It just depends on the particular ...

Scoville: But I'm not aware of any evidence for that, but my feeling is the evidence would be wiped out.

I. Mirabel: I wanted to comment on the previous question: 4×10^{10} solar masses in a molecular cloud of molecular gas is unreasonable. No. If the original galaxies that are colliding are giant spiral galaxies, we know from HI observations that there are several giant spiral galaxies with 3×10^{10} $M_\odot$ in atomic gas. Therefore, if you have as a consequence of the merging of two giant spiral galaxies at high efficiency of formation of molecular gas from the atomic gas, because of the high densities that you are getting, it is very reasonable the numbers that Nick mentioned. You can get 6×10^{10} $M_\odot$ of molecular gas out of the collision of two giant spiral galaxies. So, my conclusion is that they must be giant spiral galaxies. It's very difficult to explain these large amounts of interstellar gas in that way. Other way to confirm that these numbers might not be very much off, is that in Arp 220 we get 10^{22} atoms cm^{-2} in absorption when we look in HI. That means that the column density of HI is tremendous. We know that there is a lot of cool HI in this galaxy that we know is always associated with the formation of molecular clouds. The last comment I want to make that has not been mentioned, is that a way to probe the kinematics of the molecular gas, the very high resolution in space and velocity is with megamasers. More than 90% of these galaxies are OH megamasers and we can probe the kinematics of the inner region with ground-based interferometry.

Scoville: Good point. I think what you're suggesting is that the gas content which is generally much more distributed and being HI but that's been compressed and brought into the inner disk of the galaxy.

G. Miley: Nick, the picture you showed of NGC1068, the bar, am I right in saying that it's in the same direction as the radio?

Scoville: Yes.

G. Miley: So one person's bar is another person's jet. And I'm asking you, doesn't this ...

Scoville: Yes, but just think. The nearest red bar is longer than the radiojet. The radiojet extends out about half the radius of the ...

G. Miley: But the fact that it's in the same direction I think implies that it's not just a bar phenomena. And I'm saying that maybe this is more evidence that there's a connection between star formation and radiojets.

Scoville: Yeah. Well, maybe Andrew Wilson may have something ...

A. Wilson: I was just going to comment that I don't think that it really implies that the bar has anything to do with the radiojet directly. I think you can have a situation in which the bar forms a gas disk with its axis along the bar in the inner nucleus and it's the gas disk which collimates the radiojet. So there's an indirect relation.

J. Young: Also, with reference to Pete Stockman's question, I think there may be two processes going on that lead to the high H_2 masses and the ultraluminous galaxies and one is, that there are plenty of galaxies, NGC 7479 is an example, a spiral relatively cool disk, with star formation extended over the whole galaxy, and 2×10^{10} $M_\odot$ of gas. It's like the Milky Way in its star formation efficiency now, but if two of those galaxies collided, you'd have enough gas to make the Arp 220 and the concentration in the middle could easily lead to the high efficiency. But we also see in those, not necessarily in the most ultraluminous sample, but in sort of the more intermediate luminosity sample, that the molecular to atomic gas ratio is higher in the interacting galaxies than it is in Scs so there may also be the enhanced conversion of atomic to molecular gas that's leading to the higher molecular masses, although it could also be the removal of HI that's leading to that result.

Scoville: The issue is I think that there basically don't appear to be many galaxies down in this area of the diagram which could jump up into this area, but these galaxies are selected on the basis of far infrared luminosity or some optical criteria, but Young and Devereaux do have a survey which is in some sense distance limited; presumably, one could derive a mass function for the normal galaxies.

J. Young: We actually have more galaxies along your lower line there that go up to 10^{10} $M_\odot$ of gas at low efficiency.

STELLAR CONTENT OF STARBURST GALAXIES

Robert Joseph
University of Hawaii
Inst. for Astronomy
2680 Woodlawn Drive
Honolulu, HI 96822
USA

1. INTRODUCTION

The infrared spectral region provides a number of unique and powerful tools for gaining insight into the stellar content of galaxies. The infrared perspective on this subject is not widely appreciated, partly because infrared astronomy is itself a young subject, and also because it is only relatively recently that technical improvements in infrared spectrometers available on 4 m class telescopes made it possible to do serious (but still relatively low resolution) extragalactic observations at all. Thus, extragalactic infrared spectroscopy is still largely in its initial exploratory phase, and it is possible to read all the papers in the field in a few sittings (cf. Joseph 1989, Moorwood 1989).

I have therefore emphasized the infrared approach to investigating the stellar content of starburst galaxies, partly because it is what I know best, but partly also in reaction to a number comments and tacit assumptions made at this conference that one can only learn about massive stars in galaxies by making ultraviolet observations, and that the inherent limitations in ultraviolet studies pretty much define what we will ever be able to learn about the upper end of the initial mass function. I hope that the papers by Rieke, Gatley, and myself in these proceedings will illustrate the advantages of a more catholic approach to this subject, and the unique contributions that the infrared perspective can provide.

For a complementary review which also incorporates results from optical and ultraviolet diagnostics, I recommend the excellent recent by John Scalo, "Top-Heavy IMF's in Starburst Galaxies" (Scalo, 1989).

2. INFRARED TECHNIQUES FOR MEASURING EXTINCTION

Starbursts are located in the central few hundred parsecs of galaxies. Since these are also the regions of highest extinction in spiral galaxies, optical and ultraviolet observations may be severely limited; indeed, as I shall illustrate below, optical and ultraviolet spectroscopy may not be probing this region at all in many galaxies.

The first point to note is that infrared diagnostics are relatively insensitive to extinction. This is important for observations of the central regions of most galaxies, but becomes crucial for the ultra-luminous infrared galaxies (cf. Joseph and Wright 1985, Sanders et al. 1988) which apparently are powered by the most powerful starbursts known, with luminosities equalling those of quasars (Soifer et al. 1986).

Moreover, the extinction can be measured directly using infrared spectroscopy in several relatively model-independent ways. Firstly, one can use two infrared lines in which the transitions are from the same upper level. In this case the relative intensities are given by the ratio of the Einstein A-coefficients. Comparing this ratio with the observed ratio give the extinction at the relevant wavelength. One example is in the vibration-rotation spectrum of H_2. These quadrupole lines appear in the infrared spectra of many starburst galaxies and in virtually all the interacting galaxies we have observed (Joseph 1989). The ratio of the 1-0Q(3) line to the 1-0S(1) line should be 0.70. Comparison with the observed line ratio gives the extinction at $\sim$ 2.2 μm. There are also several infrared lines due to [FeII] which are prominent in galaxy spectra. The lines at 1.257 and 1.644 μm also share the same upper level, and the extinction-free ratio of these is 1.358. This ratio provides a measure of the extinction at $\sim$ 1.5 μm.

One can also used standard recombination theory for atomic hydrogen. Then the ratio of infrared recombination lines like Brackett α to Brackett γ should be 2.83 with no extinction, and this measurement provides the extinction between 2 and 4 μm.

Finally one can use solid-state dust features in absorption, like the 9.7 μm "silicate" feature to estimate the extinction. This is a little dicey because attempts to calibrate the relation between visual extinction, A_v, and the optical depth in this feature vary by about a factor of two, but roughly, $A_v/\tau(9.7\ \mu m) \sim$ 10–25 (Telesco 1988).

Lester et al. (1990) used their infrared spectra of M82 to compare the visual extinction using these techniques. Both the [FeII] and the ratios of Brackett lines in the 1-2 μm region gave A_v 5.4–5.7, whereas the ratio of Brackett α (at 4 μm) to Brackett γ gave $A_v \sim$ 15. This latter measurement is consistent with the extinction inferred from the silicate absorption, which implies that $A_v >$ 15. Nagata et al. (1989) used the H_2 quadrupole lines to measure the extinction to a more powerful starburst galaxy, NGC 3690, and they found from the 2 μm measurements an inferred $A_v \sim$ 14.

The two points I wish to emphasize from these examples (in addition to the general point that one really can use infrared spectroscopy in this way) are that, the longer the wavelength, the more extinction is inferred, and that the corresponding visual extinctions are large. Even the near-infrared measurements at 1-2 μm may not be probing as deeply into the galactic nuclei as those at longer wavelength. Having indicated why I think that observations at shorter wavelengths may be of limited value, I should now like to outline some of the infrared diagnostics which may be used to investigate the stellar populations in galaxies. In Section 6 I will illustrate how one can apply these tools, using our observations of the luminous starburst galaxy NGC 3256. Since the light from a starburst is the integrated light from a composite stellar population up and down the main sequence, one must compare the measured diagnostic quantities outlined above with the predictions of a starburst model, in order to derive inferences from the measurements about the stellar population. However, there are several free parameters in a starburst model, and therefore one must use the information from several of these diagnostics to constrain the model. The classic study of this type is that of Rieke et al. (1980) for M82 and NGC 253. Rieke has updated this analysis in these proceedings, and concludes that the starburst in M82 is a) biased against stars $<$ 3 $M_\odot$, and b) biased against stars of mass > 30 $M_\odot$, compared to the local IMF.

In the following I shall take a similar approach, but I will use a simple analytical starburst model, rather than a large computer code. I will first illustrate how several different infrared diagnostics may be used to gain insight into the stellar populations in starburst galaxies. Finally, I will use data from our infrared observations of the southern merging galaxy NGC 3256, combining all the infrared constraints on the stellar population in this very powerful ($L_{IR} \sim 3 \times 10^{11}\ L_\odot$) infrared galaxy. It will be particularly interesting to compare the inferred properties of the stellar population in this galaxy with those Rieke et al. find (these proceedings) for the dwarf spiral M82.

3. STARBURST MODEL

Telesco and Gatley (1984) developed a very simple analytical model for a starburst which we have found to be very useful. It has the distinct advantage that the assumptions on which it is based are transparently clear, and it is easy to change any of the numerical values in the model and see how a particular result is affected. Another virtue is the fact that solutions are analytical, and so observational results can be analyzed in the context of this model with only a pocket calculator.

Telesco and Gatley assume that the initial mass function, the luminosity of a star on the main sequence, the lifetime of a star on the main sequence, and the number of ionizing photons/sec, are all power-law functions of the mass of the star in a starburst:

$$\Psi(m)dm = Km^\gamma dm$$
$$L(m) = Am^\alpha$$
$$\tau(m) = Bm^\beta$$
$$N_i(m) = Dm^\delta$$

In general the constants γ, A, α, B, β, etc. change from low to middle to high mass ranges, and typical values are given by Telesco and Gatley. The constant K sets the scale of the starburst, and is determined by the observational data.

One can then integrate analytically expressions for the total mass, luminosity, etc. of a starburst of a given age. For example, for a starburst of age T_o, where T_o is older than the main sequence lifetime of the most massive stars, the total mass of the starburst is given by:

$$M(T_o) = \int_{m_l}^{m_u} \Psi(m) m T_o dm + \int_{m_l}^{m_u} \Psi(m) m \tau(m) dm$$

and the luminosity of the starburst by:

$$L(T_o) = \int_{m_l}^{m_u} L(m)\Psi(m) T_o dm + \int_{m_l}^{m_u} L(m)\Psi(m)\tau(m) dm$$

Thus, M/L will be a function of the starburst age, T_o, m_l, and m_u.

4. THE LOWER MASS CUTOFF

Most of the luminosity of normal and starburst galaxies is in the infrared; integrating the infrared continuum energy distribution provides a good approximation to the bolometric luminosity. If one has a rotation curve, one can then obtain the dynamical mass in the central region where there is a putative starburst, and hence the mass/luminosity ratio. Since most of the starburst mass is in low mass stars, the M/L ratio is a sensitive function of m_l, the lower mass cutoff of the starburst initial mass function (IMF). Of course, one must assume a slope for the IMF, but in practice astrophysically interesting values of m_l are found for a range of plausible IMF slopes (cf. Rieke et al. 1980).

As mentioned earlier, the lower mass cutoff, m_l, is a sensitive function of the mass/luminosity ratio, since most of the mass of a starburst is in low mass stars. Wright et al. (1988) used this approach to investigate the lower mass cutoff in nine interacting and merging galaxies for which they had both small aperture ($\sim$ 5 arcsec) infrared photometry and rotation curves from which the mass in this same volume could be estimated. They assumed a starburst age of 10 million years, an upper mass cutoff of 60 $M_\odot$, and a Miller-Scalo slope to the IMF, and they required that the accumulated starburst mass to be no more than $\sim$10% of the total (dynamical) mass. From the mass/luminosity ratios of these nine galaxies they inferred that

$$m_l \sim 3 - 6 M_\odot.$$

If m_l were to be as low as 0.1 $M_\odot$, then the accumulated starburst mass would generally greater than the total dynamical mass.

Scalo (1989) has pursued this result further, adjusting the parameters to get a smaller m_l while staying within the total mass constraint. He finds that to make $m_l \sim 0.1$ $M_\odot$, requires that the slope of the IMF at the high-mass end be flatter than -1.5 and $m_u > 100$ $M_\odot$, and this still gives a starburst mass which is half the dynamical mass.

Prestwich (1989) has worked in the context of the starburst model outlined above, but used constraints derived from both the total infrared fluxes and from Brackett γ H recombination line spectroscopy. She defines a parameter, Q, which is the ratio of the star formation rate inferred from the infrared luminosity to the star formation rate inferred from the Lyman continuum flux. She goes on to show that Q is not sensitive to either m_l or m_u for a Zero Age Main Sequence starburst. However, for an evolved starburst, with the OBA star formation rate equal to the rate of evolution off the main sequence, this parameter is very sensitive to the lower mass cutoff. Comparing the values of Q she calculated for both ZAMS and evolved starbursts with different lower mass cutoffs with the data for several luminous infrared galaxies, Prestwich clearly shows that the starbursts in some of these galaxies are very young, whereas in others the starburst is probably evolved, and in this case the lower mass cutoff is ~ 5 $M_\odot$.

5. THE UPPER MASS CUTOFF

The Lyman continuum flux, inferred from H recombination lines in the Brackett series provides information on the population of massive stars in starbursts that is far more reliable, because it probes much more deeply than Hα measurements can.

Since the ionization potential of He is twice that of H, any He/H recombination line ratio should provide information on the shape of the ultraviolet continuum and therefore on the stellar population at the upper end of the main sequence. There is a strong HeI line at 2.058 μm, close to Brackett γ at 2.166 μm. The HeI/Brackett γ ratio is therefore quite insensitive to extinction, but very sensitive to effective temperature or spectral type. Using a starburst model, Doyon and Puxley (1990) have calculated the HeI/Brackett γ ratio for a starburst with a Miller-Scalo IMF slope. They show that the ratio is insensitive to starburst age, but it is a strong function of m_u, and is insensitive to m_l. This line ratio is thus a very powerful tool for exploring the possible existence of high mass stars in the dusty centers of galaxies which are unavailable to ultraviolet or optical spectroscopy.

Ho, Beck and Turner (1990) observed the Brackett α and Brackett γ lines in seven infrared-bright galaxies. From these measurements they derived the corresponding Lyman continuum luminosity and the total luminosity in OBA stars. They assume a Miller-Scalo IMF, and ask how old the starburst can be and have the observed infrared luminosity equal the luminosity in young stars. To obtain agreement they find that the upper mass cutoffs must be ~ 30–$60\ M_\odot$.

The analysis by Prestwich et al. (1989) referred to above also provides evidence for the existence of a population of massive stars in starbursts. Those galaxies which fit her ZAMS model must still have stars up to $60\ M_\odot$ present.

The other diagnostic for determining the upper mass cutoff, the HeI/Brackett γ ratio, has only been applied to one galaxy, NGC 3256. Discussion of this galaxy is deferred until Section 6 below.

6. POPULATION OF EVOLVED MASSIVE STARS

There is a very prominent CO absorption band at $\sim 2.3\ \mu$m in the spectra of late-type stars. The band is weak in dwarfs, stronger in giants, and strongest in supergiants. Thus, a deep CO feature indicates the presence of a significant population of red supergiants, and therefore it provides information on the massive stellar population from which these supergiants have recently evolved. However, it has generally been used only qualitatively in studies of starburst galaxies, e.g., a deep CO feature indicates the presence of population of supergiant stars.

In practice, the quantitative use of this feature for starburst galaxies has not yet provided any very convincing results, although it has been used very successfully in studies of stellar populations in early-type galaxies and the bulges of spirals (Frogel et al. 1978, Frogel 1985). Determining a reliable estimate of the continuum level, in order to get a good measurement of the equivalent width of the feature, is difficult, since the long wavelength end of the feature runs into the (ragged) edge of the atmospheric transmission window. Correcting the shape of the continuum for extinction and "warm" dust emission are therefore crucial, using JHKL colors for example, since the difference between the equivalent width of the feature for giants and super-giants is not large. Additional complications are that the feature is metallicity-dependent, and there is often strong emission due to the 1-0Q-branch of H_2 in the feature. The correct way to use this tool is to do stellar population synthesis, and to my knowledge this has not yet been attempted.

Doyon et al. (1989) investigated the CO absorption feature in a sample of five putative starburst galaxies. The spectra were mostly obtained in a 20 arcsec aperture. They attempted to correct the continuum for reddening using JHK colors, and then ratioed the interacting galaxy spectra by the spectra of a "normal" galaxy, NGC 4565. The result is that the CO depth is about the same for the starburst and normal galaxies. This is probably not surprising, since the starburst would be severely diluted in such a large aperture (corresponding to 10–14 kpc for the three most distant galaxies in the sample). The one spectrum which did show a residual CO absorption after ratioing with the normal galaxy spectrum was that for Arp220 taken in an 8 arcsec aperture. The qualitative result is that there does seem to be a large population of giant and

7. NGC 3256

I should like now to focus attention now on NGC 3256. This southern galaxy is the most nearby ($\sim$ 40 Mpc) example of a classic luminous merger. It has the two tidal tails indicative of a recent merger between two disc galaxies of about equal mass. A kinematic study of the ionized gas has revealed the presence of non-circular motions, probably resulting from it violent past history (Feast and Robertson 1978). It has infrared luminosity of $\sim 3 \times 10^{11} L_\odot$ (Sargent, Sanders and Phillips 1989). It is a powerful 10 μm source (Graham et al. 1984, 1987) and half of this emission is extended outside the central kiloparsec. We have concentrated considerable effort on this galaxy, and we have the best infrared spectroscopy on it of any galaxy we have observed.

Using this data and the infrared diagnostics discussed above, I should like to make as comprehensive statement as possible on the stellar population in this powerful starburst galaxy. The following analysis is discussed in detail in Rene Doyon's thesis (Doyon 1990).

Firstly, Doyon has determined the extinction over the central 3.5 arcsec ($\sim$ 650 pc). Using Brackett series line ratios, the ratios of the [FeII] lines, and the JHK colors, we get virtually the same result, $A_v = 6$.

Secondly, he has quantitatively examined the depth of the CO feature. The continuum is first corrected for warm dust emission, based on the K-L color, which suggests that about 5% of the 2 μm continuum is due to dust emission. The resulting CO equivalent width is then found to be almost 50% deeper than that for a KIII giant, and we infer that at least 60% of the 2 μm continuum emission is due to red supergiants. This then implies that the starburst age is about 12 million years.

Thirdly, Wright et al. (1988) have used the starburst model discussed above and the mass/luminosity ratio in the central kiloparsec to place constraints on the lower mass cutoff to the IMF. The $M/L_{IR} \sim 0.01$ in solar units. Using a Miller-Scalo slope for the IMF, and a requirement that no more than 10–20% of the dynamical mass be in the starburst requires that the lower mass cutoff be > 3–6 $M_\odot$. Even if the starburst mass were 50% of the dynamical mass, this would require a lower mass cutoff of about 1 $M_\odot$. Doyon (1990) has pursued this question in more detail, calculating a variety of models with various combinations of IMF slopes and starburst ages. The qualitative result does not change. Clearly, the starburst in NGC 3256 is biased against formation of low-mass stars.

Finally, Doyon (1990) has used the HeI 2.06 μm/Brackett γ line ratio to investigate the upper end of the main sequence in this galaxy. This line ratio, 0.35, implies an effective temperature of about 35,000 K, i.e., about equivalent to that of an O8 star. To find a constraint on the upper mass cutoff, Doyon has used a starburst model with a Miller-Scalo slope for the IMF. HeI/Brackett γ line ratios were calculated for different upper mass cutoffs for both an unevolved and an evolved starburst. The result is that the ratio is insensitive to age, and has a very strong dependence on m_u. For NGC 3256, the ratio is 0.35, and this implies that the upper mass cutoff to the IMF is $\sim$ 25–30 $M_\odot$.

In summary, then, using these infrared diagnostic tools to investigate the stellar population in the powerful starburst galaxy NGC 3256, we find:

$m_l \sim 3 - 6 M_\odot,$

$m_u \sim 25 - 30 M_\odot,$

the starburst age ~ 12 million years and,

red supergiants dominate the 2 μm continuum.

It is indeed striking that these results are almost identical to those presented by Rieke (elsewhere in this volume) for the starburst in M82. It is important to note that the diagnostic tools and methods of analysis Rieke used for M82 are not the same as those we have used for NGC 3256, although some of the physical ideas are the same. Moreover, M82 is a gas-rich dwarf, and it might very well have been argued that it is quite untypical of a starburst galaxy. The fact that these results are so similar, for two very different types of galaxies, suggests that these features may have more general relevance for describing the stellar populations in starburst galaxies.

8. CONCLUSIONS

i) Several independent lines of argument and studies of several different samples of spiral galaxies are all tending support the view that the stellar populations in starbursts are biased against formation of low mass stars. The IMF's in starburst galaxies are "top-heavy."

ii) There is some evidence that there is a small upper mass cutoff at $\sim 30\ M_\odot$. Although this result is relatively firm for a few galaxies, much more data and analysis is required before one can begin to think it might be a common feature of starbursts.

iii) The excellent agreement between the results presented for the starbursts in M82 and for NGC 3256 is quite remarkable. However, whether it is coincidental or indicative of more universal characteristics of starbursts requires further studies.

iv) There is a new generation of infrared spectrometers under development for various telescopes. These spectrometers are designed to take full advantage of the increased detections sensitivity, spectral resolution, and spatial coverage made possible by the advent of recent technological advances in infrared detector arrays. I hope I have shown convincingly the power of infrared diagnostics, and especially of infrared spectroscopy, for gaining insight into the stellar content of starburst galaxies.

REFERENCES

Doyon, R., 1990, Ph.D. Thesis, University of London.

Doyon, R., Joseph, R. D. and Wright, G. S., 1989, in *Infrared Spectroscopy in Astronomy*, 22nd ESLAB Symposium, ed. B. H. Kaldeich, European Space Agency, Noordwijk, The Netherlands, p. 477.

Feast, M. W. and Robertson, B. S. C., 1978, *M.N.R.A.S.*, **185**, 31.

Frogel, J. A., 1985, *Ap. J.*, **298**, 528.

Frogel, J. A., Persoon, S. E., Aaronson, M. and Matthews, K., 1978, *Ap. J.*, **220**, 75.

Graham, J. R. Wright, G. S., Meikle, W. P. S., Joseph, R. D. and Bode, M. F., 1984, *Nature*, , 310, 2313.

Graham, J. R., Wright, G. S., Joseph, R. D., Frogel, J. A., Phillips, M. M. and Meikle, W. P. S., 1987, in *Star Formation in Galaxies*, ed. C. J. Persson, US Government Printing Office, Washington, D. C., p. 517.

Joseph, R. D., 1989, in *Infrared Spectroscopy in Astronomy*, 22nd ESLAB Symposium, ed. B. H. Kaldeich, European Space Agency, Noordwijk, The Netherlands, p. 439.

Joseph, R. D. and Wright, G. S., 1985, *M.N.R.A.S.*, **214**, 87.

Lester, D. F., Carr, J. S., Joy, M. and Gaffney, N., 1990, *Ap. J.*, **352**, 544.

Moorwood, A. F. M., 1989, in *Infrared Spectroscopy in Astronomy*, 22nd ESLAB Symposium, ed. B. H. Kaldeich, European Space Agency, Noordwijk, The Netherlands, p. 431.

Prestwich, A. H., 1989, Ph.D. Thesis, University of London.

Rieke, G. H., Lebofsky, M. J., Thompson, R. I., Low, F. J. and Tokunaga, A. T., 1980, *Ap. J.*, **238**, 24.

Sanders, D. B., Soifer, B. R., Elias, J. H., Madore, B. F., Matthews, K., Neugebauer, G., and Scoville, N. Z., 1988, *Ap. J.*, **325**, 74.

Sargent, A. I., Sanders, D. B. and Phillips, T. G., 1989, *Ap. J.*, **346**, L9.

Scalo, J. M., 1989, in *Windows on Galaxies*, eds. A. Renzini, G. Gabbiano, and J. S. Gallagher, Kluwer, in press.

Soifer, B. T., Sanders, D. B., Neugebauer, G., Danielson, G. E., Lonsdale, C. J., Madore, B. F. and Persson, S. E., 1986, *Ap. J. (Letters)*, **303**, L41.

Telesco, C. M., 1988, *Ann. Rev. Astr. Ap.*, **26**, 343.

Telesco, C. M. and Gatley, I., 1984, *Ap. J.*, **284**, 557.

Wright, G. S., Joseph, R. D., Robertson, N. A., James, P. A. and Meikle, W. P. S., 1988, *M.N.R.A.S.*, **233**, 1.

DISCUSSION

Anon: Now let me just talk about something which I think it's important to distinguish when we talk about upper mass cutoffs between a present day one, which is what you're talking about here, and the initial one, which is what George Rieke talked about yesterday. In fact there're two methods to get the two different ones and I want to say that if you're looking at the initial upper limit, what George used yesterday was the oxygen yield constraint and, as was mentioned by several people that really has got to be completely redone with the new yield models. But that wasn't what you were using?

Joseph: No.

Anon: You're talking about the present day upper limit and in fact, I think one is on pretty safe ground there when you're talking about hydrogen and helium recombination lines which is also what you were using but if you're using oxygen lines, I just want to remind everyone about this. The new models of Kudritzki may have a really significant effect on that. Nobody knows yet because they're just come out and nobody's really done the analysis.

Notice, there's a little bit of an inconsistency with what you ended up with because you had some O stars on the one hand and on the other hand you have an age of a little over 10^7 years. And O stars don't last that long. So they must be still being made now. So it might be that if there were more massive stars there, that have come and gone, when this burst is no longer making it ...

Joseph: I don't think that's an inconsistency. The starburst model is assuming an evolved population; it's a starburst that's gone on long enough that the earliest stars have evolved off the main sequence but one has new stars.

R. Kudritzki: My crucial constraint comes from your observation of say the helium to hydrogen ratio. But I'm wondering in a situation of strong extinction, how do we treat the extinction in the far ultraviolet? You have an enormous amount of extinction, so I'm wondering how reliable the real number is that you can get out of this ratio.

Joseph: Well it's going to be the differential. That's a very hard one, isn't it, because one is talking about the whole Lyman continuum compared to the UV at 584 which is what pumps that 2 singlet P transition which is the top level from which the helium line goes. There may be a real question about absolute extinction, but it's the relative extinction between some effective temperature for somewhere in the Lyman continuum compared to 584 and what Renee Duron has done, I'm sure, is just assume the same extinction.

F. Bruhweiler: It's unfortunate you've had such bad luck using IUE because if you treat IUE data as an image rather than a spectrum you can get an immensely improved data quality. I like to comment on the extinction NGC 1068. From what I can tell, if you look at the visual extinction $E(B-V)$, you get something out of the order .4 to .6 on the basis of the Balmer decrement. If you look in the ultraviolet you get something like .1. This is what you have using good UV data. You look at the UV flux distribution you get something like B0 supergiants. In order to reproduce that extinction ... equivalent to that is the LMC or SMC, you get a spectral slope that is much too high and the only fit that you can do is something that's much lower.

N. Devereux: I think the CO absorption feature may run into some complications because Judy Young and I have been trying to use it to determine the stellar population in starburst galaxies. The CO equivalent widths for two of the objects are the lowest equivalent widths ever observed in galaxies and we think that this is because there's some emission. It's traditionally assumed that the emission is hot dust. And you correct the CO index by assuming the shape of the energy distribution, but actually it looks more like its line emission from CO and until we firmly establish this, I think that the interpretation of the CO index is going to be very, very tricky.

N. Scoville: So that in your plot of the CO index versus time you chose the rising part and I was wondering why you don't equally well choose the falling part.

Joseph: Well it didn't fall on this graph but it will fall obviously at some point. That's a good point, Nick. I'm not sure that I could get consistency with the accumulated masses and so on if I did that, that in the starburst model that's, roughly 10 million years is roughly the age that I've used. If I actually went to more like a hundred million years I don't think I could get consistency with the starburst model in terms of mass to light ratios over that, another factor of ten in time, but that's an interesting point to pursue.

K. Garmany: Let me ask you a naive question. Yesterday we heard many references to the standard IMF. You referred to the Scalo 1986 IMF. In my talk I used the Scalo's quote saying we know essentially nothing about the IMF, so I would hope that if you are using the number here that there's at least an error bar on it. What I'm curious about is here you've done this using say a slope of -1.5 and -2 and fold that in and see how that affects your upper mass cutoff and all these other factors.

Joseph: Yes, if you make a flatter slope, in fact, maybe I have that on a viewgraph and didn't show it. Yes, I did have it, but I think I didn't say it. If you take the starburst models using the mass to light ratio and go to a slope of -1.5, you also need to take the upper mass cutoff to something like more than a 100 $M_\odot$s. But if you do those two things, you can get rough consistency in the dynamical mass and the accumulated starburst mass with a lower mass cutoff of somewhere between 1 and .1. You have to push it very hard to try to get the lower mass cutoff below one solar mass but in principle, you can do that. That's right. But as far as I understood from your talk as well as some others, well one doesn't know these things; an IMF of a slope of -2 looked pretty good.

K. Garmany: It looked pretty good but I don't think it's a standard model.

C. Leitherer: I'd like to follow up on the question of the upper mass cutoff. Now we've heard from several talks that obviously the preferred value for the upper cutoff mass is something like 30 solar masses or so, which translate into a main sequence spectral type of maybe O6, O7 or so. So what you're saying is that during these bursts of star formation, stars of spectral types O3, O4, O5, O6, possibly O7 are not observed. Now my question is: How well constrained is this prediction of your model because if we look at the best observed starburst galaxy, which is the LMC, in particular, 30 Doradus, we see there are lots of high mass stars, O3, O4, O5, O6 stars and to use Nolan's words, that's a Rosetta Stone of starburst galaxies. Now are you saying that starburst galaxies talk a language which is not on this Rosetta Stone.

Joseph: Well, I think that's a good question. What, I'm not really saying there are no O3's, what I'm saying is if I take an IMF, then, it's this power law, I'm continuing to extend to 60 solar masses or a 100 solar masses, the relative proportion of O3's to O8's or O7's is very small compared to that IMF. That's really all I can say. It's not that there are zero of them, I guess the relative number is less.

C. Leitherer: It's different from 30 Doradus. That's what you are saying.

Joseph: Yes, that's true.

T. Heckman: I think this is a really an important point and I wanted to follow up something that Peter Conti said, because both George Rieke's argument and your argument for upper mass cutoff really hinges on the ionization state of the gas. And so what I want to ask is the people who make model atmospheres or massive stars particularly if you allow for the possibility that the metal abundances of these stars could be several times solar, how convincing you find the evidence for an upper mass cutoff based on ionization state of the IMS gases?

R. Kudritzki: From the calculations I have shown on Tuesday, there is not the big influence of metallicity on the ratio of helium to hydrogen ionizing photons, neutral helium ionizing photons, etc. I think that the other points that I have just raised the point of the extinction role in the extreme ultraviolet is a very important thing, because it is obvious that due to dust part of these photons are absorbed, and this has to enter into this relative numbers of photons. You have to adopt something for the absorption in the ultraviolet and I don't know what your observations tell you about this kind of absorption. So there's another possibility that could cause problems.

Joseph: Can I say something. From an observational point of view, I have been very uneasy about this ratio; it was actually Phil Puxley and Renee Duron who came up with the idea, and what we started doing was testing whether that calculation applies to H II regions. So we've looked at this; ratio is for a variety of H II regions both huge H II regions like W51 in the galaxy for which there's data in the literature and we just had an observing run at UKIRT about 2 weeks ago looking at the variety of H II regions that are small enough that we have the whole H II region in the beam. And in the cases where you know what the effective temperature of the star is and it's more than 38,000 degrees, we're getting virtually identical results within the errors, to this ratio of about 1.1, so whatever the theoretical arguments and the concerns about extinction, if you look at an H II region in the galaxy and we've now done more than 10, I think, about 3 or 4 more from the literature, we get a Helium to Brackett γ ratio of about 1.1. So I think there's some observational support for this kind of ratio.

R. Kennicutt: Obviously I don't work on stellar atmospheres but I model nebulae a lot. And there is a very serious problem with these forbidden line indicators. The ionization of the nebula is extremely sensitive to something called the ionization parameter which is basically related to the density, the ionizing flux and clumpiness of the gas. You can put an O5 star in a bit of ionized gas, and by varying the density by maybe a factor of 100 I can produce any value of [O III] to $H\beta$ or fine structure O III ratios. It's a very difficult task, in fact, the problem is exacerbated for the fine structure lines because the densities in these high density environments, the densities are comparable to the critical densities for collisional de-excitation and it's just tricky business. I wanted to ask another very general question: As much as I admire George Rieke's evaluation of the parameters, when one starts talking about these lower mass cutoffs of 5, 6, or 7 solar masses in all of the starbursts, I'm going to get very worried. I'm going to give two very simple qualitative reasons. One is, 40 years ago Morgan developed the spectral classification of galaxies and he classified these galaxies as A Type. They're dominated in the optical by main sequence of A star absorption lines which I guess are a few solar masses and so somewhere those have to be made. More fundamentally, Colin has termed these the paradigm for galaxy formation and certainly if these results are right, his statement can't be correct because in the bulge of our galaxy, the most high density environments on our own galaxy and of the bulge, moreover, globular clusters were dominated by low mass stars. In fact, both Ken Freeman and Mario Mateo found that the globular clusters of Magellanic Clouds, in fact, have a beautifully smooth scale IMF of 15 solar masses down to 1. The question I'd like to turn around is where are the starbursts that formed the Milky Way and the globular clusters. Are you not seeing them, are they hiding somewhere in the present day universe, because they've got to be taking place today.

Joseph: Sure, I entirely agree Rob, I just don't know what to do about it. That's the data, but this is an extreme kind of star formation event and I think an idea of bimodal star formation with different physical environments, different parameters, it's quite reasonable that one could get in the early universe different properties for star formation activity. But I'm not an expert on that. I have no idea what the answer is.

N. Walborn: Two compact questions. It's not clear to me if the starburst is twelve million years old. How is it that you rule out a large number of 100 solar mass stars twelve million years ago which disappeared nine million years ago.

Joseph: Because my model is assuming an equilibrium between star birth and star death. And so it's the second integral in those integrals I wrote up. So you have stars that were formed in the starburst and are still sitting on the main sequence and an assumption that there's a uniform star formation rate from zero to twelve million years.

N. Walborn: Is there an observational discriminant between that assumption and the other possibility.

Joseph: No, that's simple analytically to do.

N. Walborn: Well, that could be the explanation right there. If you have a twelve million year old starburst. And maybe your assumptions are wrong.

Joseph: No, my constraints are based on a uniform rate of star formation. I'm creating new stars right up the IMF to the upper mass cutoff and they're in equilibrium between birth and death. That is the physical assumption that went into the constraint that gave the lower mass cutoff. The observational constraint on the upper mass cutoff was this Helium to Brackett γ ratio.

N. Walborn: That's presently observed, not 12 million years ago. But let me ask about the present situation. The same question I asked Dr. Rieke privately after his talk yesterday. In this galaxy, which I am not familiar with, do you observe giant H II regions or not, and if you do, there are almost certainly 50 to 100 solar mass stars there, and if you don't there almost certainly not.

Joseph: About the first comment I didn't really understand. One has an observational constraint on the present day content of upper mass, of high mass stars from the helium to Brackett line. I mean there may be a problem with the physical assumptions that went into that. But the ratio of the helium line to the Brackett γ line is .35. And if you take reasonable assumptions about the stellar atmospheres and calculate what you expect, that gives an O8 or 35,000 K is a temperature.

MODELS OF STARBURST GALAXIES

Colin A. Norman
Department of Physics and Astronomy,
Johns Hopkins University
and
Space Telescope Science Institute
3700 San Martin Drive
Baltimore, MD 21218
USA

Abstract. Dynamical studies are discussed showing how gas moves to the center of a merging and interacting galaxy on timescales on the order of the dynamical timescale. The effects of the massive stars formed in a starburst on the interstellar medium of such a galaxy are outlined including the formation of superbubbles, chimneys, and galactic winds. The problem of understanding initial mass functions that are skewed towards massive stars is presented and various possible theoretical solutions that have been proposed are reviewed. Current ideas on what makes the bursts and why they are important for galaxy formation and evolution is sketched.

1. INTRODUCTION

The phenomenon of starburst galaxies was brought sharply into focus by the IRAS results on luminous infrared galaxies. It soon became apparent that the most luminous galaxies all seemed to be associated with violent interactions and mergers, and that even at lower luminosities the incidence of such interactions was high (Joseph Chapter 15, Scoville Chapter 14). The inferred star formation rates were so extreme that standard estimates of the consumption time for the gas were very short $\sim 10^7 yr$. This estimated lifetime is too short to account for such large numbers of starbursts that are observed (Rieke Chapter 12, Weedman Chapter 18). The solution that is generally accepted is to assume that only massive stars are formed in these systems. For galaxies with normal mass functions the luminosity is given by the upper part of the IMF and the mass all resides at the lower mass end. For the only massive stars hypothesis both the luminosity and the mass are at the upper mass end and the estimated lifetimes are at least an order of magnitude longer. One should carefully note that there is no good determination of the IMF for starburst galaxies directly inferred from observations and until such time as the IMF is sufficiently well established, this massive star hypothesis should be taken to be just that—a good working hypothesis (Gatley Chapter 7, Conti

Chapter 2, Leitherer Chapter 1, Walborn Chapter 8, Bernlöhr 1990).

Most of the luminous infrared galaxies are optically undistinguished even though they are inferred to have star formation rates that are close to those of a galaxy forming itself into stars on a galactic dynamical timescale. These prodigious star formation rates make the starburst galaxies we observe into what is sometimes referred to as a good laboratory for studies of low redshift galaxy formation. In this sense we can observe relatively locally the feedback processes from massive star formation that may have a significant influence in determining the overall structure and characteristics of galaxies in the general dissipative collapse (Rees and Ostriker 1977, Silk 1977, Ikeuchi and Norman 1990, Norman and Ikeuchi 1990).

With the powerful new millimeter arrays, the CO line and other molecular lines such as CS can be detected and mapped in starburst galaxies. Assuming the standard CO/H_2 ratio large amounts of molecular gas $\sim 10^8 - 10^{10} M_\odot$ are found in the very central regions ($\leq 1 kpc$). It is a most interesting dynamical question as to how the gas got there. This is more or less equivalent to putting the entire interstellar component of two merging gas rich galaxies into the center of the merger remnant in a dynamical time or so. Models of how this might actually happen are discussed in the next section. Generally speaking these are associated with large scale dynamical asymmetries such as bar instabilities.

The interstellar medium in these galaxies has an energy balance dominated by the energy input from many spatially and temporally contiguous supernovae. Even in normal disk galaxies the manifestation of the collective energy input from Type II supernovae is to create superbubbles, chimneys, and winds (c.f. Norman and Ikeuchi 1989). These extreme starbursting environments, vast chimneys and superwinds are formed in the disk and emanate from the central regions of the disk along the polar axis. The energy and momentum input into the circumgalactic medium in these winds is significant (Heckman et al. 1990). The photoionization balance in these central regions with the very high internal UV flux from the massive star population can result in a fraction, of order unity, of the gas mass being in the form of photodissociated molecular gas as inferred from observations of the [CII] line at 157μ. The [CII] is produced in photodissociation regions (PDRs) at the edges of molecular clouds that are subject to an intense UV radiation field. FIR observations with the Fabry Perot spectrometer on board the KAO indicate that this is indeed the case (Stacey et al. 1990).

The formation of stars in such extreme environments as are found in central starbursting regions is quite plausibly different from those in normal Galactic star forming regions. The massive star hypothesis indicates that these extreme conditions of density, temperature, pressure and collective mechanical energy input from supernovae all lead to the formation of high mass stars and inhibit the formation of low mass stars. There is no accepted theoretical basis for this statement at present since the physics of both high and low mass star formation for single stars and binaries is the subject of intensive current research. It is interesting to note that in similar high pressure environments found in cooling flows the star formation is often claimed to be restricted to the very low mass end since the Jeans mass is inversely proportional to the square root of the ambient pressure.

There have been a number of interesting attempts to see if the burst phenomenon is a generic property of the structure of the interstellar medium of disk galaxies (Habe et al. 1984, Scalo and Struck-Marcell 1984, Vasquez and Scalo 1989, Li and Ikeuchi 1989). Time delays in the flow of mass, energy, magnetic field etc. can change the ISM system from a stable steady state to an unstable one by changing the feedback loop into a positive feedback system rather than a negative feedback one. In a number of cases the equations describing the system are similar to those of competing populations that are known to be unstable. Other ideas for bursts that have their foundation in observations are that collisions between clouds are seen to trigger massive star formation in our own Galaxy (Scoville et al. 1986) and other galaxies such as M51 (Rand and Tilanus 1990). This effect is hinted at in the HI observations of blue compact dwarf galaxies where there is a hint of a correlation between the blue knots and the boundaries of colliding HI clouds (Brinks 1990). Clearly in merging and interacting systems the cloud-cloud collision rate can be induced to rise dramatically.

Starburst galaxies are important as yet another paradigm in nature for the concept that evolution by a set of catastrophic events can dominate the slow but steady mode of evolution. Galaxy formation may not just be an initial burst followed by a slow steady evolution but may be rather more like a series of small bursts each one responsible for a significant step towards the formation of the galaxy with its various components. The detailed understanding of these small but significant steps may be most useful in the development of this paradigm for galaxy formation.

This paper is organized as follows. In the next section we will review dynamical studies of starburst galaxies and then move on in section 3 to considerations of the ISM under these extreme conditions. The next two sections (4 and 5) will address the formation of stars under these extreme conditions and the various models proposed for bursts. Section 6 will examine the post starburst phase and 7 will discuss briefly the question of face-on starbursts. Finally (8) the connection between models for galaxy formation at low redshift and models at high redshift will be described.

2. DYNAMICAL STUDIES

The general dynamical ingredients to set up the physics of initiating and fuelling a nuclear starburst are a merger or interaction event or a strong dynamical asymmetry such as a central and possibly transient bar (Byrd et al. 1986, Lin et al. 1988, Shlosman et al. 1989, Devereux 1987, Ishizuki et al. 1990, Telesco et al. 1990, Noguchi 1990, but see however Pompea and Rieke 1990). The ISM in starbursts is obviously very gas rich and therefore realistic models of the dynamics of starbursts must include the gas component and in particular the molecular phase. This has been achieved by Hernquist and Barnes in their excellent simulations of merging gas rich disks to form a residual elliptical system. Hernquist (1989) showed how a companion falling into a cold gas-rich disk galaxy can trigger global dynamical instabilities in the gas that can drive rapidly a large fraction of the gas into the central regions of the disk galaxy. The angular momentum loss from the gas to the background stars is strong continuous and monotonic. The tidal torques formed by the asymmetric infall of the companion are initially the most important component of the angular momentum loss. After the gas has become sufficiently concentrated to be self-gravitating in massive blobs, the angular momentum loss can become driven by the dynamical friction against the background halo of stars.

Following the pioneering work of Negroponte and White (1983), Hernquist and Barnes (Barnes 1990) have used an SPH (single-particle-hydrodynamics) code to simulate major mergers between equal mass gas rich galaxies (c.f. Olson and Kwan 1990a, b). Of great interest is the effect in the central regions of the merging galaxies which exhibit strong shocks and strong bars, which initially fuel gas into the centers of both the galaxies. The final nucleus is formed by an eventual merger of the cores. Very significant fractions of the entire gas mass end up in the central region. This result is extremely interesting in the context of the starburst problem since the models seem consistent with both the large observed central gas masses and the rapid process of gas accumulation necessary to account for the observed frequency of starbursts.

A qualitative way to understand the likely dynamical evolution of starbursts has been indicated earlier (Norman 1987, 1988a,b). There are essentially three phases:

I. *Normal Starburst Phase.* Interaction and merging generates bars and waves that act to give angular momentum transfer outward and consequently mass transfer to the center. The central surface density rises and the star formation rate increases. The timescale is $10^8 - 10^9 yr$.

II. *Very Luminous Starburst Phase.* The gas mass accumulates sufficiently in the center so that the gas itself develops appreciable self gravity and then subsequently becomes bar unstable. The angular momentum loss is consequently very rapid.

III. *Ultra Luminous Starburst Phase.* The gas is now so centrally concentrated that it becomes completely self gravitating. A massive central star cluster and central object such as a black hole may result.

Depending on the strength of the interaction and the gas masses involved the starburst systems may find themselves only able to achieve stage I in the case of weak bursts with low gas masses or to complete phase III in the case of strong bursts with high gas masses.

In phase I where waves can be induced in the disk by interactions with companions or with merging systems the general form for the timescale on which gas will flow inwards toward the central regions is given by

$$\tau_{in} \sim \frac{1}{2m^2\gamma} \left(\frac{\Omega}{\Omega - \Omega_p}\right) \left(\frac{E_{total}}{E_{wave}}\right) \quad (1)$$

where $\Omega(r)$ is the angular velocity of the stellar and gaseous disk, m is the number of arms of the induced wave pattern, Ω_p is the pattern velocity of the induced wave pattern, E_{wave}/E_{total} is the ratio of wave energy to total energy, and γ parameterizes the dissipation process due to cloud-cloud collisions or shocks. Thus the angular momentum losses due to this process dominate the purely viscous ones if $E_{wave}/E_{total} \geq (1/\sqrt{3})(h/mr)$ where h and r are the scale height and radius of the disk. A crude but useful formula is

$$\tau_{in}(10 \text{ kpc}) \sim 10^8 \left(\frac{E_{total}}{E_{wave}}\right) \text{ yr} \quad (2)$$

so that for normal disks $\tau_{inflow} \sim 10^{10} yr$. However for systems with large distortions $\sim 30\%$ due to merging and interaction timescales as short as $\sim 3 \times 10^8 yr$ can be obtained. It is important to note that (Lubow 1987) even for a 10–15% gas to star ratio by mass all the response is in the gas not the stars.

When phase II is reached we are at the stage where the gas mass interior to radius r is a significant fraction of the total interior mass. This leads us to the almost inevitable conclusion that a bar instability will form in the gas. In the absence of strong star a robust criterion to assess this effect is the Ostriker-Peebles condition $T/W \sim 0.14$ (Ostriker and Peebles 1973, Efstathiou et al. 1983, Sellwood 1989, White 1988) where T and W are the rotational and potential energy respectively. It follows that the observed central gas masses are inferred to be bar unstable if $M_{gas}/M_{dynamical} \geq 10\text{--}30\%$. When the bar does form in the gas the time scale for further contraction is short (Sellwood 1980, Weinberg and Tremaine 1983)

$$\tau_{dec} \sim 1 - 10 \left(\frac{M_{halo}}{M_{bar}}\right) t_{dyn} \sim 3 \times 10^6 - 10^7 \text{ yr} \qquad (3)$$

After a time delay of order $\sim 10^7$ yr the gas will then have become concentrated by a factor of 3–10. Then a significant fraction of the central gas mass will become totally self gravitating. This can lead to the formation of a central massive black hole and starburst and should be associated with the phase III- the ultra-luminous phase. We have analyzed the fate of a massive central star cluster of $\sim 10^8 - 10^9 M_\odot$ within a galactocentric radius of 10 pc (Scoville and Norman 1987, Norman and Scoville 1987). We have assumed efficient star formation and analyzed the evolution of the star cluster for a range of initial mass functions and upper and lower mass cut-offs. Specific calculations involve the production of red giants, the mass loss from the stars, the fate of the lost mass most of which goes to feed a central black hole and the role of stellar collisions. Interesting effects result from the influence of the strong central continuum radiation source on the envelopes of the red giants giving a strong externally induced mass loss. The ionization structure, density and size for these stellar envelopes is consistent with those for the broad line clouds. At later times the covering factor of the red giants is of order 10%. A black hole mass of $10^8 - 10^9 M_\odot$ has now grown and the orbital velocity of the clouds is $\sim 5,000$ km s^{-1}. Much early work is related to this (Zel'dovich and Novikov 1964, Spitzer 1971, Begelman and Rees 1978, Bailey 1980, Shull 1983, Mathews 1986) but here at last we have good initial conditions for the active galactic nuclei and black hole scenarios. In this version dense central star clusters should be associated with AGNs and in the immediately post active phase dense stellar cores with A-type spectra should be observable.

A complementary approach to the Barnes and Hernquist studies has been followed by Pfenniger and Norman (1990) who incorporated the essential ingredients of a massive central starburst namely a barred system, a central mass concentration, and dissipative molecular clouds orbiting in the combined potential of the bar and central mass concentration. The orbits of the clouds were modelled as test particles with a drag component that effectively incorporates the dissipative effects of shocks on gas streamlines or of dissipative collisions. Some very interesting effects were found. In barred systems with sufficiently concentrated central masses, strong radial and vertical resonances develop. Any dissipative particle close to a radial resonance will lose rapidly its radial action, and consequently its angular momentum. It is clear from the results of this work that rapid angular momentum loss occurs in jumps as the dissipative clouds move through the resonance regions. The net effect of this is to very significantly reduce the timescale on which gas clouds can flow into the center.

An additional effect that is clear in the results of many test particle runs is that the vertical resonance structure allows significant vertical heating from the disk to the bulge or halo. Stars inside the vertical resonance region enter the chaotic regime where all the integrals of motion, except the Jacobi integral, vanish. The shape of the Jacobi integral inside a rotating bar bounds the motion of stars within concentric, nearly spheroidal shells. In a most interesting simulation of a rotating bar with a live halo Raha et al. (1991) have found a fire-hose instability that produces a strong vertical heating. The central density is also raised. This instability has the advantage that it is non-resonant. Bulge formation and disk thickening may be significantly affected by this process. In fact, barred Sc galaxies with central mass concentrations should be forming bulges at present.

This process of inflow is clearly self enhancing. The inflowing gas increases the central mass concentration which in turn increases the resonances which has the effect of enhancing the rate of angular momentum loss thus feeding gas into the nucleus at an increasing rate. The limit of this process is discussed by Hasan and Norman (1990) who showed that the fundamental family of orbits that support the bar will be dissolved if the central mass concentration grows to between 10–20% of the total dynamical mass.

A crude but self-consistent approach exhibiting this limiting process has been calculated by Friedli and Pfenniger (1990) who ran an N-body calculation where each of the test particles had some dissipative drag so that an initially barred system slowly grew a central mass concentration which then dissolved the bar and turned the system into a hot triaxial bulge or lens.

The fundamental obstacle to further progress in modelling these important secular processes is that the self-consistent N-body codes do not have sufficient orbital integration accuracy to treat reliably these longer term evolutionary processes. A technical breakthrough is needed here.

3. INTERSTELLAR MEDIUM

The starburst galaxies have a most interesting interstellar medium. For the powerful systems the medium most resembles that of Galactic star forming regions (Stacey et al. 1990) where much of the gas is photodissociated molecular gas in the strong radiation field from the massive stars making up the starburst. These systems are at the extreme end of the chimney model of the interstellar medium as the massive central starburst seems to act coherently to blow a huge chimney or superwind out of the central regions (Chevalier and Clegg 1985, Tomisaka and Ikeuchi 1988, Heckman et al. 1990, Waller 1990). The inferred mass outflow rates are 1–100 $M_\odot$ yr^{-1} at speeds of 100s km s^{-1}. In milder, but closer, versions of this process chimney-like structures emanating from systems such as NGC 3079 are clearly visible in narrow band filter images originating from both the central regions and the disk. These structures can be modelled in the context of the chimney models where a relatively normal disk galaxy is put into the starburst mode by increasing the star formation rate in the disk (Norman and Ikeuchi 1989). With sufficient energy release the entire ISM could be driven out in a galactic wind but this is not observed in general and the energy release is more via localized chimneys. An examination of figures 3 and 4 in Norman and Ikeuchi can show the likely states of the halo of a starburst (confined halo, chimney or wind) as a function of two global parameters the energy input and the associated mass input.

Apart from exhibiting these extreme aspects of chimneys and outflows the issues in the study of the interstellar medium of starburst galaxies are concerned with complex questions such as:

(1) The basic multiphase structure of such a dense region with a total mass of order $10^{10} M_\odot$ and a mean density of order 100 cm^{-3}. The nature of molecular clouds needs to be studied in such a medium where the intra-cloud medium may be mainly in the form of PDRs with [CII] as the dominant coolant (Stacey et al. 1990).

(2) The evolution of supernovae remnants needs to be modelled accurately in such an environment. It may just be possible to observe them in the K-band (van Buren and Norman 1989).

(3) The general photoionization structures associated with the massive OB associations need to be modelled to make predictions of the various line ratios that can be observed above the disk (c.f. Norman 1990, Norman and Panagia 1991, Carlstrom and Kronberg 1991).

(4) It would be most useful to have an accurate map of the temperature distribution of the clouds as well as some idea of the cloud size distribution (Eckart et al. 1990).

(5) Diagnostics are needed that can constrain the fascinating pumping mechanisms for the OH and H_2O megamasers and the origin of the powerful H_2 line at 2.2 μm (Draine and Woods 1990)

(6) Quantitative estimates of the cloud-cloud collision rates would be useful for the analysis of the models where cloud collisions lead to massive star formation (Scoville et al. 1986).

4. STAR FORMATION

A detailed knowledge of star formation is probably necessary to really understand starburst galaxies. For a recent review of IMF physics see Larson (1990). The current theories divide into models of low mass star formation and high mass star formation. The low mass star formation mode is thought to be due to a slow accretion onto a central core (Shu et al. 1988) where the magnetic field is removed from the gas by ambipolar diffusion. A rough estimate for the timescale for ambipolar diffusion is $\tau_d \sim 10^{14} x$ yr where x is the ionization in the cloud. The star formation process for these low mass stars can be self-regulated by processes due to other low mass stars such as winds (Norman and Silk 1980) or the ionization due to the X-ray emission from the low mass T-Tauri component (Norman and Silk 1983). The high mass stars through their ionization effect could self-regulate the ambipolar diffusion process in the clouds including the formation of the low mass stars (McKee 1989). This latter point may be important for starburst galaxies in that a burst of high mass stars may be able to completely inhibit the slow ambipolar diffusion driven process of low mass star formation and therefore give a plausible basis for a mass function that is truncated at the low mass end. Both Wolfire et al. (1990) and Stacey et al. (1990) have shown that the central region of M82 is very similar to the massive star forming regions in our own Galaxy.

The high mass star formation mode seems to be triggered in associations correlated with spiral arms in disk galaxies and is generally correlated with the giant molecular cloud complexes. The high mass star formation mode is thought to be associated with the magnetic Jeans mass. For a given flux Φ the magnetic Jeans mass is given by

$$M_{J,mag} \sim \left(\frac{\Phi}{G}\right)^{1/2} \qquad (4)$$

which in a given pressure environment in rough equipartition can be written as

$$M_{J,mag} \sim \frac{v_A^4}{p^{1/2}} \sim \frac{\sigma_t^4}{p^{1/2}} \sim \frac{T^2}{p^{1/2}} \qquad (5)$$

where p is the pressure, v_A is the Alfven velocity, σ_t is the turbulent velocity dispersion, and T is the cloud temperature. High masses can occur if the temperature is large and this can occur in a self consistent way in regions of high stellar luminosity due to already existing massive stars.

If the central turbulent velocity dispersion is large as is the case in a medium stirred up by the energy and momentum injection from massive stars the magnetized Jeans mass can also be very large. Such large turbulent velocities are observed in starburst regions and also in galactic regions of massive star formation.

Therefore it seems possible that starbursts might operate in a way that suppresses the low mass star formation mode. In addition, the presence of existing massive stars and their associated winds and supernovae may act to keep the star formation at the high mass end.

5. MAKING THE BURST

A number of general concepts and models have been set down to understand how the burst phenomenon can be understood (Scalo and Struck-Marcell 1983, Vasquez and Scalo 1990, and Larson 1987). The general framework is as follows:

I. There is a density increase due, for example, to accretion, collapse, compression, or cooling. Star formation ensues.

II. There is an increase in the internal energy of the gas due to the energy and momentum input from the star formation process.

III. Clouds are then shredded and the star formation drops.

IV. The gas reservoir builds up.

V. A trigger occurs in a situation where the built up preexisting clouds are now well above the Jeans mass applied by the trigger due to a density compression, collapse, etc. and the cycle continues.

Very interesting work on multiphase models of the ISM and their fundamental burst properties has been done by Habe et al. (1984), Li and Ikeuchi (1990), and Hensler and Burkert (1990).

Without going into details the general property invoked is that there is a mass exchange between the phases such as between the cold, hot and warm phases. There should also be a property such as the circulation time between the disk and the halo. This introduces into the system the possibility of time delays in a feedback loop and can lead to a burst.

6. THE POST-STARBURST PHASE

The conventional starburst lifetime is of order $\sim 10^8$ yr for the powerful starburst galaxies. These lifetime estimates are based on mass exhaustion times that are calculated assuming truncated IMFs etc but, as yet, are not backed up by direct measurements using supernovae rates, Wolf-Rayet stars (Conti 1990), counting of Brackett line photons etc. Let us make the reasonable assumption that the starburst to post-starburst phase must satisfy a number continuity equation. Let us further characterize the post starburst phase by that for which it will still exhibit characteristic A-type spectra. The characteristic lifetime for this phase is given by $\tau_A \sim 5 \times 10^8$ yr.

The continuity equation can be written

$$\frac{N_{post}}{N_{burst}} \sim \frac{\tau_{post}}{\tau_{burst}} \sim 5 \qquad (6)$$

so that $N_{burst}/N_{post} \sim 5$. Account must be made of fading of the burst from $10^8 - 5 \times 10^8 yr$ and conventional synthesis models for single bursts give about 1.5 magnitudes (a factor of four; Charlot and Bruzual 1991) for this effect in the B-band. Therefore there should be approximately two times as many post starbursts as starbursts. This large a number of post-starburst galaxies does not seem to be observed (Zepf 1991) and this potentially poses the question of significant revisions to the starburst models (*e.g.*, Thronson *et al.* 1990) including possibly their age estimates (could they be longer lived?), and the massive star content (perhaps there are no stars of sufficiently low mass for there to be an A-star component?). Note that for a continuous burst lasting $\sim 10^8$ yr the fading from 10^8–5×10^8 yr is as much as 2.5^m (S. Charlot, private communication). This factor of 10 reduction in flux gives a value of $N_{post}/N_{burst} \sim 1/2$ and thus may account for some of the dearth of post starburst systems. The essential difference between the single burst and the continuous burst models is that in a continuous burst the peak of the luminosity occurs at $\sim 10^8$ yr due to the accumulation of massive stars whereas in a single burst such stars are absent at $\sim 10^8$ yr.

There is a second interesting issue associated with the gas content of post starburst galaxies. It is unlikely that star formation is 100% efficient and therefore the amount of gas left behind after the burst should be similar to the amount that went into stars and that was observed in the bursting phase. This gas could have been moved out of the galaxy in outflows Heckman *et al.* (1990) have shown that starburst galaxies have such energetic outflows of gas. This ejection of the gas which is left over from the central burst of star formation is necessary if mergers of gas-rich spirals are to make gas-poor elliptical galaxies. The outflowing material observed by Heckman *et al.* is moving with velocities of 200–300 km/s, which means that in several $\times 10^8$ yrs it only moves ~ 50 kpc. Therefore, in post-starburst galaxies this gas should be observable, either in soft x-rays and optical line emission if it is still cooling, or in HI and molecular gas if the cooling time is short. This prediction should be tested with deep HI maps and ROSAT observations of prototypical post-starburst galaxies such as NGC 7252 and NGC 5102.

Observations of QSO absorption lines using HST near starbursts and post-starbursts may reveal this left over gas that could have been more widely diffused than simple estimates initially predict and that may only therefore be of such low density that it may only be observable in absorption line spectra. The constraints from this will, in any case, be most interesting.

7. FACE-ON STARBURSTS

A luminosity of 10^{45} erg s^{-1} from the inner 100 pc of a galaxy should be a very spectacular object. Even M82, if unshrouded, should be one of the most visibly luminous objects in the local universe. The standard response is that we see no such objects and therefore they are all shrouded in gas and dust whatever the viewing angle. This is a reasonable point of view but the huge outflows seen normal to the disk of starbursts could possibly clear out a cone surrounding the nucleus. If the viewing angle of the starburst were such that one was observing the starburst down the cone then the pure starburst nucleus could be observed. The class of objects that immediately come to mind are the class of AGNs called Seyfert IIs. The problems here are that the Seyferts and starbursts can apparently be distinguished on the basis of their infrared colors and that it is not obvious that the Seyfert IIs are predominantly face on although this could be checked fairly readily. The Markarian galaxies could be partially shrouded versions of the starburst galaxies. Here, again, the inclination effects could be analyzed. The pure starburst models for a burst of 10^9 $M_\odot$ formed into stars with a normal Salpeter IMF predict that for a K band luminosity of -25^m that the underlying component of the pure starburst in the B band would have a luminosity of -23^m. The extreme point of view, emphasized by R. Terlevich, that all AGNs and quasars are pure starbursts with no central component such as a black hole is not pursued in the chapter but is reviewed by Heckman, in Chapter 17.

8. GALAXY FORMATION

The relation between the starbursting systems that we can observe directly and the process of galaxy formation may be fairly close. If the starburst scenario is correct then an appreciable fraction of the central regions of a starburst galaxy are being freshly minted in a burst. This is at least partial galaxy formation! Those high redshift objects associated with radio sources that may be associated with forming galaxies (Chambers et al. 1990) have inferred star formation rates similar to those of ultra luminous starbursts. Assuming this is so, some interesting progress can be made by using what we know about the physical conditions in starbursts to extrapolate to physical conditions appropriate to galaxy formation models. Feedback considerations from the star formation process in galaxy formation and the effect on the galaxy collapse have been studied by Ikeuchi and Norman (1990) who used a simple star formation model based on forming massive stars in cloud collisions (c.f. White 1990). The energy input is taken to be due to supernovae. Characteristic masses, radii, and binding energy of the galaxies were calculated in terms of fundamental constants. In addition the basic cloud parameters were inferred in terms of fundamental constants as were the properties of the hot component of the interstellar medium of the protogalactic cloud.

In pursuing the feedback idea further it is possible that the bursts in the high redshift objects are being triggered by the high internal pressure of the radio lobe. In general triggering of starbursts can occur due to the action of a central source such as a AGN, radio source or quasar. In analyzing this process Norman and Ikeuchi (1990) found that such triggering was likely as a central energy source turned on and began to interact with the multi-phase protogalactic medium. One very interesting result that came out of this work is that the binding energy of galaxies and the amount of injected energy from the central source are roughly equal from general stability considerations of the collapsing gaseous protogalaxy. Furthermore for canonical parameters for central luminosities and the speed of the energy input, the binding energy of galaxies turns out to be of order $10^{60} - 10^{61} erg$ which is roughly correct for massive galaxies.

It is a pleasure to thank Stephan Charlot, Tim Heckman, Satoru Ikeuchi, Claus Leitherer, Daniel Pfenniger, Jim Pringle, Jerry Sellwood, Nolan Walborn and Steve Zepf for stimulating discussions.

REFERENCES

Bailey, M. E. 1980, *M.N.R.A.S.*, **191**, 195.
Barnes, J. 1990, in *Dynamics and Interactions of Galaxies*, ed. R. Wielen, p. 186 (Springer-Verlag).
Begelman, M. C. and Rees, M. J. 1978, *M.N.R.A.S.*, **185**, 847.
Bernlöhr, K. 1990, in *Paired and Interacting Galaxies*, IAU Colloquium 124, ed. W. Keel and J. Sulentic, in press.
Brinks, E. 1990, in *Dynamics and Interaction of Galaxies*, ed. R. Wielen, p. 146 (Springer-Verlag).
Burkert, A. and Hensler, G. 1990, in *Chemical and Dynamical Evolution of Galaxies*, ed. F. Matteucci (Elba Workshop).
Byrd, G. G., Valtoneu, M., Sundeleus, B. and Valtaja, L. 1986, *Astr. Ap.*, **166**, 75.
Carlstrom, J. E. and Kronberg, P. P. 1990, *Ap. J.*, in press.
Chambers, K. C., Miley, G. K., and van Breugel, W. J. M. 1990, *Ap. J.*, **363**, 21.

Charlot, S. and Bruzual, G. 1991, *Ap. J.*, in press.
Chevalier, R. A. and Clegg, A. W. 1985, *Ap. J. (Letters)*, **300**, L107.
Conti, P. S., 1990, preprint.
Devereux, N. 1987, *Ap. J.*, **323**, 91.
Draine, B. T. and Woods, D. T. 1990, *Ap. J.*, **363**, 464.
Eckart, A., Downes, D., Genzel, R., Harris, A. I., Jaffe, D. T. and Wild, W. 1990, *Astr. Ap.*, in press.
Efstathiou, G., Lake, G. and Negroponte, J. 1982, *M.N.R.A.S.*, **199**, 1069.
Friedli, D. and Pfenniger, D. 1990, in press.
Habe, A., Ikeuchi, S. and Tanaka, Y. D. 1981, *PASJ*, 33, 23.
Hasan, H. and Norman, C. A. 1990, *Ap. J.*, **361**, 69.
Heckman, T. 1990, in *Paired and Interacting Galaxies*, IAU Colloquium *124*, ed. W. Keel and J. Sulentic, in press.
Heckman, T. Armus, L. and Miley, G. K. 1990, *Ap. J. Suppl.*, in press.
Hernquist, L. 1989, *Nature*, **340**, 687.
Ikeuchi, S. and Norman, C. A. 1990, *Ap. J.*, in press.
Ishizuki, S., Kawabe, R., Ishiguro, M., Okumuva, S. K., Morita, K. I., Chikada, Y., Kasuga, T. and Doi, M. 1990, *Ap. J.*, **355**, 436.
Larson, R. B. 1987, in *Starbursts and Galaxy Evolution XXII Moriond Meeting*, ed. T. X. Thuan, T. Montmerle, J. T. Tran, p. 467 (Editions Frontières).
_____. 1990, *Pub. A.S.P.*, **102**, 709.
_____. 1990, in *Fragmentation of Molecular Clouds and Star Formation, IAU Symposium 147*, ed. E. Falgarone, in press.
Li, F. and Ikeuchi, S. 1989, *PASJ*, **41**, 221.
Lin, D. N. C., Pringle, J. E. and Rees, M. J. 1988, *Ap. J.*, **328**, 103.
Lubow, S. 1987, private communication.
Mathews, W. C. 1986, *Ap. J.*, **305**, 187.
McKee, C. F. 1989, *Ap. J.*, **345**, 782.
Negroponte, J. and White, S. D. M. 1983, *M.N.R.A.S.*, **205**, 1009.
Noguchi, M. 1990, in *The Interstellar Medium in Galaxies*, ed. H. A. Thronson and J. M. Shull, p. 323, (Kluwer Academic).
Norman, C. A., 1987 in *Starbursts and Galaxy Evolution XXII Moriond Meeting*, ed. T. X. Thuan, T. Montmerle, J. T. Tran, p. 483 (Editions Frontières).
_____. 1988a, in *Galactic and Extragalactic Star Formation*, ed. R. Pudritz and M. Fich, (Dordrecht: Reidel), p. 495.
_____. 1988b, in *Comets to Cosmology*, ed. A. Lawrence (Berlin: Springer), p. 177.
Norman, C. A. 1990 in *The Disk-Halo Interaction, IAU Symposium 144*, ed. H. Bloemen, in press.
Norman, C. A. and Ikeuchi, S. 1989, *Ap. J.*, **395**, 372.
_____. 1990, *Ap. J.*, submitted.
Norman, C. A. and Panagia, N. 1991, *Ap. J.*, in preparation.
Norman, C. A. and Silk, J. 1980, *Ap. J.*, **238**, 158.
_____. 1983, *Ap. J. (Letters)*, **272**, L49.
Norman, C. A. and Scoville, N. Z. 1988, *Ap. J.*, **332**, 124.
Olson, K. M. and Kwan, J. 1990a, *Ap. J.*, **349**, 480.
_____. 1990b, *Ap. J.*, **361**, 426.

Ostriker, J. P. and Peebles, P. J. E. 1973, *Ap. J.*, **186**, 467.
Pfenniger, D. E. and Norman, C. A. 1990, *Ap. J.*, **363**, 391.
Pompea, S. and Rieke, G. H. 1990, *Ap. J.*, **356**, 416.
Raha, N., Sellwood, J. A., James, R. A. and Kahn, F. D. 1991, *Nature*, submitted.
Rand, R. and Telamns R. 1990, in *The Interstellar Medium in Galaxies*, ed. H. A. Thronson and J. M. Shull, p. 525 (Kluwer Academic).
Rees, M. J. and Ostriker, J. P. 1977, *M.N.R.A.S.*, **179**, 541.
Scalo, J. and Struck-Marcell, C. 1984, *Ap. J.*, **276**, 60.
Scoville, N. Z. and Norman, C. A. 1988, *Ap. J.*, **332**, 163.
Scoville, N. Z., Sanders, D. B. and Clemens, D. P. 1986, *Ap. J. (Letters)*, **310**, L77.
Sellwood, J. A. 1980, *Astr. Ap.*, **89**, 296.
_____. 1989, *M.N.R.A.S.*, **238**, 115.
Shlosman, I., Frank, J. and Begelman, M. C. 1989, *Nature*, **338**, 45.
Shu, F. H., Adams, F. C. and Lizano, S. 1987, *Ann. Rev. Astr. Ap.*, **25**, 23.
Shull, J. M. 1983, *Ap. J.*, **264**, 446.
Silk, J. 1977, *Ap. J.*, **211**, 638.
Spitzer, L. 1981 in *Galactic Nuclei*, ed. D. O'Connell, North Holland, p. 443.
Stacey, G. J., Geis, H., Genzel, R., Lugten, J. B., Poglitsch, A., Sternberg, A. and Townes, C. H. 1990, *Ap. J.*, in press.
Telesco, C. M., Campins, H., Joy, M., Dietz, K. and Decher, R. 1990, *Ap. J.*, in press.
Thronson, H. A., Majewski, S., Descartes, L. and Hereld, M. 1990, *Ap. J.*, in press.
Tomisaka, K. and Ikeuchi, S. 1988, *Ap. J.*, **330**, 695.
van Buren, D. and Norman, C. A. 1989, *Ap. J. (Letters)*, **336**, L67.
Vasquez, and Scalo, J. 1989, *Ap. J.*, **343**, 644.
Waller, W. H. 1991, *Ap. J.*, in press.
Weinberg, M. D. and Tremaine, S. 1983, *Ap. J.*, **264**, 364.
White, R. L. 1988, *Ap. J.*, **330**, 26.
White, S. D. M. 1990, in *The Interstellar Medium in Galaxies*, ed. H. A. Thronson and J. M. Shull, p. 371, (Kluwer Academic).
Wolfire, M. G. L., Hollenbach, D. and Tielens, A. G. G. M. 1989, *Ap. J.*, **344**, 770.
Zel'dovich, Y. B. and Novikov, I. D. 1964, *Dokl. Acad. Nauk SSSR*, **158**, 811.
Zepf, S. 1991, Ph.D. Dissertation, Johns Hopkins University, in preparation.

DISCUSSION

A. Kinney: I was puzzled by the simulation. There seems to be a segregation between the gas and stars in those simulations and if there is just a phase difference you'd expect the gas to be leading and the stars to be trailing or vice versa. Is that just the way the data is displayed?

Norman: It may be the 3-dimensional. If you look in 3-D it may be resolved.

A. Kinney: It looked as if the stars were coating the gas.

__Norman__: Well, I haven't noticed that. So I'm not sure. It may be the way it's displayed.

__P. Appleton__: I have some real worries about this gas star separation. The worries go back to some work that Rod Davis and myself did a few years ago on an interacting system which showed very clearly the separation or the peeling off of these tidal filaments in H1 relative to the stars. It has worried me ever since because I think it's becoming very much more clear that the gas-gas interactions between clouds are really extremely dominant, probably one of the most important aspects of the starburst phenomenon rather than say the stellar density wave kind of picture. What worries me is that if this separation between gas and stars can happen quite early on in terms of the dynamical interaction that one then is worrying a lot about trying to model systems in which the gas and stars are really being separated, and behaving as a dynamically different nuance. Given our poor understanding of how the interstellar medium even works in normal galaxies, irrespective of what happens to individual clouds which is somehow pulled out of galaxies and behaves differently, so I worry that such gas can provide a lot of angular momentum input into a disc and I wonder how you feel about modeling gas in discs given that we have such a big uncertainty about how gas clouds interact and so forth.

__Norman__: Well I think that is the major problem at the moment. The physics of the interstellar medium in the Hernquist-Barnes models is very crudely modeled. Basically, if clouds get too massive, the code stops them growing, and fixes them at a certain temperature and therefore pressure. They did that in order to do the calculations which I think are intrinsically very interesting but there may be some serious modifications if we incorporate reasonable models of the interstellar medium.

__T. Heckman__: I just wanted to show a viewgraph that's relative to Colin's question of face-on starbursts. I'm not sure whether there is a problem. I was going to show this later in my talk but I think we better show it now. This is just this famous luminosity function of everything in the universe and the thing to look at here is to concentrate on the solid red line. This is bolometric luminosity versus space density and then these things that I've circled here are the Markarian starburst galaxies and they actually don't look all that different within a factor of two or something. Therefore the Markarian sample of starburst galaxies may correspond to something more face on.

__Norman__: But if you really look down the center of an M82 it would be even more obvious, wouldn't it?

__G. Miley__: There is a point I want to make about the difference between starbursts and AGN Type objects. Seyferts Type II spectra have a very big difference in 60 to 25 micron spectral indices from starbursts.

__Norman__: Well we discussed this last night. In any good theory, you're allowed to throw away one thing and that's what I did.

__J. Krolik__: I would like to introduce two complications of the face-on starbursts and whether or not you should be able to see an active difference.

One of these complications is that in the small number of cases where we have some evidence about the orientation of obscuring stuff in the center, it's often tilted by a large angle with respect to the plane of the outer galaxy and so the face-on starbursts may not be particularly the right population to search in. They may not be, on average, more crucial places to look than any orientation.

The second is a suggestion due to Begelman, Shlosman and Frank, which is that there may be a relation between starbursts and active galaxies, but it could be a two step process in which you need, in order to get an active nucleus, you need both steps to be activated, whereas in order to get a starburst, you may need only the first. Looking at those space densities that Tim just showed, it's clear that the active galaxies are sufficiently rarer that these effects may weaken that test you proposed.

G. Hensler: This concerns the self-organization of starbursts. I think that there's no question that interactions can trigger star formation. We have done calculations for the evolution of galaxies by taking the multi-phase interstellar medium into account and also very complicated interactions between the stars and the interstellar medium. That's necessary because we know that stars are losing mass to the hot, or to the warm phase, but stars can only be formed by the cool phase, so one needs at least two phases of interstellar medium. And one needs also metal enrichment in order to look for processes which are metal dependent. I want to show you one-dimensional calculations. We could show that with these calculations, starbursts occur. They are dependent on the initial density and on the initial mass of the galaxy.

We try to explain the observations and to look back what the initial conditions for the galaxies have to be. And we have now the chance to explain the star formation bursts from the initial conditions and we hope to compare them statistically with the observations.

N. Panagia: A very simple question. In the simulation of a merger that you've shown, I have the impression that the post-merger galaxy is spatially much bigger than the two galaxies that we are merging. If this is true, then we should expect to see in the sky unusually big galaxies with normal mass, or twice as big as normal.

Norman: Well, I'm not sure whether your impression is right, but the canonical wisdom is that they are massive galaxies called ellipticals.

I. Mirabel: I hope to show a spectacular galaxy like this on images obtained with the NTT. This object is half a megaparsec in size. I will show a case where you have a correlation between two giant spiral galaxies with antennae that spread over half a megaparsec, namely five times the classical antennae. So they really do exist—we have plenty of these systems.

N. Panagia: My point was, after the merger, the clearly visible effect is gone, where there is still a much bigger galaxy, so what one sees in clusters a few huge galaxies, and other ones which are normal ones. Is there any population of unusually big spiral galaxies, for instance?

Norman: Yes, there is an ultraluminous infrared galaxy from Sanders' sample of very warm galaxies which is a CD galaxy.

J. Bland Hawthorn: I wanted to make the comment that in NGC 6240, which is a classically irregular system to look at, the kinematics are actually very regular. There is a poster on that in collaboration with Andrew Wilson. The kinematics are extremely well ordered and I was talking to Hernquist and Barnes about their simulations and they were saying in fact that you do see well-ordered, separate components all the way up until almost the final merger. So that's actually observed and NGC 6240 may well be the best case where you see 2 resolved kinematic disks.

Norman: You should quote your central mass if you want.

J. Bland Hawthorn: The central masses of both systems—there is a caveat if anyone's interested—but the central masses of both systems are something like $4 \times 10^{10}\ M_\odot$ in both disks uncorrected for inclination, and that's over a scale of a few kiloparsecs in both cases. In one disk, you see classical flat rotation and the other disk there are two forms of rotation you could be seeing: one is Keplerian in that you have a very dense central mass and a rapid fall-off and unfortunately that might be some tidal influence from the internal forces. We have very limited velocity coverage—you may be picking up the blue N II line, so therefore that could be flat rotation by moving those points up by 750 kilometers per second, just to give you the wavelength difference. But there again you do not remove the central mass distribution, you just change the nature of the rotation.

Anon: I have another question for you, and that is, being a clever theorist, can you think of a way that starbursts can be triggered that doesn't involve some external perturbation. Are there secular processes entirely internal to a galaxy, that just sitting in isolation somewhere that would lead it to simply have a burst.

Norman: Well, that's a good question. Most disks you see are almost sitting at marginal stability. They're right on the threshold. And you know, I think you need some perturbation, but it may not have to be as large as—it could be something like the Magellanic Clouds doing something to an Sc galaxy—not our Galaxy probably but something a bit more gas-rich—and that could produce some perturbation. Now, Steve Lubow has worked on what happens when that occurs, that is, when waves get launched into the center, they're action is in fact conserved—that's basically the amplitude over the characteristic frequency. The characteristic frequency goes up as you go to the center and because the action's conserved, the amplitude rises and shocks can result. Something like that could happen on a small scale, but probably would not actually explain ultraluminous starbursts, but may explain the weaker ones.

N. Scoville: One way to identify these post starburst galaxies would be to look for $R^{1/4}$ law—elliptical distributions of young stars. Are there galaxies like that?

Norman: Of the A type spectra? That's a good idea.

N. Scoville: But the thing that bothers me is it should be a large number of post starbursts.

Norman: That's right. My calculation was fairly independent of everything. You know, it's the age of A stars over what we infer to be the lifetime of the burst. We should see that population. There should be at least as many post-starburst A type spectra as there are starburst galaxies. And either we're missing something or ...

N. Scoville: ... if you say the burst has anything to do with dynamics, there's no dynamic mass ...

Norman: Right.

G. Rieke: Just a comment. It seems to me the lack of this population of A0 dominated stars is actually what you would predict if there's a lower mass cutoff to the IMF, so I'm not saying that proves it ... keep it in mind.

Norman: No, no, that's it exactly—it has to be fairly high lower mass cutoff.

G. Rieke: That's right. It would obviously vary from one galaxy to another so it's got to be seen at some level.

Norman: That's a very interesting point.

R. Kennicutt: Not to disagree not too strongly with my colleague and boss, but one thing to bear in mind is that these ultraluminous galaxies are exceedingly rare and the space densities are very low compared to normal galaxies. And I think Allison Campbell's comment has to be borne in mind. Schweizer has obtained long-slit spectra of several of these $R^{1/4}$ looking post-merger galaxies and finds A stars in them. I've done the same thing—I think it's quite a frequent feature so I think this maybe not be a problem.

Norman: Can I then ask: Is there any gas in these systems? See, there is the other question which I didn't talk about, which is how you blow the gas away so fast. You've got all this gas, right. You go from gas-rich starburst to gas-poor post-starburst.

R. Kennicutt: Well, some of these things where you look for A star spectra show optical emission lines so that's a little more evidence for gas. I think one problem that does exist is if you look back at a $z \sim 1/2$ in rich clusters, these sorts of galaxies are extremely commonly found. I think 1/4 of the blue galaxies are these E+As and certainly the current space density is much, much lower but maybe that has something to do with mergers.

Norman: Yes, I think the E+As are definitely post-starburst systems. The local ones were what I wanted to discover.

D. Weedman: It's always easy for observers to confuse the issue, but one of the most outstanding puzzles around is Markarian 231 which is the most super of the superluminous IRAS sources, and as it's name implies, it was discovered by Markarian and the reason is, because it has an A star spectrum. And we've never understood why it has an A star spectrum, so it has a remnant of an old superluminous starburst and a current ongoing superluminous starburst plus it's full of gas and it's full of dust.

Norman: Very good. Yes, it's at least 5×10^8 years old.

G. Hensler: One question concerning the interaction of the inflow of satellites towards stars. Would one expect that already a density rate which is also shown in your viewgraphs could already trigger star formation, but not in the nucleus because it's not necessary in this case to fuel the nucleus, but already the density rate would lead to enhancement of the cooling and then already in radial sites further out of the nucleus the starburst would already be triggered.

Norman: That's stage one of what I've been thinking about, but there's not that much evidence that that's happening. I think the point that Lubow was making was that you really get the effect as you go to the center, because of the steepening of the wave that's being propagated inwards. I don't know whether anyone's studied these.

Anon: I just have a question about pre-main sequence stars. We're dealing with young objects. Is it not possible that pre-main sequence stars, because they're over-luminous for their mass, should not confuse the issue of determination of the lower mass cutoff.

Norman: Yes, quite possible.

Anon: How serious a problem is it?

Norman: It seems to me that if you had a lot of pre-main sequence stars, they're infrared luminosity per unit mass is enormous of course. They may last for some significant time like 10^7 years. It may confuse the issue.

THE STARBURST–AGN CONNECTION

Timothy M. Heckman
Department of Physics and Astronomy
The Johns Hopkins University and
The Space Telescope Science Institute
3700 San Martin Drive
Baltimore, Maryland 21218
USA

Abstract. I review the possibility that there is a link between starbursts and Active Galactic Nuclei (AGN's). First, I estimate that AGN's and starbursts are of comparable energetic significance in the local universe and that AGN's and massive stars have produced similar amounts of total radiant energy over the history of the universe. I then discuss whether and how we can discriminate between phenomena powered by massive stars and those demanding a more exotic energy source such as a supermassive black hole. This is discussed in the contexts of both the model that Terlevich and his collaborators have proposed for AGN's ('AGN's Without Black Holes') and the controversy regarding the energy source in powerful infrared galaxies (starburst vs. 'buried' quasar). I then address whether there might be an indirect or evolutionary connection between AGN's and starbursts. I review the evidence that AGN's may reside in galaxies with unusually high present or recent star-formation-rates. This involves comparing the stellar population and ISM in galaxies with and without AGN's.

1. INTRODUCTION

The discovery three decades ago that the nuclei of galaxies can be the sites of incredible violence fundamentally altered our view of a serene, nearly immutable extragalactic universe. More recently, the vision of a stately 'clockwork' universe has been further eroded by the discovery of starburst galaxies—galaxies that are currently forming stars at a rate significantly greater than could be sustained for a Hubble time. This philosophical link between the AGN and starburst phenomena—together with the plenitude of powerful AGN's in the early universe (presumably in young or proto-galaxies), the nuclear or circum-nuclear location of starbursts, and the apparent association of both starbursts and AGN's with interacting galaxies—has naturally led to the suspicion that the two phenomena are somehow related. Indeed, in §2 I will show that massive stars and AGN's have produced similar amounts of radiant energy over the history of the universe (further evidence of a possible starburst–AGN connection).

The boundary between the AGN and starburst phenomena is very clear according to conventional prejudices: one is powered by a 'monster' (supermassive black hole?) while the other is powered by massive stars and their corpses. However, the phenomenologically defined boundary has never been sharp and seems to be growing ever more indistinct. On the one hand, there is an emerging consensus that most of the continuum radiation (from soft X-rays through the far-IR) in most AGN's (excluding only the relatively rare 'violently variable' AGN's like BL Lac's) is thermal emission (cf. Sanders et al. 1989). On the other hand, the 'ultraluminous' IR galaxies—believed by many to be powered by massive stars—have bolometric luminosities rivaling those of quasars. Both massive stars (as confirmed by direct observation) and supermassive black holes (as fervently attested to by hosts of eminent theorists) are capable of producing thermal X-rays, ionizing radiation, a 'featureless, blue' optical continuum, thermal IR emission, nonthermal radio emission, and high velocity gas motions. Thus, perhaps the most crucial questions regarding the 'Starburst-AGN Connection' are: 1. Are the two phenomena fundamentally the same thing? and 2. (How) can we discriminate observationally between phenomena powered by massive stars and those requiring a 'monster'? I will address these two questions in §3 and §4 below in two specific contexts: Terlevich's 'AGN's without Black Holes' model and the on-going controversy concerning the nature of powerful IR galaxies.

Even if AGN's are not powered by massive stars, a crucial evolutionary link to starbursts may still exist. For example, Weedman (1983) and Norman and Scoville (1988) have argued that a nuclear starburst can lead directly to the formation of the mythical monster, and so starbursts are the evolutionary precursors of AGN's (see also Rees, 1984). Conversely, it is also quite plausible that the energy released by an AGN could trigger star formation in the surrounding galaxy (Sanders and Bania 1976; Ikeuchi 1981; DeYoung 1981,1989; Begelman and Cioffi 1989; Rees 1989; Daly 1990). Even if AGN's and starbursts are not related in an evolutionary sense, they might still be indirectly related because they have a common fueling mechanism like galaxy interactions (see Heckman 1990 for a recent review). In §5 I will therefore address a third basic question: do AGN's reside in galaxies with unusually high present or recent star formation rates? I will attempt to answer this question by comparing the stellar population and ISM in galaxies with AGN's to those without.

2. THE ENERGY INPUT FROM MASSIVE STARS VS. AGN'S IN THE UNIVERSE

Nucleosynthesis in massive stars and accretion onto supermassive black holes are two very distinct processes. It is likely *a priori* that one process will dominate the other in terms of overall energetic significance, and I believe that most astronomers would guess that massive stars dominate AGN's. A census of bona fide 'monsters' (quasars and type 1 Seyfert nuclei) and of unambiguous starbursts (*e.g.*, Markarian or other UV-selected starburst galaxies) in our local part of the universe, shows that the two types of 'activity' actually seem to have similar bolometric luminosity functions over the relatively narrow range between $L_{BOL} \approx 2 \times 10^{10}$ to 5×10^{11} $L_\odot$ where both functions are well-determined (Soifer et al. 1987). Of course, not all massive stars are formed in starburst galaxies.

A global comparison of massive stars and AGN's is more instructive. The classic calculation of Soltan (1982), recently updated by Padovani, Burg, and Edelson (1990), shows that the total amount of radiant energy produced by AGN's over the history of the universe is equivalent to a co-moving density of energy production of $\approx 1.2 \times 10^{-15}$ erg cm^{-3} (independent of H_0 and q_0). This estimate is strictly a lower limit since it was based only on quasars with $m_B \leq 22.5$ and $z \leq 2.2$. Extrapolating to $m_B = 25$ and $z = 4$ based on our present understanding of quasar counts might increase this by a factor of a few. This estimate is also based on optically-selected quasar samples and so does not include the contribution of either dust-shrouded 'buried quasars' that may populate the IRAS catalog (see §4 below) or the high-luminosity type 2 Seyfert galaxies recently discovered by DeGrijp, Miley, and Keel (1990). Cowie (1988) has estimated that the total space density of metals in the present universe is $3.6 \times 10^{-34} h^2$ gm cm^{-3} inside galaxies and at most $1.4 \times 10^{-33} h^{0.5}$ gm cm^{-3} outside galaxies (the upper limit applying if the Coma Cluster is typical of the global ratio of metals outside galaxies to blue light in the universe). The energy associated with the nucleosynthetic production of these metals by massive stars implies that the radiant energy density produced by these stars was $2 \times 10^{-15} h^2$ erg cm^{-3} and $9 \times 10^{-15} h^{0.5}$ erg cm^{-3} for the metals inside and outside galaxies respectively. Cowie also shows that these numbers are consistent with the light produced by the known population of faint blue galaxies (*e.g.*, Tyson and Seitzer 1988).

I conclude that starbursts and AGN's have a similar energetic significance in the local universe, and that AGN's and massive stars have produced comparable amounts of radiant energy over the history of the universe. Does this rough equality imply some fundamental link between 'true' AGN's and starbursts?

3. AGN'S WITHOUT BLACK HOLES

In a series of provocative and imaginative papers, Roberto Terlevich and his colleagues have argued that the great majority of AGN's are powered by massive stars and supernovae, albeit somewhat unusual stars in a highly unusual environment. I would like to begin this section by briefly summarizing how this model attempts to account for the puzzling properties of AGN's. These arguments are given in more detail in Terlevich and Melnick (1985—hereafter TM85), Terlevich and Melnick (1987, 1988), Terlevich, Melnick, and Moles (1987), Terlevich (1989), and Terlevich (1990a,b). I will then make what I believe to be fairly robust arguments concerning potential problems with this model.

3.1. The Model and Supporting Evidence

The starting point for this model is a starburst occurring in the nucleus of an early-type galaxy. The duration of the episode of star-formation is taken to be no greater than the main sequence lifetime of a massive star (a few million years). The observational manifestations of this nuclear starburst are hypothesized to differ from those of ordinary regions of high-mass star-formation (*i.e.*, the giant H II regions described elsewhere in this volume by Kennicutt) for two basic reasons: the metal abundance of the stars is taken to be several times solar (typical of the stars and gas in the nuclei of early-type galaxies—cf. Pagel and Edmunds 1981 and references therein) and the stars are immersed in an exceptionally dense ISM.

For the first several million years (during the main sequence lifetime of the massive stars) the nuclear starburst will look like a rather ordinary giant H II region (although it will be extremely luminous). Specifically, the emission-lines will be relatively narrow (line-width ≈ orbital velocity ≈ few hundred km s^{-1}) and the relative emission-line intensities will be consistent with photoionization by relatively soft radiation from O stars. There will be little X-ray or nonthermal radio emission, and the optical/UV continuum will be dominated by the light from massive main sequence stars. Strong reprocessed thermal IR emission from dust can also be present.

After several million years, the starburst enters the 'Warmer' phase. 'Warmers' are the name given by TM85 to post-main-sequence 'bare-core' stars with $T_{eff} > 10^5$ K, which they identify as WC4, WC5, and WO Wolf-Rayet stars. They argue that the relative importance of such stars would be greatly enhanced in a starburst with trans-solar metallicity because of the strong metallicity-dependence of the mass-loss rate in massive stars. TM85 take the hard input spectra predicted for Warmers and show that they naturally account for the 'anomalous' relative emission-line intensities seen in the Narrow-Line Region of Seyfert galaxies and LINER's.

Do 'Warmers' exist? Several gaseous nebulae apparently photoionized by a WO star are now known (Davidson and Kinman 1982; Dopita et al. 1990; Garnett et al. 1990). The latter authors use the flux ratio of the nebular He II λ4686 and Hβ lines to derive an Zanstra-like 'color temperature' of ≈75000K for at least one WO star. Similarly, Pakull (1990) has found two WN stars in the Magellanic Clouds whose nebulae have strong He II emission. He classifies these as 'WN1 stars' and derives temperatures of 80000K and 95000K for the two stars. These temperatures may be compared with typical O star values of $T_{eff} \approx 45000$K.

After about 10^7 years the starburst enters the phase in which it is dominated by emission from supernovae and SNR's. Depending on the strength of the starburst, this phase is supposed to produce type 1 Seyfert nuclei or radio-quiet quasars. The X-ray, UV, and radio continua are produced by a combination of supernovae and SNR's which interact with a very dense ISM. The optical continuum is produced by supernovae and supergiants and the IR continuum is reprocessed thermal emission. Terlevich (1990a) gives semi-empirical estimates of the relative luminosities of the different components of the continuum, and claims that the overall spectral energy distribution is quite similar to that of a typical type 1 Seyfert or quasar. The strong emission-lines from the 'Broad Line Region' in AGN's are hypothesized to be produced by individual supernovae. Indeed, Terlevich (1989) emphasizes that two extragalactic supernovae discussed by Filippenko (1989a)—SN1987F and SN1988I—have emission-line spectra that are remarkably similar to those of type 1 Seyferts and quasars.

During this phase, variability of the lines and continuum will arise on two timescales with two total energies: 'flares' of total energy 10^{49-50} ergs lasting a few weeks and associated with individual supernovae and longer term variations with energies of 10^{51-52} ergs and times scales of order a year that are driven by individual SNR's interacting with the dense ISM. Terlevich (1989; 1990b) argues that these variability times and energies are consistent with individual optical events in Seyfert nuclei and with a large data-base on the long-term optical variability of radio-quiet quasars.

Terlevich, Diaz, and Terlevich (1990) have provided some very intriguing indirect evidence that the optical/near-IR continuum in at least some Seyfert nuclei is produced by massive stars. They show that the near-IR Ca II triplet photospheric absorption features are strong in the nuclei of several Seyfert nuclei, even when the photospheric MgI absorption feature at 5174Å is very weak. This is consistent with the optical continuum being produced by OB stars plus blue and red supergiants and the near-IR being produced by red supergiants. In contrast, this result is difficult to understand in the conventional picture where the optical and near-IR continua are the sum of the emission from an old stellar population and a 'featureless' continuum produced by the monster. It must be noted however that many type 1 Seyfert nuclei and quasars have very strong Ca II triplet *emission* (not absorption) lines (cf. Persson 1988). The generality of the Terlevich, Diaz, and Terlevich result is therefore unclear for Seyfert 1's and quasars.

3.2. Consequences and Possible Problems

Neither Terlevich's model nor the standard supermassive black hole model for AGN's can be expected to account in a simple way for the plethora of detailed phenomenological properties of AGN's. Thus, a point-by-point comparison of the two models in such a fashion is probably not very instructive. Instead, it is important to stick to the basics and keep clearly focused on what it was that originally made the supermassive black hole model so attractive (*e.g.*, Rees 1984).

Rapid variability is often cited as one of the cornerstone pieces of evidence supporting the standard AGN construct. The rapidly and violently variable 'Blazar' class of AGN's (including BL Lac's and the relatively rare OVV quasars) clearly can not be accommodated within the Terlevich model (and indeed he makes no claims to explain them). Turning to more normal AGN's, Terlevich (1989; 1990b) shows that the observed optical variability of most quasars, which occurs over typical timescales of weeks or longer, is in fact compatible with his model. The recent discovery of extremely rapid (several minutes), large amplitude X-ray variability in several type 1 Seyfert galaxies (cf. Matsuoka *et al.* 1990 and references therein) is therefore extremely important. It must be stressed that this rapidly fluctuating X-ray emission represents a significant fraction of the bolometric luminosity of these nuclei ($\approx 10\%$), so this can not be ascribed to some process with little bearing on the overall nature of the AGN energy source. Terlevich (1990a) responds to this data by pointing out that such rapid, strong variability is not readily understood in the context of the standard supermassive black hole model either. Still, many will find this variability rather persuasive evidence that something other than an ensemble of massive stars is needed in these AGN's at least.

The copious production of high energy photons by AGN's is another seemingly robust argument against massive stars as the power source. Indeed, to zeroth order type 1 Seyfert nuclei and quasars produce roughly equal power per decade of frequency between the far-IR (few $\times 10^{12}$ Hz) and hard X-ray (few $\times 10^{19}$ Hz) spectral regions (*e.g.*, Urry 1989). Terlevich (1990a) has explained in general terms how a massive population of stars might produce such a spectral energy distribution (SED), and I would now like to carefully consider this argument and its far-reaching implications.

Compared to an AGN, photospheric emission from a massive star has a much more 'narrow-band' SED that extends only to energies of a few Rydberg (≈ 50 eV). To explain the high energy radiation and the nonthermal radio emission from AGN's, Terlevich appeals to the kinetic energy produced by massive stars in the form of supernova ejecta and high speed winds. If such high velocity (several thousand km s^{-1}) gas is completely thermalized via shocks, the implied post-shock temperatures are one-to-several $\times 10^8$ K. An equivalent maximum temperature may be estimated by using 10^{51} erg in kinetic energy (typical supernova kinetic energy) to heat ≈ 10 $M_\odot$ of ejecta (typical supernova ejected mass). Thus, tapping the kinetic energy of massive star ejecta could (in principle) produce a copious supply of thermal photons up to energies of several tens of keV ($\nu \approx 10^{19}$ Hz). While radioactive decay during the early expansion phase of a supernova explosion or nonthermal processes associated with an SNR could produce

Gamma-Rays, only about 10^{49} ergs per supernova at most could be produced in this form ($\approx 1\%$ of the total available energy).

One clear prediction of Terlevich's model then is that AGN spectral energy distributions should exhibit a dramatic high energy 'cut-off' at energies of several-tens-of-keV (and surely by 100 keV). Unfortunately, data on AGN's at such high energies are fragmentary. Rothschild et al. (1983) showed that a dozen X-ray bright AGN's (mostly type 1 Seyferts) have spectral energy distributions that are still slowly rising ($\nu P_\nu \propto \nu^{0.3}$) at $h\nu = 100$ keV. Data like these severely strain a Terlevich-type model, but may not yet exclude it. The data on AGN's at still higher energies are almost non-existent. There are five AGN's that have been detected as Gamma-Ray sources according to Bignami and Mereghetti (1989). Three of these are radio-loud AGN's, and so are beyond the scope of Terlevich's model (at least in its present form). The two detected radio-quiet AGN's are the type 1 Seyfert galaxies NGC 4151 and MCG 8-11-11. Emission in the 1-10 MeV range comprises a significant (NGC 4151) or even dominant (MCG 8-11-11) fraction of the bolometric luminosity of these AGN's. If this turns out to be typical of type 1 Seyfert nuclei, it is hard to see how massive stars can power such AGN's all by themselves.

Integrated over its lifetime (including the supernova phase), the kinetic energy produced by a massive star is only a few percent of the radiative (mostly UV) energy it emits. Thus (as Terlevich himself shows), only a 'post-starburst' population (with an age of one-to-several $\times 10^7$ years) can generate a flat 'AGN-type' SED. This is because the supernova rate is then maximized while the optical and UV luminosities (and hence the thermal IR luminosity) have all dropped significantly compared to their initial values. This has several important consequences for Terlevich's model.

First, let me consider a light-to-mass argument. In principle, an ensemble of very high mass stars (because they are supported by radiation pressure) could have nearly as large a light-to-mass ratio (L/M) as a Black Hole accreting at the Eddington limit (L/M $\approx 5 \times 10^4$ in solar units). However, because the quasar/Seyfert 1 phase occurs during the post-starburst phase (when the bolometric luminosity of the nucleus has declined significantly relative to $t = 0$), Terlevich estimates a value of only L/M ≈ 30 for this phase. Is such a low L/M consistent with the quasars and type 1 Seyfert nuclei?

The nucleus of the proto-typical type 1 Seyfert galaxy NGC 4151 has a bolometric luminosity of $\approx 10^{44}$ erg s^{-1} (about 30% of which is at $h\nu > 500$eV). For L/M ≈ 30, the implied mass of the post-starburst is then $\approx 10^9$ $M_\odot$. From direct observation we know that this luminosity is produced within a region with a radius < 5 pc (Neugebauer et al. 1990; Ebstein, Carleton, and Papaliolios 1989). Terlevich, Diaz, and Terlevich (1990) measure a line-of-sight stellar velocity dispersion of $\sigma_{1D} = 170$ km s^{-1} for the nucleus of NGC 4151 in a 1.5 $\times$ 2.1 arcsec aperture. For a stellar number density law $N(r) \propto r^{-2}$ (singular isothermal sphere) in the nucleus, the half-light radius of the region whose velocity dispersion they measured is 0.44 arcsec (28 pc). The predicted orbital velocity at this radius for $M = 10^9$ $M_\odot$ is ≈ 400 km s^{-1}, equivalent to $\sigma_{1D} \approx 280$ km s^{-1}. This value is not terribly discrepant with the measured value, especially allowing for the possibility of projection effects (e.g., rotation viewed pole-on). However, their value for σ_{1D} was measured using the near-IR Ca II triplet line which—according to them—is produced by red supergiants within the compact (< 10 pc diameter) Seyfert 1 'post-starburst' nucleus (rather than by an ordinary old-bulge population). Thus, the true half-light radius for their σ_{1D} measurement should have been considerably smaller than 28 pc, and σ_{1D} well in excess of 300 km s^{-1} would have been expected (e.g., $\sigma_{1D} \approx 700$ km s^{-1} at $r = 5$ pc).

The prototypical high-luminosity quasar 3C273 has a bolometric luminosity of $\approx 10^{47}$ erg s^{-1}, and the implied post-starburst mass is then $\approx 10^{12}$ $M_\odot$. We know that the region producing this energy is smaller than an arcsec, and the implied orbital velocity at $r = 0.5$ arcsec (≈ 1.2 kpc) is therefore ≈ 2000 km s^{-1} for L/M $\approx$ 30. While there are no measures of the stellar dynamics in the nucleus of 3C273, the arcsec-scale region of ionized gas in 3C273 (the 'Narrow-Line Region'—NLR) has a velocity dispersion of $\sigma_{1D} \approx 500$ km s^{-1} (Hooimeyer, 1990). To generalize this result, I have calculated the orbital velocity predicted by Terlevich's model at a radius of 0.5 arcsec for 25 bright type 1 Seyfert nuclei and low-z quasars whose bolometric luminosities could be estimated. I then find typical values ranging from 300 to 700 km s^{-1}. These values are inconsistent with the typical values $\sigma_{1D} \approx 100$ to 300 km s^{-1} for the emission-line profiles produced in the arcsec-scale NLR of type 1 Seyferts and low-z quasars (cf. Heckman, Miley, and Green 1984).

A related argument can be made for the AGN population as a whole. As noted previously in §2, Soltan (1982) and Padovani, Burg, and Edelson (1990) have shown that (integrated out to redshifts of several) AGN's have produced a total amount of luminous energy equivalent to $\approx 5 \times 10^{60}$ h^3 erg per present-day typical (L_*) galaxy. Suppose that all this energy has been produced by Terlevich-type AGN's. About half of this luminous energy is at $E \gg 100$eV, and according to Terlevich was produced from the kinetic energy supplied by massive stars. The models for supernova explosions summarized by Woosley and Weaver (1986) imply that the production of $\approx 2 \times 10^{60}$ erg in kinetic energy will leave behind a remnant mass $M_{rem} \approx 10^{10}$ $M_\odot$ in neutron stars and black holes per galaxy.

Since we do not know what fraction of present-day galaxies were once powerful quasars, it is possible that (for example) 1% of today's galaxies have all the remnants ($M_{rem} = 10^{12}$ $M_\odot$ per galaxy). Let's assume instead that they are evenly spread among all bright galaxies, since that makes them the most difficult to detect. Since a typical galaxy rotation velocity of 200 km s^{-1} implies an enclosed mass within a radius r of $M(<r) \approx 10^{10}$ (r/kpc) $M_\odot$, it is clear that the size of the region containing these hypothetical remnants must have $r >$ several kpc if the remnants are not to produce pronounced dynamical effects in galaxies. Because of the strong cosmological evolution of the quasar population, most of the luminous energy I have used to estimate M_{rem} was produced by high-redshift quasars ($z > 1$). Based on our direct images of high-z quasars (Heckman et al. 1990; see §5.3 below) we know that the diameter of the region producing the bulk of the optical and UV continuum can be no larger than about 0.5 arcsec (and so has a radius < 2 kpc). This size would imply that about half the mass within $r = 2$ kpc in a typical spiral galaxy is contained in neutron stars and black holes, and that the mass-to-light ratio would rise by a factor of two inside $r \approx 2$ kpc. It is hard to believe that such effects would have gone unnoticed (cf. Kent 1988). Of course, much more restrictive limits can be placed by HST measurements of the angular size of high-z quasars.

Finally, if quasars are indeed descendants of starbursts, where are the starbursts that spawned them? A population of massive stars produces about 30 times more energy in the form of UV photons than it produces in mechanical energy. Thus, for every erg of high-energy radiation produced during the quasar phase by tapping the kinetic energy of massive stars, there should have been about 30 ergs of UV radiation, most of which would have been radiated during the starburst phase. It seems hard to avoid the conclusion that the starburst phase would have been far brighter (especially in the UV) than the quasar phase. Why then don't objective prism/grism surveys turn up many more such objects (out to large redshifts) than quasars?

One plausible possibility is that the starburst phase is shrouded in dust, and that this dust is blown away or destroyed by the time the quasar phase begins (cf. Sanders et al. 1988b). Indeed the IRAS galaxy and quasar bolometric luminosity functions imply that powerful IRAS galaxies 'outshine' quasars in the local universe by a factor of ≈ 5 (Soifer et al. 1987). Possibly these IRAS galaxies are the starburst precursors to the quasars predicted by Terlevich. Let me now consider this possibility more globally. Following Soltan (1982) and Padovani, Burg, and Edelson (1990), the mean bolometric surface brightness of the sky due to the integrated effect of all AGN's is at least $\approx 10^{-6}$ erg cm^{-2} s^{-1} ster^{-1}. In Terlevich's model, about half the total radiation (that at $E > 50$eV or so) was powered by the mechanical energy of massive stars. Again, taking the UV output from these stars to be ≈ 30 times their mechanical energy and assuming this is all converted to thermal IR radiation, the implied all-sky IR surface brightness of the dust-shrouded precursors to the quasars would be at least $\approx 1.5 \times 10^{-5}$ erg cm^{-2} s^{-1} ster^{-1}. Such a far-IR/sub-mm background (primarily in the wavelength range ≈ 0.1–0.5 mm, based on the SED's of local starburst galaxies and the known redshift dependence of the quasar population) should be readily detectable by COBE. The metals produced by these hypothetical quasar precursors would lead to a present-day metal density of $\approx 3 \times 10^{-33}$ gm cm^{-3}. This can be compared to Cowie's (1988) estimates of $3.6 \times 10^{-34} h^2$ gm cm^{-3} and $1.4 \times 10^{-33} h^{0.5}$ gm cm^{-3} for the minimum amount of metals inside galaxies and the maximum amount outside galaxies respectively.

So far, I have only discussed the implications and possible problems of Terlevich's model for type 1 Seyfert nuclei and quasars. None of the arguments I have given above can be convincingly applied to type 2 Seyfert nuclei, because these are not known (at present) to generally be strong producers of high energy photons (cf. Lawrence and Elvis 1982). In fact, the overall SED's of type 2 Seyfert galaxies from radio to X-ray wavelengths resemble the SED's of IR-starburst galaxies more strongly than they resemble the SED's of type 1 Seyfert nuclei or quasars. Traditionally, the main evidence that type 2 Seyfert nuclei contained something other than massive stars was the presence of strong, high-ionization emission-lines (e.g., Osterbrock 1989). If Warmers really exist in abundance in galactic nuclei, TM85 argued that they could produce these high-ionization lines. It is very significant then that several type 2 Seyfert nuclei have recently been shown to contain 'hidden' type 1 Seyfert nuclei, visible in reflected (polarized) light (Antonucci and Miller 1985; Miller and Goodrich 1990). If type 1 nuclei are hidden by obscuration in type 2 Seyferts (cf. Krolik and Begelman 1988), then observations at high enough energies (e.g., above the Klein-Nishina limit) should show type 1 and type 2 Seyferts to be equally bright. In addition, X-ray spectroscopic measurements of Fe Kα line emission have yielded indirect (but persuasive) evidence for an intrinsically-strong 'hidden' X-ray continuum source in the prototypical type 2 Seyfert NGC 1068 (Koyama et al. 1989). If type 2 Seyfert nuclei generally turn out to be intrinsically strong sources of high energy radiation, the arguments given above concerning type 1 Seyferts and quasars can then be applied to them.

To summarize, I believe that present data severely strain, but do not decisively rule out, a Terlevich-type model in which a population of massive stars is the power source for type 1 Seyfert nuclei and quasars. Measurements in the near future that could provide definitive tests include: 1) Determination of the SED of typical AGN's/quasars at $E \gg 100$ keV (with GRO?). 2) High angular resolution measurements of the stellar dynamics in the nuclei of bright type 1 Seyferts or quasars using the Ca II triplet lines (this may require the second-generation instrument STIS on HST). 3) High angular

4. FAR-INFRARED GALAXIES: STARBURSTS OR 'BURIED' QUASARS?

I hope that the arguments given above (if nothing else) convince my audience that it is not easy to *unambiguously* discriminate between massive stars and more exotic energy sources in galactic nuclei. It should therefore come as no surprise that there is considerable controversy concerning the energy source in the dust-shrouded class of galaxies discovered by IRAS (cf. Soifer, Houck, and Neugebauer 1987). Everyone agrees that thermal emission from dust grains comprises the great bulk of the bolometric luminosity of such galaxies. However, the heating source for these grains has been variously argued to be a powerful AGN (*e.g.*, Sanders *et al.* 1988b; Becklin and Wynn-Williams 1987), a starburst (*e.g.*, Rieke *et al.* 1985), an old stellar population (Thronson *et al.* 1990), or the kinetic energy of violently-colliding galaxies (Harwit *et al.* 1987). The problem of course is that the only 'windows' through which one can potentially see directly into the region of energy production lie in the hard X-ray, IR, and microwave spectral regions. These are not the optimal places to look for the unambiguous signature of a population of young, massive stars. In what follows I will summarize the evidence as to whether Far-IR Galaxies (FIRG's) are powered by starbursts or by AGN's/quasars. Since my charge was to review the "Starburst-AGN Connection", I will not consider further the interesting alternatives proposed by Harwit *et al.* (1987) or Thronson *et al.* (1990).

4.1. UV, Optical, and Near-IR Continua

The relatively few powerful ($L_{IR} \geq 10^{11}$ $L_\odot$) FIRG's with published UV spectra do show strong C IV and Si IV absorption lines, presumably indicative of a substantial population of hot, young stars (NGC 1068—see the poster paper contributed by Bruhweiler at this conference; NGC 3256—Joseph, Wright, and Prestwich, 1986; NGC 3690—Augarde and Lequeux, 1985 and Sekiguchi and Anderson 1987). There are considerably more IUE data available for lower-IR-power galaxies, especially the Blue Compact Dwarfs: see Leitherer and Lamers (1990) and Thuan's contribution to this volume for a summary of these data and the relevant references.

Armus, Heckman, and Miley (1989) have recently analysed the properties of the optical continua in a sample of powerful FIRG's with 'starburst-type' IR SED's. The photospheric absorption-lines seen in normal (old) stellar populations (*e.g.*, the Mg I $\lambda 5174$ line) are very weak in these galaxies, and the region in which they are weak is spatially-extended in all the cases for which such a measurement is possible. They argue then that the optical continuum is produced by young stars rather than an AGN. The detection of Wolf-Rayet emission lines in some of these galaxies (see also Armus, Heckman, and Miley 1988) and of moderately strong TiO absorption in many others is taken as evidence for a major contribution by massive post-main-sequence stars to the integrated visual light (cf. Bica, Alloin, and Schmidt 1990).

While these data provide fairly direct evidence that massive stars dominate the

UV/optical continuum in FIRG's, they do not prove that these young stars are the major power-source for the far-IR emission. This is because the optical and (especially) the UV continua suffer heavy extinction by dust that is not distributed as a simple uniform screen (*i.e.*, it is probably very patchy and inter-mixed with the stars). It is therefore very difficult to correct for extinction and thereby determine the intrinsic luminosity of the stellar population and compare it to the far-IR luminosity. While we could in principle be seeing an energetically-unimportant starburst surrounding an optically-invisible quasar, Armus, Heckman, and Miley (1989) argue that the luminosities of young stars in the FIRG's in their sample are a significant fraction of the IR luminosity, even if the average extinction suffered by the starlight is as low as that implied by the Balmer decrement ($A_V \approx$ several magnitudes).

Near-IR data are promising, because we can then penetrate through the dust-shroud. However, the available near-IR data have not yielded a definitive answer either. The near-IR colors of the central regions of most (but not all) FIRG's are consistent with a reddened stellar population, with perhaps some contribution from hot dust at $\lambda > 2$ μm (*e.g.*, Carico et al. 1988). The strong CO and H_2O absorption features detected spectroscopically in several well-studied FIRG's confirms that the near-IR continuum must be mostly starlight (cf. Rieke et al. 1985; Lester, Harvey, and Carr 1988). The extinction-corrected near-IR luminosity is much smaller than ($\approx 1\%$ of) the far-IR luminosity in the majority of the FIRG's in Carico et al. (1988). This is true even for many of the 'ultraluminous' ($L_{IR} \geq 10^{12} L_\odot$) FIRG's in Sanders et al. (1988b). Starbursts (which should radiate relatively little of their energy in the near-IR), are consistent with this result, but normal quasars (which radiate roughly equal power per decade of frequency from the X-ray through the IR) are not. Thus, if FIRG's are typically powered by a quasar, the quasar is still almost completely invisible even in the near-IR. Counter-examples—FIRG's with quasar-like near-IR colors—include the well-known peculiar Seyfert 1/quasar Mrk231 and IRAS 05189-2524 (Sanders et al. 1988b). Significantly, these two FIRG's also have much higher near-IR luminosities than typical FIRG's ($L_{NIR} \approx 20\% L_{FIR}$).

While it appears that the near-IR continuum is usually starlight, the nature of these stars remains controversial (as Joseph and Rieke both discuss elsewhere in this volume). The strengths of the near-IR CO and H_2O absorption features are luminosity-sensitive (see the R. Humphries contribution elsewhere in this volume), but various groups have disagreed about whether it is possible to discriminate between ordinary M giants (as expected in a normal galactic nucleus) or M supergiants (as expected in a starburst). Further discussion of this issue can be found in (for example) Rieke et al. (1985) and Lester et al. (1988). Eales et al. (1990) argue that the high near-IR surface brightnesses of the nuclei of FIRG's favor the supergiant/starburst interpretation (see also Devereux 1989).

4.2. Optical and Near-IR Properties of the Emission-Line Gas

The same arguments made above about the optical continuum apply to the use of optical emission-lines as diagnostics of AGN's vs. starbursts: because it is difficult to accurately correct the optical emission-lines for extinction, it is hard to prove that the ionization source for these emission-lines is energetically significant compared to the observed IR continuum luminosity. That is, the gas we see in the optical may be photoionized by hot stars, but an 'invisible' and much more powerful AGN may power most of the IR (or vice versa). However, as I will discuss further in §4.3 below, the FIRG's whose optical emission-lines are AGN-like (starburst-like) also seem to have AGN-like (starburst-like) radio continuum properties (Norris, 1989). This suggests that the optical emission-lines may be a surprisingly reliable guide to the energy source in FIRG's.

Several large optical spectroscopic surveys of FIRG's have reached a qualitatively similar conclusion that the fraction of such galaxies having 'AGN-like' optical emission-line properties increases as a function of IR luminosity (Armus, Heckman, and Miley 1989; Leech et al. 1989; Sanders et al. 1988b). By 'AGN-like' I mean that characteristic emission-line ratios like [OI]6300/Hα or [S II]6716,6731/Hα are significantly larger than in ordinary giant H II regions (cf. Baldwin, Phillips, Terlevich 1981; Veilleux and Osterbrock 1987) and/or that the emission-lines are relatively broad (FWHM > 300 km s^{-1}). At modest IR luminosities ($L_{IR} < 10^{11}$ erg s^{-1}), the majority of FIRG's have optical emission-line properties suggesting that the gas is photoionized by an ensemble of rather normal OB stars.

At the highest luminosities ($L_{IR} > 10^{12}$ $L_\odot$), a majority of FIRG's have AGN-like optical emission-line properties. It is quite interesting however that only a minority of these 'ultraluminous' FIRG's have classical quasar or Seyfert-type spectra. Only 1 of the 10 ultra-luminous FIRG's in the Sanders et al. sample and only 1 of the 15 in the Armus, Heckman, and Miley sample are type 1 Seyferts (have a Broad-Line Region). Only 2 of the 15 in the Armus, Heckman, and Miley sample and 4 of the 10 in the Sanders et al. samples are type 2 Seyferts (have an [O III]5007/Hβ flux ratio greater than four). Instead, the majority of the ultraluminous 'AGN-like' FIRG's have LINER spectra (Low-Ionization Nuclear Emission-line Region—cf. Heckman 1980, 1987b). Low luminosity LINER's are common in the nuclei of bright, nearby galaxies and are generally thought to be low-power analogs to Seyfert nuclei in which gas is photoionized by an extremely dilute, but hard continuum produced by a feeble AGN (cf. Ferland and Netzer 1983; Filippenko 1989b). However, LINER's and old supernova remnants have very similar spectra, and indeed models of shock-heated gas can also reproduce LINER-type spectra (cf. Fosbury et al. 1978; Shull and McKee 1979). Thus, the mere presence of strong LINER-type emission in FIRG's does not—by itself—imply that a powerful AGN must be present.

As an alternative to the AGN interpretation of FIRG's having LINER-type spectra and high velocity gas motions, my colleagues and I have proposed that we are witnessing a galactic-scale 'superwind': a global outflow of shock-heated gas being driven by the kinetic energy supplied by the supernovae and stellar winds in a dust-shrouded central starburst (Heckman, Armus, and Miley 1987,1990; McCarthy, Heckman, and van Breugel, 1987; Fabbiano, Heckman, and Keel 1990; see also Chevalier and Clegg 1985; Tomisaka and Ikeuchi 1988; Fabbiano 1988; Bland and Tully 1988).

Near-IR spectroscopy of FIRG's shows that they are generally strong sources of recombination-line radiation. DePoy (1988) and Ho, Beck, and Turner (1990) find that $L_{IR} \approx 10^{4\pm1}$ $L_{Br\alpha}$ in FIRG's, rather similar to the relation obeyed by normal H II regions. This result is really consistent with either the starburst or AGN interpretation, since it only provides a constraint on the ratio of the ionizing and total luminosity. For example, an AGN with $S_\nu \propto \nu^{-1}$ will have $L_{BOL} \approx 10^4$ $L_{Br\alpha}$ for a Case B, ionization-bounded situation. Arp 220 is the most noteworthy example of a FIRG that is underluminous in Brα ($L_{IR} \approx 10^5$ $L_{Br\alpha}$). While underluminous IR recombination lines are often taken as evidence for an AGN (cf. DePoy 1988; Ho, Beck, and Turner 1990), they can also be readily understood as the by-product of a star-formation rate that declines with time or of an IMF that is deficient in very massive stars (cf. Rieke et al. 1980; 1985; Rieke, this volume). DePoy (1988) also finds that the Brα line in Arp 220 is quite broad (FWHM ≈ 1300 km s^{-1}). This is certainly suggestive of the presence of an AGN, but such a broad line could also arise in a starburst-driven 'superwind', since typical supernova and stellar wind outflow velocities are thousands of km s^{-1}.

There are now about two dozen galactic nuclei in which the [Fe II] line at 1.64 μm and the H$_2$ S(1) vibration-rotation line at 2.12 μm have been detected (e.g., Kawara et al. 1990; Mouri et al. 1989,1990; Moorwood and Oliva 1989). These lines are strongly enhanced relative to Brγ in supernova remnants compared to H II regions, and could perhaps be used to distinguish between the AGN and starburst models for FIRG's. Most of the galactic nuclei (including well-known FIRG's like M82, NGC 253, Arp 220, and NGC 3690) have [Fe II]/Brγ and H$_2$/Brγ ratios that are intermediate between H II regions and SNR's. In contrast, NGC 6240 has exceptionally strong (SNR-like) near-IR H$_2$ and [Fe II] emission-lines. There is also a trend for the nuclei that have the more AGN-like *optical* spectra to have stronger [Fe II] and H$_2$ lines relative to Brγ.

4.3. Mid/Far IR Sizes

Since there is universal agreement that the IR emission is thermal, Stefan-Boltzmann-type arguments give a minimum size for the IR-emitting region that is typically 300–1000 pc for dust temperatures of 30 to 50 K and IR luminosities of 10^{11} to 10^{12} $L_\odot$. Thus, IR sizes significantly larger than these minimum sizes would require a distributed heating source for the dust, favoring a starburst over an AGN. Small IR sizes do not necessarily rule out either model.

Data on the size of the mid/far IR sources in powerful FIRG's ($L > 10^{11}$ $L_\odot$) are lamentably scarce. The IR source in Arp 220 is unresolved, with a maximum size of 700 pc at 10 μm and 20 μm (Becklin and Wynn-Williams 1987) and ≈3 kpc at 50 μm and 80 μm (Joy et al. 1986). On the other hand, the IR source in NGC 3256 is several kpc in extent at λ = 10 μm, and clearly requires an extended heating source (Graham et al. 1987). In NGC 1068 the detailed structure of the mid-IR emission implies that about half the IR luminosity is produced by a several-kpc-scale starburst and the other half by a compact nuclear source (Telesco 1988 and references therein).

4.4. Radio Continuum Properties

The IR luminosity of a galaxy appears to correlate better with the nonthermal radio continuum power than with any other single quantity (e.g., Soifer, Houck, and Neugebauer 1987 and references therein). This relation extends over about three orders-of-magnitude in luminosity, from normal late-type galaxies to 'ultraluminous' FIRG's. While we do not understand the physical basis of this correlation (but cf. Condon and Yin 1990), the very existence of such a good correlation is often taken as evidence that the IR and radio continua have a common power-source, and that this power-source is fundamentally the same in normal galaxies and powerful FIRG's. That is, UV radiation from massive stars heats the dust the produces the IR radiation, and the kinetic energy supplied by these stars is tapped to accelerate relativistic, synchrotron-emitting electrons.

The existence of the IR-radio correlation and its attribution to a central starburst were pointed out by Condon et al. (1982), in an important paper the predates IRAS.

They also showed that the radio sources in a radio-selected sample of disk and peculiar galaxies generally had a diffuse (rather than a jet-like) morphology. They therefore suggested that these radio sources were powered by a starburst rather than by an AGN. Their list of galaxies include such now-famous FIRG's as NGC 3690/IC694, Arp 220 = IC4553, and NGC 6240.

The radio-IR correlation has recently been re-investigated in a most instructive way by Condon et al. (J. Condon—private communication). They have used the VLA to measure fluxes and angular sizes for the nonthermal radio sources in a IR flux-limited sample of galaxies. They segregate the radio sources into two categories based on the angular size of the radio source compared to the minimum possible size of the thermal IR source (see above). The first class of galaxies have radio sources whose sizes are greater-than-or-equal-to the minimum IR size. This class then consists of galaxies in which the radio and IR sources may have similar sizes and arise in the same region. Such galaxies follow a very tight correlation between the radio and IR luminosities. The second class comprises the galaxies in which the radio sources are smaller than the minimum IR source size. These galaxies—in which the radio and IR sources can not arise in the same region—exhibit a much looser radio-IR correlation. Condon et al. then interpret the two respective classes as 'starbursts' and 'monsters'. The 'starburst' class constitutes the clear majority of the population except at the highest IR luminosities ($> 10^{12} L_\odot$).

A related technique has been effectively employed by Norris (1989). He has searched for very compact radio sources in three classes of galaxies: Seyfert galaxies, 'classical' (UV/optical-selected) starburst galaxies, and FIRG's. He finds that such sources are much more common in Seyfert nuclei (17%), than in classical starburst nuclei (4%). The FIRG's are intermediate between the Seyfert and starburst populations, with 9% having a compact radio source. Tellingly, when the FIRG's are segregated—using their optical emission-line spectra—into 'AGN' and 'starburst' classes, Norris finds that compact radio sources are much more common in the former group (45% vs. 0%). He concludes that the radio continuum and optical emission-line data can therefore be used to reliably classify FIRG's as AGN's or starbursts. Carral, Turner, and Ho (1990) have made similar arguments based on high-resolution radio maps of IR-bright spiral galaxies.

The results of these surveys strongly suggest that FIRG's are a mixed bag (some are powered by AGN's and others by starbursts). Several caveats are in order, however. First, the physics that underlies the IR-radio correlation is poorly understood at best. Second, the radio emission is energetically trivial ($L_{Radio} \approx 10^{-5} L_{IR}$). Thus (for example) a 'true believer' in starbursts could argue that while an AGN may well be present in those FIRG's with compact radio sources, the AGN might contribute only a minor fraction of the bolometric luminosity.

4.5. Molecular Gas

For a detailed discussion of the topic of molecular gas in starburst galaxies, I refer the reader to the review elsewhere in this volume by Scoville. In what follows, I will emphasize the material that is most pertinent to the 'starburst vs. quasar' controversy.

Rickard and Harvey (1984) and Young et al. (1984) showed that IR-bright galaxies exhibited a correlation between their IR continuum luminosity and the luminosity of the $J = 1-0$ line of ^{12}CO at $\nu = 115$ GHz (the most readily detectable molecular line). The availability of the massive IRAS data-set and the rapid development in mm-wave instrumental capabilities naturally led to many subsequent investigations of molecular gas in IR-bright galaxies. Single dish surveys of large samples confirmed this correlation, adding many galaxies with very high IR luminosities (*e.g.,* Sanders and Mirabel 1985; Sanders *et al.* 1986) and also encompassing ordinary spiral galaxies (*e.g.,* Kenney and Young 1988; Stark *et al.* 1986). These studies also showed that the correlation was not a linear one: the CO luminosity flattens out at the highest IR luminosities (cf. Young 1987; Tinney *et al.* 1990). The CO emission from IR galaxies has also been mapped with high spatial resolution (several arcsec) using mm-wave interferometers (Scoville *et al.* 1989,1990; Sanders *et al.* 1988a; Meixner *et al.* 1990). These maps show that the bulk of the CO emission originates in the central kpc-scale circum-nuclear region where the IR emission is thought, or in some cases known, to originate (see above).

The authors of most of the papers cited above argue that the $J = 1$–0 ^{12}CO line reliably and quantitatively traces the location of molecular gas: the CO luminosity is directly proportional to the mass of molecular gas and the surface brightness of the CO line emission is directly proportional to the surface mass density of the molecular gas. The CO-IR correlation then arises because the IR luminosity (powered by massive stars) is proportional to the star-formation rate, which is in turn related to the mass of molecular gas. If the $J = 1$–0 ^{12}CO line is a reasonable tracer of the molecular gas in FIRG's, then these data provide strong (though circumstantial) evidence for the starburst model: not only are FIRG's chock full of molecular gas, this gas is confined to the central region of the galaxy and has a surface mass density so high that a starburst seems virtually inevitable (*e.g.,* Larson 1987).

There are several caveats, however. First, the decline in the ratio of CO-to-IR luminosity at high IR luminosities has been cited as evidence that the bulk of the IR emission in the 'ultraluminous' FIRG's is powered by a shrouded AGN rather than a starburst (Sanders *et al.* 1988b). Alternatively, Young (1987) interprets this decline as meaning that the 'efficiency' of star-formation in such galaxies is unusually high (that is, they are converting their molecular gas into stars over an abnormally short time-scale and are hence producing an abnormally large amount of IR radiation per unit mass of molecular gas).

A potentially more fundamental problem is that the luminosity of the $J = 1$–0 CO line may not be proportional to the mass of molecular gas (Maloney and Black 1988 and references therein). Because the emissivity in this line is proportional to the temperature of the molecular gas, we might expect that the CO emission will be the brightest where the heating rate is the greatest. Thus, the central CO sources in FIRG's may be very bright not just because they represent regions of large molecular mass, but also because the molecular gas there has been heated to unusually high temperatures by a powerful energy source (be it an AGN or starburst).

4.6. Hard X-Rays

One of the main themes in §3 was that high energy radiation is a very important discriminant between massive stars and supermassive black holes as the energy source in galactic nuclei. This is particularly true in the case of the IR galaxies because hard X-rays are one of the few types of radiation that ought to be able to penetrate the gas and dust that shroud the energy source in these galaxies.

We know that type 1 Seyfert nuclei and quasars are strong sources of hard X-rays, with $\approx 10\%$ of the bolometric luminosity in the 2 to 10 keV energy band (*e.g.,* Lawrence and Elvis 1982). In contrast, Rieke (1988) showed that 15 high-luminosity IR galaxies were characterized by luminosities in the 2 to 10 keV band that ranged from ≈ 0.01 to $< 2 \times 10^{-4}$ of the IR ($\approx$bolometric) luminosity. These galaxies were thus deficient in hard X-rays by factors of 10 to nearly 1000 relative to quasars or type 1 Seyfert nuclei with the same bolometric luminosities. If the X-ray deficiency is due to absorption, column densities in excess of 10^{24} cm^{-2} are required, corresponding to visual and near-IR (K-band) extinctions of greater than 1000 magnitudes and 100 magnitudes respectively. These values are far higher than the extinction measured or inferred in the IR.

Rieke (1988) therefore argues that high-luminosity IR galaxies are powered by either a starburst or some type of 'X-ray Quiet AGN'. Type 2 Seyfert nuclei are good candidates for the latter, because (as I stressed above) they are both X-ray quiet and commonly believed to be powered by a 'true' AGN and not a starburst. Thus, sorting out the true nature of type 2 Seyfert nuclei is relevant to both the 'AGN's without Black Holes' and 'starburst vs. buried quasar' controversies.

4.7. Space Densities

Sanders et al. (1988b) pointed out that the luminosity functions of IR galaxies and quasars/type 1 Seyfert nuclei are similar to one-another (they have similar space densities at a given bolometric luminosity). They argue that this is unlikely to be a coincidence, and therefore represents evidence that the two phenomena share a common energy source. This is a rather elegant argument that has a ring of truth to it. However, as I emphasized in §2, the luminosity functions of IR galaxies and optically-selected ('Markarian') starburst galaxies are also similar to one-another, so an analogous argument could be made on behalf of the starburst model for IR galaxies.

4.8. Conclusions

Neither side in the debate has been able to provide the knock-out punch, though suggestive evidence exists to support both the 'starburst' and 'buried quasar' models. The fraction of IR-selected galaxies that are powered by a quasar/AGN almost certainly increases as a function of IR luminosity. However, even if we restrict our attention to the relatively rare galaxies with $L_{IR} > 10^{12}$ $L_\odot$, it seems likely that they are a mixed bag of objects united by the presence of a powerful energy source of some kind and loads of dust.

I confess that this is a rather wishy-washy conclusion that leaves me a bit unsatisfied. However, given the similar radiant energy production rate of optically-selected starbursts and quasars in the local universe, perhaps it is inevitable that both processes will make significant contributions to the powerful IR galaxy population.

5. STAR-FORMATION IN ACTIVE VERSUS NORMAL GALAXIES

As I described in §1, even if "classical" AGN's (e.g., Seyfert nuclei, quasars, radio galaxies) are powered primarily by a supermassive black hole, there may still be a strong physical link between such AGN's and the starburst phenomenon. In this portion of my review, I will therefore consider whether these types of AGN's are sited in galaxies having an unusually young stellar population. Such a discussion involves the seemingly impossible task of trying to accurately decompose the properties of active galaxies into a component that is powered by the mythical nuclear 'monster' and a component that is powered by massive stars. In this section, I will take the conservative point-of-view that the stars are innocent until proven guilty (that is, I will attribute everything to the 'monster' that can not be plausibly attributed to a normal population of young stars). A more detailed discussion of some of these issues and data may be found in Heckman (1987a).

5.1. Seyfert Galaxies

Spatially-resolved color maps of samples of Seyfert galaxies show that their off-nuclear colors are similar to the colors of galaxies of moderately early Hubble type (E to Sbc)—cf. Yee (1983), MacKenty (1986). Since Seyferts are usually early-type disk galaxies (e.g., Simkin, Su, and Schwarz 1980), these data suggest that the extra-nuclear regions of Seyferts are rather normal in terms of stellar population. The difficulty of discriminating between massive stars and an AGN as the origin the blue 'featureless' nuclear continuum has already been discussed in §3 above.

In order to use the Balmer emission-line luminosity to estimate the OB star population in Seyferts we must figure out a way to separate this gas from the gas that has been photoionized by the Seyfert nucleus. As is well known observationally and well understood theoretically, the relative strengths of the prominent optical emission-lines differ significantly between gas in the nuclei of active galaxies and ordinary giant H II regions in galaxies (e.g., Baldwin, Phillips, and Terlevich 1981; Veilleux and Osterbrock 1987). Therefore, by imaging Seyfert galaxies in the light of several different emission-lines like $H\alpha$, [O III]5007, [S II]6724, and [O I]6300, we can use two-dimensional line ratio maps to classify the emission-line gas into regions photoionized by massive stars and by the AGN.

While the ionized gas in several Seyfert galaxies has been decomposed into an 'AGN' and a 'starburst' component (e.g., Wilson et al. 1986; Baldwin, Wilson, and Whittle 1987; Keel 1987; Shields and Filippenko 1990), no systematic, fully quantitative and rigorous comparison of Seyfert and normal galaxies has been made. The best attempt so far is that of Pogge (1989), who imaged Seyferts and non-Seyfert spirals in $H\alpha$+[N II]6548,6584 and [O III]5007. Based on inspection of these images, he concluded that the type 2 Seyfert galaxies had brighter extra-nuclear emission-line gas than the type 1 Seyferts, but that there was no strong, obvious excess of a star-formation component in the Seyferts vs. the non-Seyferts. Recently, we (H. Kirkpatrick, A. Wilson, and I) have embarked on a similar survey that will refine Pogge's approach in several ways. First, we have selected the 43 known Seyfert galaxies in the *Revised Shapley Ames Catalog* (*'RSA'*) and a control sample of 43 *RSA* galaxies that match the Seyferts in Hubble type, absolute magnitude, and redshift. Given the strong dependence of the $H\alpha$ properties of galaxies on Hubble type, a good match in Seyfert and control samples is critical. Second, we are imaging both sets of galaxies in the [S II]6724 doublet (as well as $H\alpha$+[N II] and [O III]). This greatly improves our ability to discriminate between gas excited by the AGN and OB stars.

Some fraction of the nonthermal radio continuum from Seyfert galaxies is powered by the extreme Population I component of the galaxy (as in the starburst galaxies) and some is undoubtedly powered by the Seyfert nucleus. Disentangling the two components has proved difficult and has engendered considerable controversy (e.g., Condon et al. 1982; Ulvestad 1982; Wilson 1987; Heckman et al. 1989). For example, the nonthermal radio power of non-Seyfert galaxies correlates well with the strength of such indicators of extreme Population I as the luminosity of the 115 GHz CO line (*e.g.,* Rickard, Turner, and Palmer 1985). Heckman *et al.* (1989) recently showed that while Seyfert galaxies also exhibit a significant correlation between the luminosity of the CO $J = 1$–0 115 GHz line and the nonthermal radio power, they are systematically stronger nonthermal radio sources than non-Seyferts with the same CO luminosity. This supports the idea of a composite origin for the radio emission in Seyferts. One possible way to discriminate between the AGN and 'starburst' components may be the morphology of the radio emission: attribute those sources with 'linear' (jet-like, double, or triple) morphologies to the AGN and those with diffuse morphologies to regions of star-formation. This interpretation is bolstered by Wilson's (1987) discovery that Seyferts with diffuse radio emission exhibit optical and infrared evidence for extra-nuclear star-formation. We (Kirkpatrick, Wilson, and I) are following up on this idea using high and low-resolution VLA data on the *RSA* samples to compare the power of the diffuse radio emission in the Seyferts to normal spiral galaxies.

Seyfert galaxies have long been known as strong infrared sources (*e.g.,* Rieke and Lebofsky 1979). More recently, IRAS has provided a large and homogeneous body of data on the global mid and far-IR properties of Seyfert and normal galaxies. While the IR emission of normal galaxies is widely used as a measure of the star-formation rate (*e.g.,* Soifer, Houck, and Neugebauer 1987; Telesco 1988; Hunter *et al.* 1986), the origin of this emission in Seyferts is still controversial. For example, Rodriguez-Espinosa, Rudy, and Jones (1987) analysed a sample of 96 Seyferts. They concluded that the IR was powered by massive stars, and that Seyferts were characterized by abnormally high star-formation rates compared to normal galaxies. In contrast, Edelson and Malkan (1986) argued that the IR continuum in Seyferts is primarily produced by the AGN (either as nonthermal radiation or re-radiated thermal emission). Comparing the IR spectral energy distributions of Seyferts galaxies to those of quasars, Markarian starburst galaxies, and normal galaxies (*e.g.,* Miley, Neugebauer, and Soifer 1985; Ward *et al.* 1987; Rowan-Robinson 1987) shows that it is likely that the IR has a composite origin in Seyferts with the relative strength of the AGN, 'starburst', and 'normal galaxy' component varying from galaxy-to-galaxy. Indeed, Wilson (1987) showed that the Seyferts with 'starburst' type IR spectral energy distributions exhibit optical and radio evidence for circum-nuclear star-formation. It is interesting that the 'starburst' type IR component is especially prominent in the type 2 Seyferts while the AGN component is more pronounced in the type 1 Seyferts.

We (Heckman *et al.* 1989) recently published the results of a large survey of CO 115GHz line emission from Seyfert galaxies. We found that the Seyferts obeyed the same relationship between the strength of the CO line and far-IR (40–120 μm) continuum emission as normal and 'starburst' galaxies. We concluded therefore that the bulk of the far-IR emission in our sample of Seyferts (which were typically nearby galaxies with rather low-power AGN's) was probably powered by the extreme Population I component of the galaxy. Restricting our attention to the optically-selected sample of *RSA* Seyferts, we found that the type 2 Seyferts were systematically stronger CO and far-IR sources (by factors of several) than either the type 1 Seyferts or normal galaxies of the same Hubble type and absolute magnitude. We suggested therefore that the type 2 Seyferts in the *RSA* may have abnormally high star-formation rates. The survey with Kirkpatrick and Wilson will be able to assess this claim with independent data sets (optical and radio). It will also be important to compare the properties of the *nuclear* molecular gas in Seyfert and normal galaxies. Meixner *et al.* (1990) found that some (but not all) *RSA* Seyferts have very bright nuclear 115 GHz CO sources with sizes of a few hundred pc and estimated H_2 masses of $\approx 10^9$ $M_\odot$ (see also Taniguchi et al. 1990). A much more systematic investigation along these lines will soon be feasible with the new generation of mm-wave interferometers.

5.2. Radio Galaxies

Powerful radio galaxies at low redshifts ($z <$ few tenths) can be conveniently divided into two categories (Hine and Longair 1979). The 'Class A' radio galaxies have high-ionization, strong (Seyfert-like) optical emission-lines and usually have 'classical double' radio morphologies dominated by bright outer 'hotspots' (e.g., Fanaroff and Riley 1974; Miley 1980). They dominate the population at radio luminosities above $\approx 10^{42}$ erg s^{-1}. The 'Class B' radio galaxies have low-ionization, weak (LINER) optical emission-lines and usually have 'twin-jet-dominated' radio morphologies. They dominate the population with radio luminosities of $\approx 10^{40}$ to 10^{42} erg s^{-1}.

There are pronounced systematic differences between the optical properties of these two types of radio galaxy (Smith and Heckman 1990). The Class B radio galaxies have colors, morphologies, and dynamics that are normal for giant elliptical galaxies. There is no evidence that they have any surplus of massive stars. In contrast, about half of the Class A radio galaxies have *off-nuclear* colors that are significantly bluer than normal giant elliptical galaxies (they have rest-frame (B-V) < 0.8). Color maps of several nearby Class A radio galaxies show strong and complex spatial variations in color, strongly suggestive of the competing effects of dust and young stars (Smith and Heckman 1989). Golembek, Miley, and Neugebauer (1988) showed that the Class A radio galaxies are strong IR sources as well (L_{IR} typically 10^{10} to 10^{11} $L_\odot$). It is not clear whether the IR is powered primarily by massive stars or by an AGN however. It is interesting in this regard that several IR-powerful radio galaxies have been detected as strong 115 GHz CO sources (Mirabel, Sanders, and Kazes 1989). About half of the Class A radio galaxies have morphological peculiarities suggestive of a recent merger or interaction with a disk galaxy (Heckman et al. 1986; Hutchings 1987; Smith and Heckman 1989). It seems plausible that many of these galaxies are examples of an AGN-plus-starburst triggered by a galaxy interaction.

High-redshift ($z > 1$) radio galaxies seem to be a qualitatively different phenomenon (Chambers 1989; McCarthy 1988). They have enormous UV luminosities ($\approx 10^{12}$ $L_\odot$) and UV-optical spectral energy distributions that can be matched by either a 'post-starburst' population with an age of a few $\times 10^8$ years and a mass of several $\times 10^{11}$ $M_\odot$ (Chambers and Charlot 1990) or by a composite of a current starburst (age $\approx 10^8$ years and star-formation rate of a few hundred $M_\odot$ per year) and an old ($> 10^9$ years) population that provides most of the mass (Lilly 1988). They also have 100 kpc-scale gaseous nebulae with Lyα luminosities of $\approx 10^{44}$ erg s^{-1}, though the strong C IV 1549 and He II 1640 emission-lines in at least some of the nebulae imply that the gas is more likely to be photoionized by an AGN than by massive stars (McCarthy et al. 1990). Because both the continuum and line-emitting structures in these radio galaxies are well-aligned with the radio source axis, it has been argued that they are examples of titanic starbursts that have been initiated by the effect of the radio plasma on the interstellar or circum-galactic medium (Begelman and Cioffi 1989; Daly 1990; DeYoung 1989; Rees 1989).

A radically different possibility is that these are not galaxies at all, but are in effect giant bi-polar reflection nebulae. That is, the continuum is actually produced as light from a powerful quasar that is hidden from our direct view escapes along the radio axis and is then scattered off dust-grains or electrons into our line-of-sight (*e.g.,* Fabian 1989). This model gains credence from the detection of qualitatively similar phenomenon in the nuclei of some type 2 Seyfert galaxies (Antonucci and Miller 1985; Miller and Goodrich 1990) and from the discovery that the continuum light in the high-z radio galaxy 3C368 is polarized (di Serego-Alighieri *et al.* 1989; Scarrott, Rolph, and Tadhunter 1990). Chambers and McCarthy (1990) have published the spectrum of the continuum light from a composite of high-z radio galaxies, and have tentatively detected the absorption-line signature of massive stars. Certainly, the composite spectrum is very different from that of a quasar. If this is scattered quasar light, the scattering medium must be hot electrons (thereby washing out the quasar emission-lines via thermal-Doppler broadening by the electrons).

5.3. Quasars

The practical difficulty in searching for evidence for massive stars in the host galaxies of quasars can not be over-emphasized. There have been several papers that have suggested that the host galaxies of radio-loud, low-redshift quasars are significantly bluer than giant elliptical galaxies (Boroson and Oke 1984; Boroson, Oke, and Persson 1985; Hickson and Hutchings 1987; Hutchings 1987). An important early paper by Miller (1981) showed that the absorption-line signature of an old stellar population was too weak in the spectra of radio-loud quasars for the host galaxies to be normal giant elliptical galaxies. Since the host galaxies of radio-loud quasars have absolute visual magnitudes like giant elliptical galaxies (see Smith *et al.* 1986 and references therein), consistency with Miller's result implies that they have a significantly younger stellar population than a normal giant elliptical galaxy. In contrast, there is as yet no clear indication that the host galaxies of radio-quiet, low-redshift quasars are exceptionally blue (Malkan 1984; Malkan, Margon, and Chanon 1984; Boroson and Oke 1984; Boroson, Persson, and Oke 1985).

Quasars are generically strong sources of infrared radiation (*e.g.,* Neugebauer *et al.* 1986). Sanders *et al.* (1990) favor a thermal (dust re-radiation) origin for the IR, but the ultimate energy source is presumed to be the UV and optical radiation of the quasar itself rather than massive stars in the host galaxy. It is interesting that several low-redshift quasars have been detected as strong 115 GHz CO sources, and seem to roughly obey the CO vs. IR luminosity correlation defined by normal and starburst galaxies (Sanders, Scoville, and Soifer 1988; Barvainis, Alloin, and Antonucci 1989). Within the context of the standard interpretation of this correlation (see §4.5), this result would imply very high star-formation rates in these quasars.

Not surprisingly, our knowledge about the starburst-AGN connection is most incomplete for high-redshift quasars. My colleagues and I have recently obtained deep optical images of 18 radio-loud quasars at $z > 2$ (Heckman *et al.* 1990), and have just started a program to obtain similar data on high-redshift radio-quiet quasars. In 15 of the 18 cases the Lyα emission can be spatially-resolved: there is a central unresolved quasar that contributes typically 80–90% of the Lyα luminosity surrounded by a 100 kpc-scale nebula with a Lyα luminosity of $\approx 10^{44}$ erg s^{-1}. The UV continuum (rest-wavelength ≈ 1200 Å) is spatially-resolved in about half the quasars. This UV 'fuzz' has a luminosity (λP_λ at 1200 Å) of a few $\times 10^{45}$ erg s^{-1}, typically 10% that of the unresolved quasar-proper. If this UV fuzz is starlight, a star-formation rate of several hundred $M_\odot$ per year is implied. Spectra of the nebulosity show no emission-lines stronger than 5–10% of Lyα to be present. This is consistent with the nebulae being photoionized by either the quasar or the massive stars that may power the UV fuzz (though the former dominates the total UV light). Overall, the emission-line and continuum properties of the spatially-resolved structures around these high-redshift, radio-loud quasars are quite similar to the high-redshift radio galaxies discussed in §5.2 above. We *may* be witnessing the creation of massive galaxies, but we can not yet exclude the possibility that we are observing giant reflection nebulae or something even more exotic.

I would like to conclude on an intriguing, but speculative note. Turnshek (1988) has argued that the 'Broad-Absorption-Lines' (BAL's)—seen in about 10% of all high-redshift radio-quiet quasars—are produced by matter with highly super-solar metal abundances. While the BAL clouds may be accelerated up to the observed velocities of $> 10^4$ km s^{-1} by the quasar, Turnshek speculates that the source of the highly metal-rich BAL clouds may be supernovae occurring close in to the quasar. Kwan (1990) has recently challenged Turnshek's inference of very high metal abundances, however. In terms of the 'starburst-AGN connection', it is amusing (though possibly irrelevant) that the spectra of BAL quasars bear a superficial resemblance to the spectra of massive stars undergoing mass-loss!

I would like to thank Mike Fall, Julian Krolik, Colin Norman, and Meg Urry for many stimulating conversations, and especially for not hiding when I wandered into their offices with a crazy gleam in my eyes and a suspicion that everything I knew was wrong!

REFERENCES

Antonucci, R., and Miller, J. 1985, *Ap. J.*, **297**, 621.
Armus, L., Heckman, T., and Miley, G. 1988, *Ap. J. (Letters)*, **326**, L45.
_____. *Ap. J.*, **347**, 727.
Augarde, R., and Lequeux, J. 1985, *Astr. Ap.*, **147**, 273.
Baldwin, J., Phillips, M., and Terlevich, R. 1981, *Pub. A.S.P.*, **93**, 5.
Baldwin, J., Wilson, A., and Whittle, M. 1987, *Ap. J.*, **319**, 105.
Barvainis, R., Alloin, D., and Antonucci, R. 1989, *Ap. J. (Letters)*, **337**, L69.
Bignami, G. and Mereghetti, S. 1989, in *Gamma-Ray Observatory Science Workshop*, ed. W. N. Johnson (NRL: Washington, DC).
Becklin, G., and Wynn-Williams, G. 1987, in *Star-Formation in Galaxies*, ed. C. Lonsdale-Persson (NASA Conference Pub. 2466), p.643.
Begelman, M., and Cioffi, D. 1989, *Ap. J. (Letters)*, **345**, L21.
Bica, E., Alloin, D., and Scmidt, A. 1990, *M.N.R.A.S.*, **242**, 241.
Bland, J., and Tully, R. B. 1988, *Nature*, **334**, 43.
Boroson, T., and Oke, J. B. 1984, *Ap. J.*, **281**, 535.
Boroson, T, Persson, S. E., and Oke, J. B. 1985, *Ap. J.*, **293**, 120.
Bruhweiler, F. 1990, private communication.
Carico, D., Sanders, D., Soifer, B., Elias, J., Matthews, K., and Neugebauer, G. 1988, *A. J.*, **95**, 356.
Carral, P., Turner, J., and Ho, P. 1990, *Ap. J.*, **362**, 434.
Chambers, K. 1989, Ph.D. thesis, Johns Hopkins University.
Chambers, K., and Charlot, S. 1990, *Ap. J. (Letters)*, **348**, L1.
Chambers, K., and McCarthy, P. 1990, *Ap. J. (Letters)*, **354**, L9.
Chevalier, R., and Clegg, A. 1985, *Nature*, **317**, 44.

Condon, J., and Yin, Q. 1990, *Ap. J.*, **357**, 97.
Condon, J., Condon, M., Gisler, G, and Puscell, J. 1982, *Ap. J.*, **252**, 102.
Condon, J. 1990, private communication.
Cowie, L. 1988, in *The Post-Recombination Universe*, ed. N. Kaiser and A. Lasenby (NATO Advanced Science Institute Series).
Daly, R. 1990, *Ap. J.*, **355**, 416.
Davidson, K., and Kinman, T. 1982, *Pub. A.S.P.*, **94**, 634.
DeGrijp, R., Miley, G., and Keel, W. 1990, in preparation.
DePoy, D. 1988, Ph.D. Thesis, University of Hawaii.
Devereux, N. 1989, *Ap. J.*, **346**, 126.
di Serego-Alighieri, S., Fosbury, R., Quinn, P., and Tadhunter, C. 1989, *Nature*, **341**, 307.
DeYoung, D. 1981, *Nature*, **293**, 43.
_____. 1989, *Ap. J. (Letters)*, **342**, L59.
Dopita, M., Lozinskaya, T., MacGregor, P., Rawlings, S. 1990, *Ap. J.*, **351**, 563.
Eales, S., Becklin, E., Hodapp, K., Simons, D., and Wynn-Williams 1990, *Ap. J.*, in press.
Ebstein, S., Carleton, N., and Papaliolios, C. 1989, *Ap. J.*, **336**, 103.
Edelson, R., and Malkan, M. 1986, *Ap. J.*, **308**, 59.
Fabbiano, G. 1988, *Ap. J.*, **330**, 672.
Fabbiano, G., Heckman, T., and Keel, W. 1990, *Ap. J.*, **355**, 442.
Fabian, A. 1989, *M.N.R.A.S.*, **238**, 41P.
Fanaroff, B., and Riley, J. 1974, *M.N.R.A.S.*, **167**, 31P.
Ferland, G., and Netzer, H. 1983, *Ap. J.*, **264**, 105.
Filippenko, A. 1989a, *A. J.*, **97**, 726.
_____. 1989b, in *Active Galactic Nuclei* (IAU Symp. 134), ed. D. Osterbrock and J. Miller (Kluwer: Dordrecht), p. 495.
Fosbury, R., Mebold, U., Goss, W. M., and Dopita, M. 1978, *M.N.R.A.S.*, **183**, 549.
Garnett, D., Kennicutt, R., Chu, Y., and Skillman, E. 1990, preprint.
Golembek, D., Miley, G., and Neugebauer, G. 1988, *A. J.*, **95**, 26.
Graham, J., Wright, G., Joseph, R., Frogel, J., Phillips, M., and Meikle, W. 1987, in *Star-Formation in Galaxies*, ed. C. Lonsdale-Persson (NASA Conference Pub. 2466), p. 517.
Harwit, M., Houck, J., Soifer, B. T., and Palumbo, G. 1987, *Ap. J.*, **315**, 28.
Heckman, T. 1980, *Astr. Ap.*, **87**, 152.
_____. 1987a, in *Starbursts and Galaxy Evolution*, ed. T Thuan, T. Montmerle, and J. Tran Thahn Van (Editions Frontieres: Paris), p. 467.
_____. 1987b, in *Observational Evidence for Activity in Galaxies* (IAU Symp. 121), ed. E. Khachikian, J. Melnick, and K. Fricke (Reidel: Dordrecht).
_____. 1990, in *Paired and Interacting Galaxies*, ed. J. Sulentic and W. Keel (in press).
Heckman, T., Miley, G., and Green, R. 1984, *Ap. J.*, **281**, 525.
Heckman, T., Smith, E., Baum, S., van Breugel, W., Miley, G., Illingworth, G., Bothun, G., and Balick, B. 1986, *Ap. J.*, **311**, 526.
Heckman, T., Armus, L., and Miley, G. 1987, *A. J.*, **93**, 276.
Heckman, T., Blitz, L., Wilson. A., Armus, L., and Miley, G. 1989, *Ap. J.*, **342**, 735.
Heckman, T., Lehnert, M., van Breugel, W., and Miley, G. 1990, *Ap. J.*, in press.

Heckman, T., Armus, L., and Miley, G. 1990, *Ap. J. Suppl.*, in press.
Ho, P., Beck, S., and Turner, J. 1990, *Ap. J.*, **349**, 57.
Hooimeyer, J. 1990, Ph.D. Thesis, Leiden University.
Hickson, P., and Hutchings, J. 1987, *Ap. J.*, **312**, 518.
Hine, R., and Longair, M. 1979, *M.N.R.A.S.*, **188**, 111.
Humphries, R. 1990, this volume.
Hunter, D., Gillet, F., Gallagher, J., Rice, W., and Low, F. 1986, *Ap. J.*, **303**, 171.
Hutchings, J. 1987, *Ap. J.*, **320**, 122.
Ikeuchi, S. 1981, *P.A.S.J.*, **33**, 211.
Joseph, R., Wright, G., and Prestwich, A. 1986, in *New Insights in Astrophysics*, ed. E. Rolfe (ESA SP-263), p. 597.
Joseph, R. 1990, this volume.
Joy, M., Lester, D., Harvey, P., and French, M. 1986, *Ap. J.*, **307**, 110.
Kawara, K., Nishida, M., and Gregory, B. 1990, *Ap. J.*, **352**, 433.
Keel, W. 1987, in *Star-Formation in Galaxies*, ed. C. Lonsdale-Persson (NASA Conference Pub. 2466), p. 661.
Kenney, J., and Young, J. 1988, *Ap. J. Suppl.*, **66**, 261.
Kennicutt, R., this volume.
Kent, S. 1988, *A. J.*, **96**, 514.
Koyama, K. 1989, in *Active Galactic Nuclei* (IAU Symp. #134), ed. D. Osterbrock and J. Miller (Kluwer: Dordrecht), p. 167.
Krolik, J., and Begelman, M. 1988, *Ap. J.*, **309**, 702.
Kwan, J. 1990, *Ap. J.*, **353**, 123.
Lamb, S., Bushouse, H., and Towns, J. 1990, preprint.
Larson, R. 1987, in *Starbursts and Galaxy Evolution*, ed. T Thuan, T. Montmerle, and J. Tran Thahn Van (Editions Frontieres: Paris), p. 467.
Lawrence, A., and Elvis, M. 1982, *Ap. J.*, **256**, 410.
Leech, K., Penston, M., Terlevich, R., Lawrence, A., Rowan-Robinson, M., and Crawford, J. 1989, *M.N.R.A.S.*, **240**, 349.
Lehnert, M., Heckman, T., Miley, G., and van Breugel, W. 1990, submitted to *Ap. J. (Letters)*.
Leitherer, C., and Lamers, H. 1990, submitted to *Ap. J.*
Lester, D., Harvey, P., and Carr, J. 1988, *Ap. J.*, **329**, 641.
Lilly, S. 1988, *Ap. J.*, **333**, 161.
MacKenty, J. 1987, PhD. Thesis, University of Hawaii.
Malkan, M. 1984, *Ap. J.*, **287**, 555.
Malkan, M., Margon, B., and Chanon, G. 1984, *Ap. J.*, **280**, 66.
Maloney, P., and Black, J. 1988, *Ap. J.*, **325**, 389.
Matsuaoka, M., Piro, L., Yamauchi, M., and Murakami, T. 1990, *Ap. J.*, **361**, 440.
McCarthy, P. 1988, Ph.D. Thesis, University of California, Berkeley.
McCarthy, P., Heckman, T., and van Breugel, W. 1987, *A. J.*, **93**, 264.
McCarthy, P., van Breugel, W., Spinrad, H., and Djorgovski, S. 1990, *Ap. J.*, in press.
Meixner, M., Blitz, L., Puchalsky, R., Wright, M., and Heckman, T. 1990, *Ap. J.*, **354**, 158.
Miley, G. 1980, *Ann. Rev. Astr. Ap.*, **18**, 165.
Miley, G., Neugebauer, G., and Soifer, B. 1985, *Ap. J. (Letters)*, **293**, L11.
Miller, J. 1981, *P.A.S.P.*, **93**, 681.

Miller, J., and Goodrich, R. 1990, *Ap. J.*, **355**, 456.
Mirabel, I.F., Sanders, D., and Kazes, I. 1989, *Ap. J. (Letters)*, **340**, L9.
Moorwood, A., and Olivia, E. 1989, in *Active Galactic Nuclei* (IAU Symp. #134), ed. D. Osterbrock and J. Miller (Kluwer: Dordrecht), p. 365.
Mouri, H., Taniguchi, Y., Kawara, K., and Nishida, M. 1989, *Ap. J. (Letters)*, **346**, L73.
Mouri, H., Nishida, M., Taniguchi, Y., and Kawara, K. 1990, *Ap. J.*, **360**, 55.
Neugebauer, G., Miley, G., Soifer, B. T., and Clegg, P. 1986, *Ap. J.*, **308**, 815.
Neugebauer, G., Graham, J., Soifer, B. T., and Matthews, K. 1990, *A. J.*, **99**, 1456.
Norman, C., and Scoville, N. 1988, *Ap. J.*, **332**, 124.
Norris, R. 1989, in *Active Galactic Nuclei* (IAU Symp. #134), ed. D. Osterbrock and J. Miller (Kluwer: Dordrecht), p. 379.
Osterbrock, D. 1989, *Astrophysics of Gaseous Nebulae and Active Galactic Nuclei* (University Science Books: Mill Valley, CA).
Padovani, P., Burg, R., and Edelson, R. 1990, *Ap. J.*, **353**, 438.
Pagel, B., and Edmunds, M. 1981, *Ann. Rev. Astr. Ap.*, **19**, 77.
Pakull, M. 1990, preprint.
Persson, S. E. 1988, *Ap. J.*, **330**, 751.
Pogge, R. 1989, *Ap. J.*, **345**, 730.
Rees, M. 1984, *Ann. Rev. Astr. Ap.*, **22**, 471.
_____. 1989, *M.N.R.A.S.*, **239**, 1P.
Rickard, L., and Harvey, P. 1984, *A. J.*, **89**, 1520.
Rickard, L., Turner, B., and Palmer, P. 1985, *A. J.*, **90**, 1175.
Rieke, G., and Lebofsky, M. 1979, *Ann. Rev. Astr. Ap.*, **17**, 477.
Rieke, G., Lebofsky, M., Thompson, R., Low, F., and Tokunaga, A. 1980, *Ap. J.*, **238**, 24.
Rieke, G., Cutri, R., Black, J., Kailey, W., McAlary, C., Lebofsky, M., and Elston, R. 1985, *Ap. J.*, **290**, 116.
Rieke, G. 1988, *Ap. J. (Letters)*, **331**, L5.
_____. this volume.
Rodriguez-Espinosa, J., Rudy, R., and Jones, B. 1987, *Ap. J.*, **312**, 555.
Rothschild, R., Mushotzky, R., Baity, W., Gruber, D., Matteson, J., and Peterson, L. 1983, *Ap. J.*, **269**, 423.
Rowan-Robinson, M. 1987, in *Starbursts and Galaxy Evolution*, ed. T. Thuan, T. Montmerle, and J. Tran Thahn Van (Editions Frontieres: Paris), p. 235.
Sanders, D., and Mirabel, I. 1985, *Ap. J. (Letters)*, **298**, L31.
Sanders, D., Scoville, N., Young, J., Soifer, B. T., Schloerb, F., Rice, W., and Danielson, G. 1986, *Ap. J. (Letters)*, **305**, L45.
Sanders, D., Scoville, N., Sargent, A., and Soifer, B. T. 1988a, *Ap. J. (Letters)*, **324**, L55.
Sanders, D., Soifer, B. T., Elias, J., Madore, B., Matthews, K., Neugebauer, G., and Scoville, N. 1988b, *Ap. J.*, **325**, 74.
Sanders, D., Scoville, N., and Soifer, B. 1988, *Ap. J. (Letters)*, **335**, L1.
Sanders, D., Phinney, E. S., Neugebauer, G., Soifer, B. T., and Matthews, K. 1989, *Ap. J.*, **347**, 29.
Sanders, R., and Bania, T. 1976, *Ap. J.*, **204**, 341.
Scarrott, S., Rolph, C., and Tadhunter, C. 1990, *M.N.R.A.S.*, **243**, 5P.

Scoville, N., Sanders, D., Sargent, A., Soifer, B. T., and Tinney, C. 1989, *Ap. J. (Letters)*, **345**, L25.
Scoville, N., Sargent, A., Sanders, D., and Soifer, B. T., 1990, *Ap. J. (Letters)*, in press.
Scoville, N. 1990, this volume.
Sekiguchi, K., and Anderson, K. S. 1987, *A. J.*, **94**, 644.
Shields, J., and Filippenko, A. 1990, *A. J.*, **100**, 1034.
Shull, J. M., and McKee, C. 1979, *Ap. J.*, **227**, 122.
Simkin, S., Su, H., and Schwarz, M. 1980, *Ap. J.*, **237**, 404.
Smith, E., Heckman, T., Bothun, G., Romanishin, W., and B. Balick 1986, *Ap. J.*, **306**, 64.
Smith, E., and Heckman, T. 1989, *Ap. J.*, **341**, 658.
_____. 1990, in *Dynamics and Interactions of Galaxies*, ed. R. Wielen (Springer-Verlag: Berlin), p. 450.
Soifer, B. T., Sanders, D., Madore, B., Neugebauer, G., Danielson, G., Elias, J., Lonsdale, C., and Rice, W. 1987, *Ap. J.*, **320**, 238.
Soifer, B. T., Houck, J., and Neugebauer, G. 1987, *Ann. Rev. Astr. Ap.*, **25**, 187.
Soltan, A. 1982, *M.N.R.A.S.*, **200**, 115.
Stark, A., Knapp, G., Bally, J., Wilson, R., Penzias, A., Rowe, H. 1986, *Ap. J.*, **310**, 660.
Taniguchi, Y., Kameya, O., Nakai, N., and Kawara, K. 1990, *Ap. J.*, **358**, 132.
Telesco, C., 1988, *Ann. Rev. Astr. Ap.*, **26**, 343.
Terlevich, R. 1989, in *Evolutionary Phenomena in Galaxies*, ed. J. Beckman, and B. Pagel (Cambridge Univ. Press).
_____. 1990a, in *Windows on Galaxies*, ed. G. Fabbiano, J. Gallagher, and A. Renzini (Kluwer: Dordrecht), p. 87.
_____. 1990b, in *Structure and Dynamics of the Interstellar Medium*, ed. G. Tenorio-Tagle, M. Moles, and J. Melnick (Springer-Verlag: Berlin).
Terlevich, R., and Melnick, J. 1985, *M.N.R.A.S.*, **213**, 841.
_____. 1987, in *Starbursts and Galaxy Evolution*, ed. T. Thuan, T. Montmerle, and J. Tran Thahn Van (Editions Frontieres: Paris), p. 393.
_____. 1988, *Nature*, **333**, 239.
Terlevich, R., Melnick, J., and Moles, M. 1987, in *Observational Evidence for Activity in Galaxies*, ed. E. Khachikian, J. Melnick, and K. Fricke (Reidel: Dordrecht).
Terlevich, E., Diaz, A., and Terlevich, R. 1990, *M.N.R.A.S.*, **242**, 271.
Thronson, H., Majewski, S., Descartes, L., and Hereld, M. 1990, *Ap. J.*, in press.
Tinney, C., Scoville, N., Sanders, D., and Soifer, B. T. 1990, *Ap. J.*, **362**, 473.
Tomisaka, K., and Ikeuchi, S. 1988, *Ap. J.*, **330**, 695.
Turnshek, D. 1988, in *Quasar Absorption-Lines: Probing the Universe*, ed. C. Blades, D. Turnshek, and C. Norman (Cambridge Univ. Press: Cambridge).
Tyson, A., and Seitzer, P. 1988, *Ap. J.*, **335**, 552.
Ulvestad, J. 1982, *Ap. J.*, **259**, 96.
Urry, C. M. 1989, in *Multiwavelength Astrophysics*, ed. F. Cordova (Cambridge Univ. Press: Cambridge).
Veilleux, S., and Osterbrock, D. 1987, *Ap. J. Suppl.*, **63**, 295.
Ward, M., Elvis, M., Fabbiano, G., Carleton, N., Willner, S., and Lawrence, A. 1987, *Ap. J.*, **315**, 74.
Weedman, D. 1983, *Ap. J.*, **266**, 479.

Weedman, D., Feldman, F., Balzano, V., Ramsey, L., Sramek, R., and Wu, C. 1981, *Ap. J.*, **248**, 105.
Wilson, A. 1987, in *Star-Formation in Galaxies*, ed. C. Lonsdale-Persson (NASA Conference Pub. 2466), p. 675.
Wilson, A., Baldwin, J., Sun, S., and Wright, A. 1986, *Ap. J.*, **310**, 121.
Woosley, S., and Weaver, 1986, *Ann. Rev. Astr. Ap.*, **24**, 205.
Yee, H. 1983, *Ap. J.*, **272**, 473.
Young, J., Kenney, J., Lord, S., and Schloerb, F. 1984, *Ap. J. (Letters)*, **287**, L65.
Young, J. 1987, in *Star-Formation in Galaxies*, ed. C. Lonsdale-Persson (NASA Conference Pub. 2466), p. 197.

DISCUSSION

J. Bland Hawthorn: I agree that the proximity of NGC 1068 makes it an ideal laboratory to study the starburst-AGN connection. But as I showed in my talk here in January, on different scales, you see Seyfert 1, Seyfert 2, shock, LINER and starburst activity even though in the integrated spectrum, the latter dominates.

Heckman: I agree. It may mean that to sort out the starburst-AGN connection we need to make detailed multi-wavelength "decompositions" of the starburst and AGN components in large and statistically well-chosen samples of active and normal galaxies.

J. Bland Hawthorn: Can Bruhweiler comment on whether or not there are Wolf-Rayet stars in NGC 1068, as Schneider *et al.* claimed a decade ago?

F. Bruhweiler: The NGC 1068 starburst knots do not show UV features of W-R stars. The closest thing is the spectrum of knots 4, 5, and NE of the nucleus. This is due to the superposition of the very high ionization extended emission line region upon the stellar spectra. The P Cygni features of knot 1 (the brightest knot) are due to a very luminous population representative of a B0 Ia spectrum. The spectrum of knot 1 agrees amazingly well with that of a galactic B0 Ia supergiant.

R. Kennicutt: It seems to me on naive theoretical grounds that starbursts and AGNs must go together at some level. If you want to explain everything as a supermassive black hole, you have to cycle enormous amounts of gas from interstellar densities in to the Schwarzchild radius. It's hard to imagine you won't form a lot of stars along the way. On the other hand, if the "warmer" picture were correct, you'd have this enormous number of stellar remnants within a few parsecs of the nucleus. Surely you must form a black hole as a result.

Heckman: I agree!

N. Walborn: A recent paper by Dopita *et al.* in *Ap. J.*, **351**, 563 (1990) discusses a remarkable WO object in the Galaxy which has an effective temperature near 100,000

K and is interpreted as the highly evolved core of an initially very massive star. It could be regarded as an observable counterpart of a warmer.

Heckman: Yes, but it bothers me that we do not commonly see any observable Wolf-Rayet signatures in the optical spectra of Seyfert nuclei. Maybe if the metal abundance is much higher than solar, Wolf-Rayet stars no longer produce observable optical emission-lines. Do any of the Wolf-Rayet pundits care to comment?

P. Conti: I'd like to comment on the locations of known WO stars: out of the half-dozen individual ones known, 1 is in the LMC, 1 in the SMC and 1 in NGC 1613. These are all "low" metal abundance environments, somewhat at odds with the high abundance predictions of Terlevich. Also, WO stars have strong emission in the O VI lines near $\lambda 3800$Å. If such stars were present in appreciable numbers in Seyferts one might expect to see these features.

A. Campbell: I'm not sure how strong a constraint the failure to observe optical WR features in Seyfert 2 spectra puts on the "warmers" paradigm. But (and perhaps André Maeder could comment) a larger problem may be the very short lifetime of the WR phase—$\sim 10^5$ yr. If Seyfert 2s are powered by WO stars in starbursts, there should be 100–300 times more spirals with starburst nuclei than there are Seyfert 2s. Is this observed?

Heckman: I think that depends on the assumption that the burst is a delta function in time. If it's extended in time, the "warmer" phase will be in some sense a convolution of the burst duration with the lifetime of the WR phase.

A. Campbell: I also believe that if warmers exist, ROSAT should see them. That's where they put out most of their radiation.

M. Fall: The high supernova rates postulated in the Terlevich et al. scenario might be expected to drive some gas out of the starburst/AGN. Such mass loss could weaken your arguments about the central velocity dispersions of the host galaxies. If more than about half the total mass were removed impulsively ($t_{loss} \ll t_{orb}$), the host galaxy would not remain gravitationally bound. If, as is more likely, the gas were removed adiabatically ($t_{loss} \gtrsim t_{orb}$), the remaining mass, characteristic radius and velocity dispersion of the starburst/AGN region would vary together as $R \propto 1/M$ and $\sigma_v \propto M$. Thus, if 90% of the mass were lost, the characteristic radius would increase by a factor of 10 and the velocity dispersion would decrease by a factor of 10. While this example may be extreme, my general point is that some changes in the central properties of the host galaxies are possible and even likely.

Heckman: That's an interesting point in the context of my arguments about where the 10^{10} $M_\odot$ of stellar remnants per L_* galaxy associated with "dead" high-redshift QSOs might be found. It does not affect the arguments about the dynamics of the nuclei of presently-active galaxies like NGC 4151 or 3C273.

N. Scoville: Don't the rapid variations in the brightness of Seyfert nuclei like Fairall 9 rule out sizes for AGNs that are as large as Terlevich's model requires?

Heckman: I don't think so. Terlevich would say these rapid variations are a "Christmas Tree" type phenomenon with different supernovae and SNRs "popping off" on short time-scales throughout a relatively large volume. I do think that the *very* rapid X-ray variability is a problem however.

COSMOLOGICAL CONSEQUENCES OF STARBURSTS

Daniel W. Weedman
Department of Astronomy and Astrophysics
The Pennsylvania State University
University Park, PA 16802
USA

Abstract. Massive star formation in galaxies is the key process being sought observationally to find evidence for the first generations of primordial star formation. Starting with a local luminosity function for star forming galaxies determined from IRAS and radio data, various constraints defined by other observations in X-ray, ultraviolet, optical and far infrared astronomy are discussed. It is shown how limits from the X-ray background and from faint optical galaxy counts restrict the amount of evolution which could have occurred for starburst galaxies. Predictions are made for the far infrared background from star forming galaxies that should be observed by COBE, and it is shown that this background will provide the tightest constraint yet available on the nature of primordial star formation.

1. INTRODUCTION

This meeting has been highlighted by many commendable presentations from people struggling to understand the processes of massive star formation. I cannot contribute significantly to such questions, but I do wish to place the overall issue in a cosmological context and impress on you the extraordinary importance of understanding star formation before we can understand the most outstanding problems in extragalactic astronomy. I will be presenting some quantitative results, but these are meant primarily as illustrations of the subjects that must be addressed. All of these subjects will soon be affected significantly by results from extant spacecraft—HST and COBE—so I do not consider the results given herein to be in any way definitive. They are meant primarily to alert you to the major comparisons and constraints that must be accommodated when we consider star formation in the largest possible context. My procedure will be to begin with an estimate of the local luminosity function for star forming galaxies and use that to extrapolate to what would be expected if such galaxies were observed to very high redshifts. Adopting a local luminosity function in this context means local in a cosmological sense. How many star forming events are there Mpc^{-3} and how are they best observed? I will discuss luminosity functions and their applications in the X-ray, ultraviolet, infrared and radio regimes—though not in this order of wavelength.

2. LUMINOSITY FUNCTIONS FOR STAR FORMING GALAXIES

2.1. Infrared Luminosity Functions

At the present time, it seems that the best census of star formation is obtained in the infrared. It is an empirical observation that star formation is commonly found in dense clouds so dusty that the optical and ultraviolet radiation is absorbed. The reradiation from this dust makes the star forming regions strong infrared sources. The IRAS all sky survey has yielded various samples of galaxies whose infrared luminosities are considered to arise primarily from star formation. Local galaxy luminosity functions have been derived from these samples using subsequent observations of galaxy redshifts. The luminosity function currently most completely defined is the IRAS "Bright Galaxy Sample", chosen on the basis of the 60 μm flux limit (Soifer et al. 1987). This luminosity function gives a starting point for extrapolating to higher redshifts and fainter flux limits, but we should note some caveats. In extragalactic astronomy, determining a correct "local" luminosity function is difficult because of the conflict between the necessity to average over an adequately large volume while also restricting to sufficiently small distances that effects of galaxy evolution do not enter. The universe is not evenly filled with galaxies, so it is necessary to average over local inhomogeneities. But if we look to distances sufficiently great that such inhomogeneities are negligible, it is possible that the more distant galaxies are systematically different from the nearby ones. The Bright Galaxy Sample has median galaxian distances of less than 50 Mpc, and we know that the distribution of galaxies is very irregular on this scale (e.g. Geller and Huchra 1989). Probably, the Bright Galaxy Sample is enhanced over a truly smooth distribution because many of the galaxies included are members of the Local Supercluster; comparing with fainter samples indicates an overdensity of order 25%, although it is difficult to separate this from evolution effects in the fainter samples (Lonsdale et al. 1990, Hacking and Houck 1987).

Unquestionably, further observations of redshift distributions for infrared selected samples will improve the "local" luminosity function. Inhomogeneities of the order mentioned are negligible sources of uncertainty in the present paper compared to some other assumptions that must be made. The advantage of using the Bright Galaxy Sample is that it yields a luminosity function that is clearly tabulated and well defined. This provides a clear reference for making future adjustments. Consequently, all calculations that follow begin with the 60 μm luminosity function of the Bright Galaxy Sample as taken from Soifer et al. (Table 1). This and all other luminosity functions used below are normalized to a Hubble constant $H_o = 75$ km s^{-1} Mpc^{-1}. All luminosity functions have units (in the log) of the number of galaxies per unit volume brighter than the listed luminosity.

2.2. Radio Luminosity Function

As a check on using the infrared luminosity function to represent all star formation, I compare it to the radio-derived luminosity function for star forming galaxies. Although the precise mechanisms relating star formation and the radio continuum luminosity are not well understood, it is clear empirically that these closely relate (Kennicutt 1983; Helou, Soifer and Rowan-Robinson 1985; Sramek and Weedman 1986). The mechanism somehow stems from the production of a non-thermal radio continuum as a consequence of supernovae. Using the assumption that the observed continuum represented a measure of star formation activity, Condon (1989) derived the "starburst" luminosity function for 20 cm radio luminosity using a radio-derived sample of faint galaxies in a small area of the sky. Classical radio galaxies with radio sources outside the galaxy or compact active galactic nuclei were omitted. Recent observations are available giving the 20 cm fluxes for the Bright Galaxy Sample (Condon et al. 1990). This gives a mean ratio of 60 μm to 20 cm flux; $\log f(60\ \mu m)/\log f(20\ cm) = 2.3$, for both fluxes measured in Jy. Using this ratio makes it straightforward to convert the Bright Galaxy Sample 60 μm luminosity function into a radio luminosity function (Table 1). This and the independently derived starburst radio luminosity functions are compared in Figure 1. Their close agreement is an independent confirmation that the IRAS derived Bright Galaxy Sample gives a correct census of star forming galaxies.

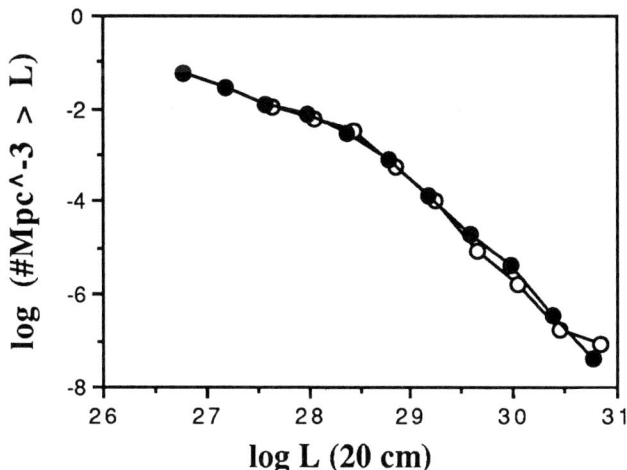

Figure 1. *Comparison of radio luminosity functions for star forming galaxies. Open circles: Condon's "starburst" luminosity function for radio selected sample. Filled circles: IRAS Bright Galaxy Sample transformed to radio using 20 cm observations. L(20 cm) has units of ergs s^{-1} Hz^{-1}*

2.3. Ultraviolet Luminosity Functions

A pressing objective for understanding the luminosity function of star forming galaxies is to address the nature and number of the faintest galaxies that will be seen in deep images with HST. Already, some preview of the extraordinary results expected is available from heroic efforts with ground based telescopes that have pushed galaxy counts to 27th magnitude (Tyson and Seitzer 1988). These images find 500,000 galaxies deg^{-2}; this number represents an upper limit to the number of faint star forming galaxies. It is the number that must be explained by extrapolating the local luminosity function to high redshifts, although future HST results will be crucial to improving the interpretation of the nature of these faint galaxies. Most of them have blue colors, and their occasional presence as arcs gravitationally lensed by intervening clusters implies that we are observing the ultraviolet spectra of galaxies with redshifts exceeding unity (Tyson, Valdes and Wenk 1990). It is necessary to know the intrinsic ultraviolet luminosity function of local galaxies to compare with these distant observations. Determining that function will be an early objective for HST. For now, I make a very approximate estimate using the limited data available from previous spacecraft. I cannot overemphasize that the present results are very uncertain and sure to be improved by HST data. I estimate three independent ultraviolet luminosity functions for star forming galaxies; the results will show that even in the ultraviolet, the IRAS Bright Galaxy Sample dominates star forming galaxies. The first luminosity function is that for normal spiral galaxies, those used by Condon (1989) in deriving the radio luminosity function. The galaxy luminosities are given as absolute blue magnitudes, M. These can be converted to monochromatic luminosities at 4400 Å, the effective blue wavelength, by the relation $2.5 \log L = 51.8 - M$, for L the luminosity in ergs s^{-1} Hz^{-1}. For conversion to 1400 Å, I use the ratio $L(1400)/L(4400) = 0.033$, deduced from observations of a few spiral galaxies with the ANS satellite (Coleman, Wu, and Weedman 1980). The resulting ultraviolet luminosity function for spiral galaxies is in Figure 2.

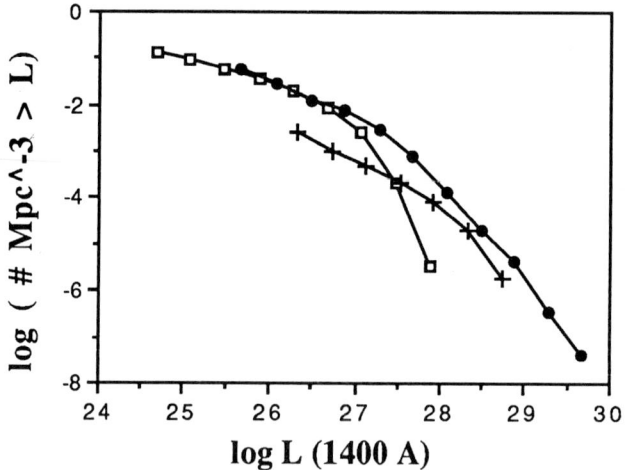

Figure 2. *Comparison of ultraviolet luminosity functions. Filled circles: IRAS Bright Galaxy Sample. Squares: Condon's spiral galaxies chosen as starburst radio sources. Crosses: Markarian galaxies. L(1400Å) has units of ergs per sec per Hz.*

The classical optical source for starburst galaxies was the Markarian survey, for which about 90% of the 1500 galaxies appear blue because of star formation. This is a second source of an ultraviolet luminosity function. To determine that, I first determined a blue magnitude luminosity function using all of the non-Seyfert galaxies in the Markarian catalog summary by Mazzarella and Balzano (1986). I transformed this to an ultraviolet luminosity function using the 30 Markarian starburst galaxies observed by the IUE and summarized in Weedman (1988). The ultraviolet luminosity function is maximized by letting it be dominated by the galaxies which have the largest ratios of $L(1400)/L(4400)$. The median for this ratio is 0.07, but 25% of the galaxies have a ratio of 0.4 or greater. For the present calculation, an upper limit to the ultraviolet luminosity function is more important, so I adopt a ratio of 0.4 but only use 25% of the real space density of Markarian galaxies, assuming that the remainder contribute negligibly in the ultraviolet. The result is also shown in Figure 2.

Finally, I try to estimate what the ultraviolet luminosity function would be for the IRAS Bright Galaxy Sample. This is uncertain because I have no ultraviolet observations of objects chosen because they were IRAS sources (obtaining such data is another

HST objective). What is available is the fact that 22 of the 30 Markarian galaxies used above with IUE spectra were also detected by IRAS. This gives a ratio of infrared to ultraviolet fluxes for this sample of starburst galaxies. But this was a sample chosen by the optical criteria of the Markarian selection; it may not be representative of a sample chosen by infrared selection. I assume that the ratio $\log L(60\,\mu m)/\log L(1400\,\text{Å}) = 3.4$, for both luminosities in ergs s^{-1} Hz^{-1}, which is the median from the Markarian sample including non-detections, is the same ratio as would be found from the entire Bright Galaxy Sample. The resulting ultraviolet luminosity function is given in Table 1 and Figure 2. A justification of the key assumptions leading to this is the fact that the IRAS Bright Galaxy Sample yields an ultraviolet luminosity function that is close to the sum of that derived from spiral galaxies plus Markarian galaxies. Many of each are included in the IRAS sample, so the final luminosity function adopted, through strictly taken from the IRAS Bright Galaxy Sample, is not very different from what would have been formulated using other samples of star forming galaxies. For attempting to explain faint galaxy counts, it will be necessary to ask what would happen if the ultraviolet continuum of star forming galaxies were never obscured by dust, so that galaxies would be brighter ultraviolet sources but not infrared sources. For an idealized starburst, in which the IMF, stellar lifetimes, and model atmospheres are assumed, there are straightforward relations between the bolometric luminosity from the massive stars and the monochromatic luminosity at 1400 Å. For a wide range of IMF, this ratio is about 1000, for monochromatic luminosity in ergs s^{-1} A^{-1} (Kunth and Weedman 1987). If all observed infrared luminosity is taken to be that amount of the bolometric luminosity from the starburst which was absorbed by dust, the fraction of ultraviolet luminosity escaping compared to the fraction absorbed can be deduced. The observed infrared flux is taken as the "FIR" flux tabulated in the IRAS catalog of extragalactic detections (IRAS 1985), and the 1400 Å monochromatic flux is as summarized in Weedman (1988). For the 30 Markarian galaxies referred to, the results of this calculation are given in Table 2 and the histogram for the fraction of luminosity escaping given in Figure 3.

TABLE 1. *Luminosity Functions**

$\log \#\text{Mpc}^{-3} > L$	$\log L\,(60\,\mu m)$	$\log L(1400\,\text{Å})$	$\log L(2\,\text{kev})$	$\log L(100\,\mu m)$
-1.36	28.2	24.8	19.9	28.5
-1.39	28.6	25.2	20.3	28.9
-1.45	29.0	25.6	20.7	29.3
-1.61	29.4	26.0	21.1	29.7
-1.82	29.8	26.4	21.5	30.1
-2.09	30.2	26.8	21.9	30.5
-2.62	30.6	27.2	22.3	30.9
-2.99	31.0	27.6	22.7	31.3
-3.73	31.4	28.0	23.1	31.7
-4.55	31.8	28.4	23.5	32.1
-5.23	32.2	28.8	23.9	32.5
-6.23	32.6	29.2	24.3	32.9
-7.19	33.0	29.6	24.7	33.3

*Luminosities L are in ergs sec^{-1} Hz^{-1}.

TABLE 2. *Ultraviolet and Infrared Properties for Markarian Starburst Galaxies**

Markarian Galaxy	1000L(1400 Å)/ L(FIR)	% UV Escaping
0007	0.34	25
0008	0.03	2.9
0019	0.36	26
0025	0.19	16
0026	>0.23	>19
0036	>0.7	>41
0052	0.06	5.4
0054	0.42	30
0059	1.69	63
0066	0.41	29
0067	>0.09	> 8
0086	0.04	3.6
0153	>1.95	>66
0170	>0.22	18
0171	0.005	0.5
0209	>1.25	>56
0213	0.04	3.8
0288	>0.01	>10
0297	0.01	1.6
0309	0.01	1.0
0325	0.078	7.2
0357	0.29	22
0432	0.028	2.7
0450	>0.17	>15
0487	0.39	28
0496	0.032	3.1
0538	0.09	8.2
0691	0.066	6.2
0789	0.016	1.6
1027	0.023	2.3

*Ratio in second column would be 1.0 if the amount of ultraviolet radiation which is absorbed and reradiated in the infrared is equal to the amount of ultraviolet radiation which escapes, so that fraction escaping would be 50%.

With this result, we can realistically derive a lower limit to the luminosity function for "clean" star forming galaxies. The distribution of emitted fractions in Figure 3 can be folded into the ultraviolet luminosity function for the Bright Galaxy Sample to produce a "cleaned" luminosity function. The correction is dominated by the approximately 50% of IRAS galaxies for which less than 10% of the ultraviolet escapes. Removing all dust from these galaxies would make half of the galaxies in the tabulated ultraviolet luminosity function brighter by a factor of 10. This "dust-free" luminosity function as it will be used later is given in Table 3.

2.4. X-Ray Luminosity Function

Optical observations to faint magnitudes will not be the only relevant or important check on star forming galaxies in the distant universe. As will be discussed, X-ray observations already provide meaningful constraints, so we will need to have an X-ray luminosity function for local star forming galaxies. This is best derived at an energy of 2 kev, for which there are the greatest number of observations with the Einstein Observatory. In order to compare with the star forming activity revealed by IRAS observations, Griffiths and Padovani (1990) have compiled all star forming galaxies with both X-ray and infrared observations. They derive a median ratio for $\log L(60 \mu m)/\log L(2 \text{ kev}) = 8.3$, for both luminosities in ergs s^{-1} Hz^{-1}. Using this result with the Bright Galaxy Luminosity Function yields the X-ray luminosity function in Table 1.

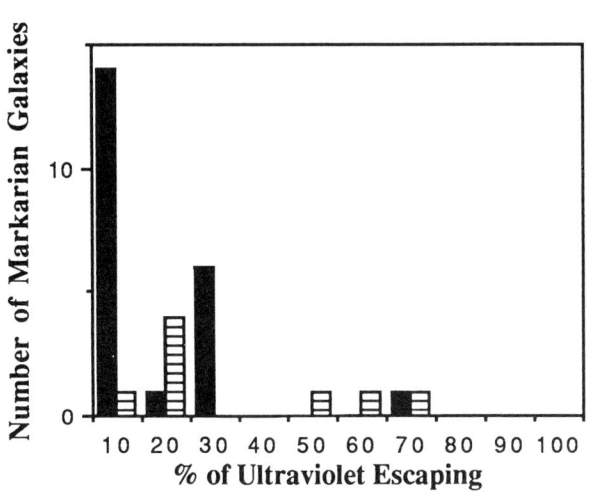

Figure 3. *Number of Markarian galaxies having different fractions of ultraviolet light escaping without dust absorption. Solid bars: galaxies in Table 2 with infrared detections. Hatched bars: galaxies in Table 2 with no infrared detections so % given is lower limit to escaping ultraviolet.*

2.5. Far Infrared Luminosity Function

The most exciting observational cosmological data of this millennium is now coming in from the Cosmic Background Explorer. The COBE results, when compared to extrapolations from the local luminosity function, will seriously constrain the number of star forming galaxies in the early universe. For relevant calculations to be given below, a luminosity function needs to be adopted at the longest wavelength observed by both IRAS and COBE. This is 100 μm, at which the luminosity ratio for galaxies in the Bright Galaxy Sample is $\log[L(100\ \mu m)/L(60\ \mu m)] = 0.31$, yielding the luminosity function in Table 1.

TABLE 3. *Dust Free Luminosity Function*

$\log L(1400\ \text{Å})$	26.6	27.0	27.4	27.8	28.2	28.6	29.0	29.4	29.8	30.2	30.6
$\log \#\ \text{Mpc}^{-3} > L$	-1.40	-1.75	-2.05	-2.40	-2.85	-3.35	-4.10	-4.90	-5.60	-6.55	-7.50

3. PREDICTIONS FOR LIMITING OBSERVATIONS

3.1. Faint Optical Galaxy Counts

To predict what should be seen from faint reaches of the universe, it is necessary to integrate the local luminosity function over all volumes examined. For the present calculations, the redshift limit for this integration is taken at $z = 4$, because an optical CCD observation at 6500 Å for this redshift would see the intrinsic spectrum shortward of Lyman α. Below that limit, the continuous spectrum will be greatly diminished by hydrogen absorption by intervening material (the "Lyman α forest"), so it is unlikely that any optical counts will be significantly affected by objects above this redshift. For consistency, X-ray and COBE background calculations are also cut off at $z = 4$.

The relevant cosmological equations used are given in Weedman (1986a), using $q_o = 0.1$ and $H_o = 75$ km s^{-1} Mpc^{-1}. Using these equations to integrate a luminosity function over successively distant volume shells requires application of a K-correction, which is most easily applied by using a single parameter to describe the spectrum shape. For a spectrum described as a power law such that the luminosity Hz^{-1} is proportional to frequency to a power alpha, this alpha is included explicitly in the equations used. Because we need here only the continuous spectrum of the starburst, and then only in the intrinsic ultraviolet, a reasonable approximation for this alpha that matches the observed ultraviolet continua of starburst galaxies from the ultraviolet through the blue is a value of -1 for alpha. (For this value, the flux per unit wavelength, which is what one sees in an IUE spectrum, would increase inversely with wavelength. For an alpha of -2, the flux per unit wavelength would be the same with wavelength, or an IUE spectrum that looked flat. Real starburst spectra from IUE seem to be between these extremes of flat or slightly rising to the ultraviolet.) Starting with the ultraviolet luminosity function in Table 1, the results for what should be seen using this function to $z = 4$ are given in Table 4. The result greatly underpredicts the number of faint galaxies actually observed by Tyson and Seitzer. What is the explanation? Clearly, we have to account for more galaxies than would be predicted from the local luminosity function of star forming galaxies alone. Actually, the observed faint counts are strictly only upper limits to the number of star forming galaxies. This is because the counts can also include galaxies of old stars. Such galaxies would be important only at low redshifts, because for galaxies in which the intrinsic ultraviolet spectrum is seen, old star populations are negligible luminosity sources. How the mix of old and young stars changes to affect galaxy counts, therefore, is a complicated function of redshift, and of the magnitude limit of the counts since fainter limits go to higher redshifts. That the limits given by the observed counts are close to the actual numbers of star forming galaxies is indicated by the blue colors of the faint galaxies, but better quantitative statements must wait on improved HST measures of these colors. We proceed by considering the counts to show that some mechanism allows star forming galaxies to be seen more easily at high redshift than would have been expected from their local properties.

TABLE 4. *Counts of faint galaxies in B magnitude**

Magnitude limit	No Change in Luminosity Function #deg^{-2}	Dust Free Luminosity Function #deg^{-2}	$z > 1$	$z > 2$	Evolution with $(1+z)^{2.5}$ #deg^{-2}	$z > 1$	$z > 2$
25	13×10^3	73×10^3	0.30	0.07	98×10^3	0.73	0.41
27	66×10^3	38×10^4	0.35	0.13	44×10^4	0.86	0.58
29	23×10^4	19×10^5	0.35	0.13	93×10^4	0.91	0.66

*Comparison target from the observations of Tyson and Seitzer (1988) is 50×10^4 galaxies deg^{-2} to magnitude 27. Fractions of total counts are given for $z > 1$ and $z > 2$.

This circumstance is similar to what also happens for quasars, and the solution invoked in that case is to apply evolution to the quasars so that they were either systematically more luminous or more numerous in the past. When discussing evolution for extragalactic sources, there are two extremes for parameterizing evolution. One extreme is "pure luminosity evolution", which assumes that the luminosity of each galaxy in the luminosity function changes systematically with redshift in some way; that is, all galaxies considered were more luminous in the past by the same factor. The most common way to express this luminosity enhancement is to allow it to scale as $(1+z)^n$, for n the evolution parameter that needs to be found to match the observations. Any form incorporating redshift or cosmic look-back time could be used; putting it in terms of $(1+z)$ is the simplest because powers of $(1+z)$ enter in various other parts of cosmological equations. A similar scaling could also be used for the other extreme of evolution, which is "pure density evolution". This assumes that the characteristic luminosities of individual galaxies do not change but that the number of galaxies involved increases with some evolutionary scaling. For star forming galaxies, pure density evolution is not a realistic option, because the luminosity functions used already incorporate the majority of all galaxies. Asking for the number of galaxies to increase significantly at high redshifts would invoke galaxies that existed then but not now. Asking only how much galaxies would have to scale up in luminosity with increasing redshift to explain the observations by scaling up the local ultraviolet luminosity function results in a value of 2.5 for the evolution parameter n. This gives the results in Table 4, results which come close to matching the number counts observed at 27th magnitude. This form of evolution would mean that the representative luminosity of a star forming galaxy was greater by a factor of 16 at $z = 2$ than it is in the local universe. There is, however, an alternative explanation that requires no evolution for star forming galaxies. Remember that most of the ultraviolet luminosity is obscured in the star forming galaxies used to derive the local ultraviolet luminosity function. What if such galaxies are less dusty at high redshifts? In the extreme, we can ask what would happen if the "dust free" luminosity function in Table 3 were extrapolated all the way to $z = 4$. Most of the galaxies would appear 10 times more luminous in the ultraviolet, so we would expect the number of galaxies seen via their intrinsic ultraviolet spectrum to be much greater. Quantitatively, the results are in Table 4, where we see that this dust free luminosity function comes close to the observations without invoking any evolution. These two alternatives, luminosity evolution vs. less dust, are considered further in the context of X-ray and COBE observations, but for the sake of further optical checks, note in Table 4 that the two choices predict different forms for galaxy counts and for galaxy redshift distributions.

3.2. The Diffuse Extragalactic X-ray Background

Optical observations to faint magnitudes are not the only relevant or important check on the nature of star formation at high redshifts. An absolute constraint on the amount of star formation in the universe is set by the requirement that the integrated flux from the galaxies in X-rays not exceed the extragalactic X-ray background. There are good reasons to believe that a substantial fraction of the 2 kev background, perhaps the majority, does arise from star forming galaxies (Weedman 1986b, Griffiths and Padovani 1990). It is also known that quasars and active galactic nuclei must contribute significantly to the background (Persic et al. 1989). The difficulty in disentangling the sources of the background arises from lack of knowledge of the X-ray spectra for the contributing sources; the spectrum of the background is known and is too hard (or flat) to be explained predominantly by quasars (Boldt 1987). This is the main reason why star forming galaxies are an important optional source, because these may prove to have flatter X-ray spectra, by analogy to the spectra of massive X-ray binaries (Nagese 1989). Regardless of the spectral issues, the monochromatic 2 kev flux sets a firm upper limit on the integrated 2 kev flux from any set of sources. Using the 2 kev luminosity function in Table 1, I calculate the X-ray background that would be produced in the "evolution" and "dust-free" options from above. (This calculation assumes an X-ray power law spectral index alpha of -0.4, chosen to match the observed index of the background, although the results for the 2 kev background are insensitive to the actual value of alpha.) The results are to be compared to the observed background of 1.2×10^{-6} Jy deg^{-2} (Schwartz 1979). The dust free model is simply the local luminosity function extrapolated with no evolution to $z = 4$. Implicit in this is the assumption that star forming galaxies are sufficiently transparent to X-rays at 2 kev that the presence or absence of dust is not relevant to the X-ray luminosity. This predicts a 2 kev background of 1.8×10^{-7} Jy deg^{-2}, or 14% of the observed background. Of this predicted background from star forming galaxies, 88% arises from galaxies at $z < 2$; this is a general result of background calculations—that relatively nearby objects dominate the background if there is no evolution with redshift. For the evolution model, projecting the luminosity function with the luminosity evolution used to compare with the optical counts, the predicted background is precisely 100% of the observed background, and in this case 50% arises from $z < 2$ and 50% from $2 < z < 4$. As it is already known from counts of quasars and active galactic nuclei that a substantial fraction of the X-ray background must arise from these sources, the above results for star forming galaxies rule out evolution as extreme as the $(1+z)^{2.5}$ form that was assumed. Constraints from the X-ray background imply, therefore, that at least some part of the enhanced optical counts of faint star forming galaxies arises from increased ultraviolet transparency rather that pure luminosity evolution. Once adequate counts of X-ray sources and their spectra for galaxies and quasars are available, from future missions Astro-D, AXAF, and possibly BBXRT (the soft energy counts of ROSAT will not be particularly helpful in this respect), very tight constraints on the total X-ray background produced in the "nearby" universe to redshift of about 2 can be determined. This will yield well defined limits on the number of primordial star forming galaxies. Because our difficulty in most modeling efforts is to avoid overpredicting the background, the situation is already troubling, because there does not seem to be much room left in the X-ray background for primordial galaxies; yet, as will be mentioned below, we already know that there are copious amounts of primordial heavy elements.

3.3. The COBE Far Infrared Background

The most definitive predictions, least subject to assumption, have been saved for last. At the time of this writing, the Cosmic Background Explorer, thoroughly chilled by its liquid helium coolant, is determining the diffuse sky radiation at many different wavelengths. For the present comparisons, the most important results are arising from the Diffuse Infrared Background Experiment, particularly with the observing channel at 100 μm. This is the longest wavelength at which the IRAS Bright Galaxy Sample luminosity function can be determined using IRAS observations. As the luminosity function is shifted to higher redshifts, intrinsic wavelengths are observed for which the K-correction term is known using the spectrum shape from the shorter wavelength IRAS observations. This yields an alpha of -1.3 between 100 μm and 60 μm, and alpha of -2.3 from 60 μm to 25 μm. Starting with the 100 μm luminosity function in Table 1, and projecting to $z = 4$, yields total sky fluxes from star forming galaxies that should be seen by COBE. If we simply project the local 100 μm luminosity function without change in either dust content or evolution, the result is that the extragalactic background observed at 100 μm should be 38 Jy deg^{-2}. If COBE observes a fainter background than this, the explanation is that galaxies at high redshift are fainter at 100 μm than local galaxies, implying they have significantly less dust. In this case, the dust-free model would be supported. In this case of no evolution, most of the background would arise from relatively nearby galaxies, 80% from galaxies with $z < 0.5$. Actually, it would be difficult to accept that the dust content changed significantly at such low redshifts, so a COBE extragalactic background fainter than this limit would be hard to accommodate consistently with the IRAS luminosity function. If instead we use the alternative model for luminosity evolution of star forming galaxies, the predicted background at 100 μm is 89 Jy deg^{-2}, of which 55% comes from galaxies with $z > 0.5$, and 28% comes from galaxies with $z > 2$. If the DIRBE experiment observes an extragalactic background greater than this, that will be extremely exciting because it would be evidence for a population of primordial galaxies at redshifts above 4. Determining the extragalactic backgrounds will be a challenge because the amounts given are only a few percent of the total observed background (Mather et al. 1990), most of which arises from radiation by zodiacal dust and dust in our Galaxy. These earliest results from COBE will not immediately yield the extragalactic background. It will be necessary to subtract a somewhat model-dependent distribution for zodiacal and Galactic components. Nevertheless, we can expect that the COBE data will soon provide the most meaningful constraints available on a population of primordial galaxies.

3.4. Element Formation in Quasars

Before concluding, I draw your attention to a subject seemingly far removed from the topic of massive star formation. Note the spectrum of the highest redshift quasar yet known (Schneider, Schmidt and Gunn 1989) at a redshift of 4.67. Here is an object that existed when the universe was less than 20% of its present age. Even then, the emission lines from carbon, nitrogen, oxygen and silicon were as strong as in present day quasars. These elements, in the context of present knowledge, had to come from massive stars. Because the emission lines seen are resonance lines, the features are the same as the strong absorption lines seen in the winds escaping from massive stars, described extensively by R. Kudritzki at this meeting. We must assume, therefore, that the presence of these emission lines in quasars is evidence that extensive generations of massive stars had gone before. From the empirical considerations of an observer, it is also important to realize that these features in absorption in the spectra of massive stars are the dominant spectral features in the ultraviolet for star forming galaxies. If we ever demonstrate conclusively that star forming galaxies exist at high redshift, that demonstration will depend on observing these absorption line features from stellar winds.

4. SUMMARY

In the remarks above, I have attempted to address as quantitatively as possible with available data the various observational tests that allow statements about the formation of galaxies in the distant universe. Incorporating data from X-ray telescopes, HST, and COBE will provide very meaningful constraints on how much star formation occurred at redshifts greater than the highest redshifts yet observed for individual objects. Particularly if a COBE background higher than that predicted is seen, imaging observations in the infrared, with the SIRTF spacecraft, will be of top priority. Continuing efforts to find quasars with heavy element emission lines at higher and higher redshifts will push back the epoch at which star formation must have begun. I hope these remarks demonstrate that solving the fundamental mysteries of observational cosmology is intimately tied to understanding the star formation process in the early universe. That requires understanding this process locally. This is a clear statement that any distinction between the objectives of "stellar" and "extragalactic" astronomers is completely artificial.

REFERENCES

Boldt, E. A. 1987, *Physics Reports*, **146**, 216.
Coleman, G. D., Wu, C.-C., and Weedman, D. W. 1980, *Ap. J. Suppl.*, **43**, 393.
Condon, J. J. 1989, *Ap. J.*, **338**, 13.
Condon, J. J., Helou, G., Sanders, D. B., and Soifer, B. T. 1990, preprint.
Geller, M. and Huchra, J. 1989, *Science*, **246**, 897.
Griffiths, R. E. and Padovani, P. 1990, *Ap. J.*, in press.
Hacking, P. B. and Houck, J. R. 1987, *Ap. J. Suppl.*, **63**, 311.
Helou, G., Soifer, B. T., and Rowan-Robinson, M. 1985, *Ap. J. (Letters)*, **298**, L7.
IRAS 1985, Cataloged Galaxies and Quasars Observed in the IRAS Survey (Pasadena: Jet Propulsion Laboratory).
Kennicutt, R. C. 1983, *Astr. Ap.*, **120**, 219.
Kunth, D. and Weedman, D. 1987, in Exploring the Universe with the IUE Satellite, ed: Y. Kondo, (Dordrecht: Reidel) p. 623.

Lonsdale, C. J., Hacking, P. B., Conrow, T. P., and Rowan-Robinson, M. 1990, *Ap. J.*, in press.
Mazzarella, J. M. and Balzano, V. A. 1986, *Ap. J. Suppl.*, **62**, 751.
Mather, J. C. *et al.* 1990, COBE preprint 90-03.
Nagese, F. 1989, *Pub. Ast. Soc. Japan*, **41**, 1.
Persic, M., De Zotti, G., Danese, L., Palumbo, G. G. C., Franceschini, A., Boldt, E. A., and Marshall, F. E. 1989, *Ap. J.*, **344**, 125.
Schneider, D. P., Schmidt, M., and Gunn, J. E. 1989, *A. J.*, **98**, 1951.
Soifer, B. T., Sanders, D. B., Madore, B. F., Neugebauer, G., Lonsdale, C. J., Persson, S. E., and Rice, W. L. 1987, *Ap. J.*, **320**, 238.
Sramek, R. A. and Weedman, D. W. 1986, *Ap. J.*, **302**, 640.
Tyson, J. A. and Seitzer, P. 1988, *Ap. J.*, **335**, 552.
Tyson, J. A., Valdes, F., and Wenk, R. A. 1990, *Ap. J. (Letters)*, **349**, L1.
Weedman, D. W. 1986a, Quasar Astronomy (Cambridge: Cambridge University Press) p. 63.
_____. 1986b, in Star Formation in Galaxies, ed: C. Lonsdale, (NASA CP-2466) p. 351.
_____. 1988, *Ap. Lett. and Comm.*, **27**, 117.

DISCUSSION

L. Dressel: Can you explain the bimodal distribution of the percentage of ultraviolet light escaping starburst Markarian galaxies?

Weedman: No!

Anon: I would like to be reminded of the mass distribution of the Markarian galaxies used in your UV sample as I note that in optically selected starbursting interacting galaxies the lower mass galaxies ($M \lesssim 10^9$ $M_\odot$) are deficient in dust even though they are experiencing strong 'starbursts.' Is there a correlation between the masses of the galaxies you use and their UV absorption? If yes, what is that correlation?

Weedman: Markarian galaxies span the range from the intrinsically faint "extragalactic H II region" class to luminous full-sized galaxies like NGC 3690 (which is Mrk 171). Markarian galaxies are scattered throughout the lists of superluminous IRAS sources, and these are mostly full-sized galaxies. I don't have an optical luminosity function with me, so I can't answer you rigorously.

T. Heckman: Aren't you worried that you've biased your calculation of the UV luminosity function for IRAS galaxies by using Mrk galaxies (which were selected to be UV-bright)? Isn't it less uncertain to just use the observed FIR luminosity function and then "de-shroud" it?

Weedman: I selected things the way I did because I had no other choice. You are correct, I get the same answer if I use the IRAS luminosity function itself.

C. Lonsdale: I can improve your luminosity function a little now. You have probably overestimated the contributions to the background by at least 25% because the Soifer et al. luminosity function is contaminated by the local supercluster.

Weedman: That would decrease the backgrounds by the same factor and cause a slight increase in the exponent of the evolution needed to explain the faint optical counts.

Z. Wang: Galaxy distributions show large-scale-structures, such as clusters and voids. If the X-ray background is attributed to galaxies, are you expecting to see similar structures or has that already been seen?

Weedman: The limits on structure in the X-ray background are not sensitive enough to see the large scale structures shown by galaxy distributions.

B. Rocca-Volmerange: The counts of normal elliptical and spiral galaxies with luminosity evolution which fit the faint galaxy counts to $m = 27$ from Tyson do not reproduce the COBE luminosity by a large factor.

Weedman: That's an interesting calculation. I haven't done that calculation, and I'm glad to hear I don't have to.

R. Wyse: Richard Ellis and collaborators are doing a redshift survey at the AAT of the population of 'Tyson-blue galaxies.' They have a small sample, size ~ 15 galaxies, but the redshifts are all $\lesssim 1$.

Weedman: If you look at the alternative predictions I made, this result implies that galaxies are getting less dusty.

J. Bland Hawthorn: How much of the X-ray background do you want to account for? Nearby, apparently bright, low luminosity AGNs are mostly detected at 10 keV to even 1 MeV. These would also tend to overpredict the X-ray background.

Weedman: Yes, AGN and QSOs must contribute significantly to the background, so the true limit allowable to starbursts is actually much less than the whole background.

FINAL DISCUSSION

N. Walborn: The interstellar absorption lines present a significant problem for spectral synthesis of low-resolution spectrograms, which was not explicitly addressed in the relevant reviews. In giant H II regions they can have extensive multiple velocity components with a total equivalent width comparable to that of stellar features (Walborn and Hesser, *Ap. J.*, **252**, 156, 1982). For instance, a relatively strong feature near 1335 Å in the low-resolution IUE spectrogram of HDE 303308 has an equivalent width of 4.4 Å. When analyzed in the corresponding high-resolution data, it is found to actually consist of similar contributions from interstellar C I and C II, and stellar O IV. An analogous problem can affect the interpretation of Ca II H and K in optical spectra.

R. Joseph: Is it possible to make any general statement about the equivalent width of the Si IV interstellar feature, compared to what one might expect if there is a substantial population of massive early-type stars?

N. Walborn: That would depend upon the ionization and velocity dispersion in the interstellar gas, and upon the nature of the early-type stars. If the stars are on the O main sequence, the Si IV will be *entirely* interstellar. If B main sequence, stellar photospheric absorption will likely dominate. If OB supergiants, the dominant feature will be a stellar-wind profile.

G. Koenigsberger: Is the use of the Si IV and C IV lines as classification criteria as applicable to stars in low-metallicity environments as it is in the Galaxy?

N. Walborn: Not too much information is available, but the SMC supergiant spectra indicate that they are. Both features are weaker than in the Galactic counterparts, but the ratio effects are similar.

G. Koenigsberger: How valid is the use of the Sekiguchi and Anderson (1987) calibration for objects with much lower heavy element abundances, given that it is based on measurements of Galactic stars.

C. Leitherer: We studied the usefulness of the Si IV/C IV lines to infer the massive-star population in starburst galaxies (Leitherer and Lamers 1990, submitted to *Ap. J.*). Our main result is that the models published by Sekiguchi and Anderson (1987) should not be used for a *quantitative* comparison with observations. Sekiguchi and Anderson assumed in their models that Si IV/C IV is due to main-sequence stars only. However, by looking at the radial velocity of these lines in many starburst galaxies, it is evident that this assumption is not correct. We observe C IV *and* Si IV as blue-shifted absorptions—indicative of stellar winds in massive stars. However, if you take a sample of OB stars on the main sequence, say all OB stars from the solar neighborhood with low-resolution IUE spectra available, you will find that C IV is indeed blue-shifted because it is due to stellar winds from mid-O stars, whereas Si IV is *unshifted* because in main-sequence stars this line is of photospheric origin in early B stars. Main-sequence stars do *not* show a blue-shifted Si IV line because their wind densities are too low. If blue-shifted Si IV is observed, this immediately implies a significant contribution from evolved O stars which have higher $\dot{M}$, and for which blue-shifted Si IV is actually observed in individual stars.

The second problem with the Sekiguchi and Anderson models is their assumption that—apart from *direct* abundance variations—the relative strengths of Si IV and C IV are the same in extragalactic stars as in their galactic calibration stars. Most probably this assumption is wrong. On theoretical grounds we expect different $\dot{M}$ characteristics for stars with different chemical composition (see Rolf Kudritzki's talk). C IV (and Si IV if post-main-sequence stars are important) originate in stellar winds so that detailed wind models are necessary to predict their strengths in starburst galaxies. Both lines are due to ions which are not in the dominant ionization stage but are trace ions in O-star winds. This means only slight differences in the densities of O-star winds (due to metallicity effects) may have enormous consequences for the strengths of these lines. Detailed modeling is necessary before one can use the line ratio of Si IV/C IV as a tracer of the initial mass function in starburst galaxies. Let me emphasize the Sekiguchi and Anderson study still is an important paper since it led the way for this kind of analysis. However, care should be taken if these models are applied *quantitatively* to starburst galaxies.

J. Bland Hawthorn: This is a question for George Rieke. Can you give some sense of the relative importance of the various observed parameters as constraints on starburst models?

G. Rieke: I would say the dynamical mass, the derived ionizing flux, and the CO band depth. I think these three pieces of data in the case of M82 inevitably drive you to the type of model I discussed. Of course the models are still very dependent on stellar evolutionary tracks. For example, in the center of the Milky Way we see very luminous stars that can not be very hot. Peter Conti has suggested to me that there may be a phase of stellar evolution where coming off the main sequence a massive star 'hangs up' for a while at a lower temperature. Perhaps our conversion from excitation temperature to stellar mass is incorrect. This would allow stars more massive than 25 or 30 $M_\odot$ to exist in the center of the Milky Way or in M82 without producing very high-excitation nebular lines. Maybe M82 is telling us something about stellar evolution rather than the upper mass limit to the IMF.

Coming back to your question, one of the problems with these starburst models is that they are not totally determinant. That is, there is no statistically valid way of saying you've explored every possible corner of parameter space. It's a generic problem with this way of approaching science.

T. Heckman: If you can relax the *upper* mass limit (allow it to drift higher) doesn't this simultaneously loosen the constraint on the *low* mass cut-off (allow it to move lower)? I am very worried that the mass range you derived for the IMF in M82 is so narrow (3–25 $M_\odot$).

G. Rieke: Yes, that's right. If very massive stars spend a lot of time with $T \sim 30{,}000$ K you can relax the upper and lower mass limits on the IMF.

A. Maeder: I think the possibility that very massive stars have such low temperatures is in conflict with both theoretical models and the observational data shown at this conference by Garmany.

Rieke: Well, there are these candidates in the Galactic center. We still don't know what they are. Perhaps metallicity is important.

C. Leitherer: Let us assume André Maeder's tracks are applicable in starburst galaxies. Assuming Z is about $Z_\odot$ or $1/3\ Z_\odot$, then André's models predict that stars below about 25–30 $M_\odot$ should evolve into red supergiants and explode as supernovae without forming Wolf-Rayet stars. Stars above this mass limit will evolve into W-R stars. This provides an interesting observational test: If the upper cut-off mass for the IMF is 30 $M_\odot$, then no W-R stars should be observed. On the other hand, in the case of W-R galaxies, we would expect that the upper cut-off should be higher, if the IMF can be derived independently. Has such an effect been observed?

G. Rieke: In the W-R galaxy NGC 5253, the mid infrared fine structure lines indicate a very hot exciting field, much hotter than for a typical starburst. However, the test you propose has not been done systematically.

T. Heckman: While we are on the subject of Wolf-Rayet stars, I'd just like to show a spectrum of an ultra-luminous infrared galaxy (IRAS01003–2238) that has very strong spectral signatures of late-type WN stars. Lee Armus and I estimate it may have several $\times 10^5$ such stars. Wolf-Rayet lines have been detected in a handful of powerful IR galaxies, so it seems that at least some of these galaxies can make very massive stars.